AF776812

Advances in Science, Technology & Innovation

IEREK Interdisciplinary Series for Sustainable Development

Editorial Board

Anna Laura Pisello, Department of Engineering, University of Perugia, Italy
Dean Hawkes, University of Cambridge, Cambridge, UK
Hocine Bougdah, University for the Creative Arts, Farnham, UK
Federica Rosso, Sapienza University of Rome, Rome, Italy
Hassan Abdalla, University of East London, London, UK
Sofia-Natalia Boemi, Aristotle University of Thessaloniki, Greece
Nabil Mohareb, Faculty of Architecture - Design and Built Environment, Beirut Arab University, Beirut, Lebanon
Saleh Mesbah Elkaffas, Arab Academy for Science, Technology, Egypt
Emmanuel Bozonnet, University of la Rochelle, La Rochelle, France
Gloria Pignatta, University of Perugia, Italy
Yasser Mahgoub, Qatar University, Qatar
Luciano De Bonis, University of Molise, Italy
Stella Kostopoulou, Regional and Tourism Development, University of Thessaloniki, Thessaloniki, Greece
Biswajeet Pradhan, Faculty of Engineering and IT, University of Technology Sydney, Sydney, Australia
Md. Abdul Mannan, Universiti Malaysia Sarawak, Malaysia
Chaham Alalouch, Sultan Qaboos University, Muscat, Oman
Iman O. Gawad, Helwan University, Egypt
Anand Nayyar, Graduate School, Duy Tan University, Da Nang, Vietnam

Series Editor

Mourad Amer, International Experts for Research Enrichment and Knowledge Exchange (IEREK), Cairo, Egypt

Advances in Science, Technology & Innovation (ASTI) is a series of peer-reviewed books based on the best studies on emerging research that redefines existing disciplinary boundaries in science, technology and innovation (STI) in order to develop integrated concepts for sustainable development. The series is mainly based on the best research papers from various IEREK and other international conferences, and is intended to promote the creation and development of viable solutions for a sustainable future and a positive societal transformation with the help of integrated and innovative science-based approaches. Offering interdisciplinary coverage, the series presents innovative approaches and highlights how they can best support both the economic and sustainable development for the welfare of all societies. In particular, the series includes conceptual and empirical contributions from different interrelated fields of science, technology and innovation that focus on providing practical solutions to ensure food, water and energy security. It also presents new case studies offering concrete examples of how to resolve sustainable urbanization and environmental issues. The series is addressed to professionals in research and teaching, consultancies and industry, and government and international organizations. Published in collaboration with IEREK, the ASTI series will acquaint readers with essential new studies in STI for sustainable development.

More information about this series at http://www.springer.com/series/15883

Francisco Fernandes · Ana Malheiro ·
Helder I. Chaminé
Editors

Advances in Natural Hazards and Hydrological Risks: Meeting the Challenge

Proceedings of the 2nd International Workshop on Natural Hazards (NATHAZ'19), Pico Island—Azores 2019

Springer

Editors
Francisco Fernandes
Regional Civil Engineering Laboratory (LREC)
Ponta Delgada, Portugal

Ana Malheiro
Regional Civil Engineering Laboratory (LREC)
Ponta Delgada, Portugal

Helder I. Chaminé
Laboratory of Cartography and Applied Geology
Department of Geotechnical Engineering
School of Engineering (ISEP)
Polytechnic of Porto
Porto, Portugal

ISSN 2522-8714 ISSN 2522-8722 (electronic)
Advances in Science, Technology & Innovation
IEREK Interdisciplinary Series for Sustainable Development
ISBN 978-3-030-34396-5 ISBN 978-3-030-34397-2 (eBook)
https://doi.org/10.1007/978-3-030-34397-2

© Springer Nature Switzerland AG 2020
This work is subject to copyright. All rights are reserved by the Publisher, whether the whole or part of the material is concerned, specifically the rights of translation, reprinting, reuse of illustrations, recitation, broadcasting, reproduction on microfilms or in any other physical way, and transmission or information storage and retrieval, electronic adaptation, computer software, or by similar or dissimilar methodology now known or hereafter developed.
The use of general descriptive names, registered names, trademarks, service marks, etc. in this publication does not imply, even in the absence of a specific statement, that such names are exempt from the relevant protective laws and regulations and therefore free for general use.
The publisher, the authors and the editors are safe to assume that the advice and information in this book are believed to be true and accurate at the date of publication. Neither the publisher nor the authors or the editors give a warranty, expressed or implied, with respect to the material contained herein or for any errors or omissions that may have been made. The publisher remains neutral with regard to jurisdictional claims in published maps and institutional affiliations.

This Springer imprint is published by the registered company Springer Nature Switzerland AG
The registered company address is: Gewerbestrasse 11, 6330 Cham, Switzerland

Scientific Committee

Chairs

Ana Malheiro, Regional Laboratory for Civil Engineering (LREC), Ponta Delgada, Azores, Portugal
Helder I. Chaminé, Laboratory of Cartography and Applied Geology, Department of Geotechnical Engineering, School of Engineering (ISEP), Polytechnic of Porto, Porto, Portugal

Board

Alberto Sayão, Pontifical Catholic University of Rio de Janeiro (PUC–Rio), Rio de Janeiro, Brazil
Alexandre Tavares, Faculty of Sciences and Technology, University of Coimbra, Coimbra, Portugal
Ana Cláudia Teodoro, Faculty of Sciences, University of Porto, Porto, Portugal
Andreas Dix, Institute of Geography, University of Bamberg, Bamberg, Germany
Andrzej Kijko, University of Pretoria Natural Hazards, Pretoria, South Africa
Antonio Pulido-Bosch, University of Almeria, Almeria, Spain
António Roque, National Laboratory for Civil Engineering (LNEC), Lisbon, Portugal
Bruno Merz, University of Potsdam (GFZ), Potsdam, Germany
Chiara Arrighi, University of Florence, Florence, Italy
Gil Mahé, HydroSciences Montpellier, University of Montpellier, Montpellier, France
Hossein Tabari, University of Leuven, Leuven, Belgium
Humberto Varum, Faculty of Engineering, University of Porto, Porto, Portugal
Ira Didenkulova, Tallinn University of Technology, Tallinn, Estonia
Jo De Waele, University of Bologna, Bologna, Italy
João Carlos Nunes, University of Azores and INOVA, Ponta Delgada, Portugal
Johnny Douvinet, University of Avignon, Avignon, France
José Luís Zêzere, University of Lisbon, Institute of Geography and Spatial Planning (IGOT), Lisbon, Portugal
José Martins Carvalho, School of Engineering (ISEP), Polytechnic of Porto, Porto, Portugal
José Teixeira, Faculty of Arts, University of Porto, Porto, Portugal
José Virgílio Cruz, University of Azores, Ponta Delgada, Portugal
Luciano Lourenço, Faculty of Arts, University of Coimbra, Coimbra, Portugal
Luis González de Vallejo, University of Complutense of Madrid, Madrid, Spain
Luis Ribeiro e Sousa, University of Tongji, Shanghai, China
Manuel João Abrunhosa, IAH–Portuguese Chapter, Porto, Portugal
Marcela Pérez, National University of Litoral (FEHS), Santa Fe, Argentina
Marco Borga, University of Padova, Padova, Italy
Marco Petitta, La Sapienza University of Rome, Rome, Italy

Maria Teresa Brunetti, Research Institute for Geo-Hydrological Protection (IRPI-CNR), Perugia, Italy
Maurizio Barbieri, La Sapienza University of Rome, Rome, Italy
Moncho Gómez-Gesteira, University of Vigo and EPhysLAB, Ourense, Spain
Nabil Khélifi, Springer, Heidelberg, Germany
Okke Batelaan, Flinders University, Adelaide, Australia
Ozgur Kisi, School of Natural Sciences and Engineering, Ilia State University, Tbilisi, Georgia
Paulo Diogo, Faculty of Sciences and Technology, New Lisbon University, Lisbon, Portugal
Romeu Vicente, University of Aveiro, Aveiro, Portugal
Rosário Carvalho, Faculty of Sciences, University of Lisbon, Lisbon, Portugal
Rui Coutinho, University of Azores, Ponta Delgada, Portugal
Sérgio Lourenço, University of Hong Kong, Hong Kong SAR
Tew-Fik Mahdi, Montréal Technological University, Montréal, Canada
Tiago Abreu, School of Engineering (ISEP), Polytechnic of Porto, Porto, Portugal
Tommaso Caloiero, Institute for Agricultural and Forest Systems in the Mediterranean (ISAFOM | CNR), Rende, Italy

Foreword by José Virgílio Cruz

This special volume of the *Advances in Science, Technology & Innovation* series introduces a set of proceedings that draw on natural hazards and hydrological risks, selected from communications presented at the 2nd International Workshop on Natural Hazards (NATHAZ'19) that took place in Pico Island (Azores, Portugal).

The specific location of the conference, in one of the nine islands from the Azores, a volcanic archipelago with a complex geodynamic setting which explains the ongoing seismo-volcanic activity, historically described since the settlement in the fifteenth century, provided a perfect scenery for the meeting. In fact, addressing hazards in islands such as Pico, where a set of natural hazards have been described, from volcanic eruptions and earthquakes to flash floods, landslides, coastal erosion and aquifer salinization processes, supply several good examples of the interaction between natural hazards and society. It is noted that, besides the island being relatively small and having a low population, Pico attracts an increasing number of tourists to their environmentally fragile territory, increasing vulnerability to natural hazards.

Furthermore, climate change scenarios suggest that in the Azores the annual precipitation will be higher and rainy events will be more concentrated over time during winter, while in summer drier conditions will be intensified. Thus, the emergence of these climatic conditions poses new challenges in the forthcoming future, as a higher number of flash floods events are to be expected, favoured by the hydrological and geomorphological characteristics associated to much more intense precipitation. Several other examples could also be drawn from the climate change implications to natural hazards in islands, such as, among others, coastal erosion or groundwater salinization.

The volume offers a vivid overview of the current research being made worldwide on natural hazards, a subject with societal impacts of great importance. The set of proceedings presented reflect a multidisciplinary approach to the workshop subject, comprising more than thirty proceedings authored by researchers from several countries.

A first chapter presents several studies on hydrological hazards, including their relationship with hydrogeomorphology, landslides and geochemical processes in aquifers, namely groundwater contamination. A second chapter addresses hazard assessment and the water role, mainly focused on slope instability, landslides and rockfalls. Nevertheless, these proceedings also include studies on other natural hazards, such as subsidence and forest fires, as well as their relationship with land use planning and climate change impacts. A third chapter contributes with several case studies of flood events and coastal processes, as well as engineering approaches to natural hazards characterization and adaptation measures. Finally, the last chapter includes an interesting guide related to the field trip.

The overall high quality of the proceedings highlights the importance of the present volume that will be a cornerstone in what regards natural hazards research and the application of their findings towards vulnerability reduction and to civil protection enhancement. Thus, students, researchers and practitioners from several fields, from the earth and environmental sciences to engineering, will find the present volume of interest and a useful companion in their bookshelf.

June 2019

José Virgílio Cruz
Faculty of Sciences and Technology
Institute of Volcanology and Hazard Evaluation
University of the Azores
Ponta Delgada, Azores, Portugal

Foreword by Nabil Khélifi

Awareness of the need to incorporate disaster risk reduction and preparedness programmes into strategies and plans for sustainable development is increasing. These programmes include risk reduction, hazard mitigation and the lessening of vulnerability as we adapt to climate change. In the 17 UN Sustainable Development Goals (SDGs) outlined in the UN 2030 Agenda for Sustainable Development and the Sendai Framework for Disaster Risk Reduction 2015–2030, disaster risk reduction and resilience are high priorities. In 10 of these SDGs, 25 targets establish disaster risk reduction as a core development strategy involving many different sectors of development.

The scientific community has stepped up its research efforts on how to reduce the risks associated with disasters, how to build resilience in future and how to achieve the goals and targets set by both the Sendai Framework and the 2030 Agenda for Sustainable Development. In the latter, the scientific community recognizes and reaffirms the urgent need to reduce disaster risk through enhanced research activities and collaborations. These actions encourage and endorse increased political commitment and economic investment to reduce risks and take development initiatives in which disaster resilience is seen as fundamental to poverty reduction and as a key dimension of sustainable development.

Inspired by the UN's SDGs, Springer Nature's Grand Challenges Programme was launched in 2017 with an initial focus on five global challenges: climate change; global health; the food–energy–water nexus; a digitally transformed world; and sustainable cities. Now in its second year, the programme supports cutting-edge research in a wide range of disciplines, including science, engineering, social sciences and humanities. It helps practitioners to develop innovative and effective policies, programmes and technologies addressing global issues by connecting them with the latest evidence-based research. The programme has inspired collective action throughout Springer Nature and beyond. From new publishing strategies to initiatives connecting academic, business and policy leaders, this programme is part of a broad, concerted effort to tackle major social, environmental and economic challenges.

In Springer, we are inspired by and want to support the Sustainable Development Goals through Springer Nature. Several of our Springer Nature Grand Challenges speak specifically to these goals and we are working to bring our strength and expertise to those areas for the benefit of our planet and mankind. Publishing research work in this field is one of the ways we seek to complement existing regional and global actions to reduce the risk of disasters and our vulnerability to them.

This edited volume published by Springer is an important piece of scientific work which informs the efforts to reduce disaster risk and to build resilience as we strive to achieve sustainable development. It highlights the importance of undertaking hazard and risk assessment studies so that hazard mitigation can be achieved. Several case studies in this volume provide ideal examples of how the scientific research community can contribute to a more sustainable and safer environment by improving our understanding of natural hazards and how a "design with nature" concept can be followed to ensure that the way we occupy and modify the Earth is planned and designed to be compatible with nature, the environment and society.

I would like to thank the editors of this volume and the authors of the chapters, who participated in the 2nd International Workshop on Natural Hazards (NATHAZ'19), for their confidence by publishing the results of their work in this volume by Springer, the leading global publisher of academic books!

Heidelberg, Germany
June 2019

Nabil Khélifi
Senior Publishing Editor

Preface

Pico Mountain, with black clouds covering its tower and coming down the whole slope which finally looks out over the heights of Friar Matias' Cave...
Carlos Faria (Kinsella 2007)

Natural hazards are the consequence of a threat of a naturally occurring event will have a harmful outcome on persons and ecosystems. That damaging effect is often called natural disasters. That is defined by the UN—United Nations (IASC 2006) as: "the consequences of events triggered by natural hazards that overwhelm local response capacity and seriously affect the social and economic development of a region". UN (IASC 2006; Report of the Secretary General to the General Assembly 2005) highlights also the key challenges faced by international community as: "The risks and potential for disasters associated with natural hazards are largely shaped by the prevailing levels of vulnerability and the effectiveness of measures taken to prevent, mitigate and prepare for disasters". For all that the shared comprehensive knowledge is the key to understand the functioning of natural systems within climate change framework and outline guidelines and measures based on sustainable environments, socio-responsibility and ethical approach to achieve balanced and integrated management. In addition, as recently stated by the UN—United Nations (IPCC 2019; Intergovernmental Panel on Climate Change) "knowledge on risk is essential for conceiving and implementing adequate responses".

Among the natural hazards and potential disasters to be considered are earthquakes, volcanic eruptions, landslides, rockfalls, subsidence, floods, droughts and coastal erosion. In addition, anthropogenic hazards occur as a result of human interaction with the environment. They comprise technological hazards, which occur due to exposure to hazardous substances in the environment. Natural systems in different framework require a comprehensive understanding of climatology, geology and hydrology data and dynamics. Thus, it is important to perform hazard and risk assessment studies to accomplish hazard mitigation. Currently, it is vital to highlight the role of the variability and climate change in natural systems. Furthermore, an accurate understanding of the natural systems and interactions with engineering and natural resources has a vital significance to the entire socio-economic sector. That is the key landmark to achieve in any natural hazards project aiming at a sustainable design that is compatible with nature, environment and society (McHarg 1992; González de Vallejo 2010).

This volume offers an overview related to natural hazards in model regions in Asia, Europe, America and Atlantic islands. It gives new insights on characterization, assessment, protection, modelling on geological hazards, water systems, urban areas, coastal zones and engineering approaches by international researchers and professionals. Furthermore, the volume gives a general overview of current research and challenges focusing on natural hazard issues and its applications to a variety of problems worldwide but highlighting volcanic islands framework.

The volcanic origin of the Azores islands, associated with the geographic location, its archipelagic nature (nine dispersed and distant islands in Atlantic Ocean often subject to adverse weather conditions) and its geodynamic framework, make Azores highly vulnerable to natural hazards (Malheiro 2006; Malheiro and Nunes 2007; Malheiro et al. 2016; Kueppers and Beier 2018). Pico is the biggest island of the Central Azorean Group and shows the highest point of Portugal (2351 m), the Pico Mountain volcano, which is the third highest volcano of the North Atlantic Ocean (Kueppers and Beier 2018). In addition, the Azores archipelago is affected by most of the all-natural hazards, namely earthquakes, volcanic eruptions, landslides, floods, coastal erosion, damages in engineering works, etc. Some of these events have resulted in heavy damage to people and goods. Consequently, it was the perfect place where experts in these fields discussed and shared several case studies and vision aiming the advance of the natural hazard's knowledge. The workshop highlighted all aspects of natural hazards focusing in hydrological hazards and risks, including the forecasting of catastrophic events, risk assessment and management, as well as all aspects of water resources impacts, and oceanic, natural and technological hazards and disasters because all that it is extremely important the sharing of the results of 2nd International Workshop on Natural Hazards (NATHAZ'19) in an archipelago marked by natural hazards and, for that reason, with a strong geodynamic context that drives us to know more. In a sense, Azores is a real natural laboratory at the centre of the Atlantic, opened to the exchange of knowledge and experiences that enrich and prepare the region for a correct societal perception and a balance sustainable environmental practice. That approach will be another pillar of Azores socio-economic and more sustainable autonomous region within the European Union context. In fact, the impressive words of the outstanding Azorean writer, Vitorino Nemésio, are still topical: "For us, geography is just as important as history, and it is not without reason that 50% of our written memories comprise records of earthquakes and floods". (Nemésio 1932).

This book comprises the select proceedings during the 2nd International Workshop on Natural Hazards (NATHAZ'19), Lajes do Pico, Pico Island, Azores 2019. The challenges of balanced management and design with natural hazards are confirmed by the diversity of contributions to this special volume (Fig. 1). Main topics include: (i) hydrological hazards, hydrogeomorphology, groundwater and disasters; (ii) hazard assessment, spatial planning and climate change; (iii) natural hazards, hydrodynamics and engineering design; and (iv) around of Pico Island geology: meeting natural hazards.

The special volume has a core of 32 original proceedings grounded on the scientific-technical sessions (including the unforgettable field trip coordinated by João Carlos Nunes from University of Azores and INOVA) and three outstanding keynote lectures by leading experts, José Luís Zêzere (University of Lisbon, Portugal), Luís Ribeiro e Sousa (University of Tongji, China) and Giuseppe Sappa (La Sapienza University of Rome, Italy). The keynote speakers gave interesting insights from hydrological hazards focused on hydrogeomorphology and disasters, geotechnical hazards highlighting the role of water, as well as landslide risks and flooding hazards and hydraulic design. The volume gathered over 115 authors from the academy, research centres and/or state laboratories from Europe, Africa, America and Asia (Figs. 2, 3, 4).

The volume will be of interest to researchers and practitioners in the field of geosciences, hydrology, groundwater, natural hazards and geotechnics, as well as those engaged in sustainable environmental sciences, earth sciences, natural resources and engineering design. Graduate students, geoscientists, engineers and natural hazard-related professionals further research in the earth and environmental sciences will also find the book to be of value.

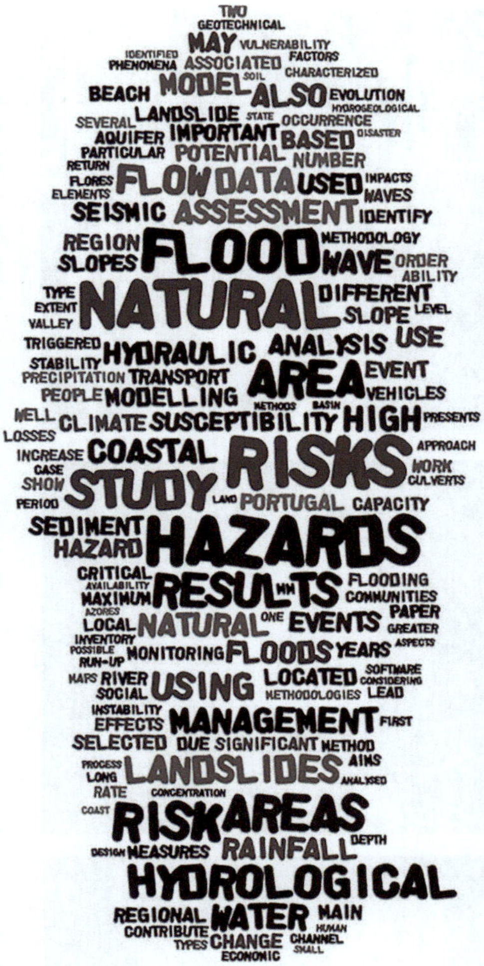

(generated by http://www.wordle.net/)

S. Miguel do Arcanjo rock fall (São Roque do Pico, Pico island Azores), june 2014 (Malheiro et al. 2016)

Fig. 1 Word cloud based on all abstracts of the special volume on "Advances in Natural Hazards and Hydrological Risks: Meeting the Challenge" (generated using http://www.wordle.net/). The image shows the recent rockfall occurred on the coastline cliff in the place of São Miguel Arcanjo, at São Roque Municipality (Pico Island, Azores). That event has affected a municipal road, the water and electricity supply system and has caused the relocation of 31 people (9 houses). The landslide occurred on a volcanic cliff. This natural hazard event took place in four stages: (i) occurred by September 2013; (ii) and (iii) on 13 June 2014; and (iv) the last happened 3 days later, on 16 June 2014. The rockfall process culminated with a global failure of a considerable segment of the cliff (ca. 100 high and 140 m length) and has triggered a retreat of the cliff of approximately 30 m (details in Malheiro et al. (2016))

Fig. 2 Overview during the sessions from the 2nd International Workshop on Natural Hazards (NATHAZ'19), Lajes do Pico, Azores (+ info: https://sites.google.com/civil-event.pt/nathaz19/homepage): opening ceremony by the President of Organizing Committee, Francisco Fernandes (LREC) and the Mayor of Lajes do Pico municipality, Roberto Silva; keynote lectures and presentations during the event

(Image: José Macedo; Design: LREC)

Fig. 3 Overview during the sessions from the 2nd International Workshop on Natural Hazards (NATHAZ'19), Lajes do Pico, Azores: presentations; a unique musical moment performed the director of Baleeiros Museum, Manuel Costa; Ana Malheiro presenting the NATHAZ'21 dedicated to Volcanic Risks and to be held in Terceira Island, Azores, 2021; closing ceremony by Scientific Committee Chairs, Ana Malheiro (LREC) and Helder I. Chaminé (ISEP)

Fig. 4 Overview during the field trip visits from the 2nd International Workshop on Natural Hazards (NATHAZ'19), Lajes do Pico, Azores, leaded by João Carlos Nunes (University of Azores and INOVA) and supported by the LREC colleagues Ana Malheiro, Francisco Fernandes, Filipe Marques, Letícia Moniz, Paulo Amaral, Roberto Dutra and Helena Brasil

The special volume comprises case studies that demonstrate the role of natural hazards understanding can contribute to a more sustainable and safer environment, also discusses the latest advances in natural hazards from diverse backgrounds, particularly highlighting the role of the variability and climate change in hydrological systems and, lastly, offers new insights on natural hazards mapping, characterization, assessment, protection and geoethics aiming a better knowledge and design with nature.

Ponta Delgada, Azores, Portugal Francisco Fernandes
Ponta Delgada, Azores, Portugal Ana Malheiro
Porto, Portugal Helder I. Chaminé
May 2019

References

González de Vallejo LI (2010) Design with geo-hazards: an integrated approach from engineering geological methods. Soils and Rocks. Int J Geot Geoenviron Eng 31(1):1–28

IASC (2006) Protecting persons affected by natural disasters. IASC Operational Guidelines on Human Rights and Natural Disasters, Washington, DC [https://www.preventionweb.net/files/1617_2006IASCNaturalDisasterGuidelines.pdf]

Kinsella JM (2007) Voices from the islands: an anthology of Azorean poetry. Gavéa-Brown Publications, Providence, Rhode Island

Kueppers U, Beier C (eds) (2018) Volcanoes of the Azores: revealing the geological secrets of the Central Northern Atlantic Islands. Springer, Berlin

Malheiro A (2006) Geological hazard in the Azores archipelago: volcanic terrain instability and human vulnerability. J Volcanol Geoth Res 156:158–171

Malheiro A, Amaral P, Marques F, Moniz L (2016) Instabilidade geomorfológica ocorrida em junho de 2014 no lugar de São Miguel Arcanjo (ilha do Pico). In Matos Fernandes M, Topa Gomes A, Couto Marques J et al (eds) Proceedings XV Congresso Nacional de Geotecnia / 8º Congresso Luso-Brasileiro de Geotecnia, FEUP, Porto, pp 1–12

Malheiro A, Nunes JC (eds) (2007) Volcanic rocks. Proceedings of the International Workshop on Volcanic Rocks, Workshop W2, 11th Congress ISRM, Ponta Delgada, Azores. Taylor & Francis Group, London, UK

McHarg IL (1992) Design with nature. 25th anniversary edition. Wiley series in sustainable design. Wiley, New York

Nemésio V (1932) Açorianidade. In Insula, Número Especial Comemorativo do V Centenário do Descobrimento dos Açores, nº 7–8, Julho-Agosto, Ponta Delgada, p 59

Report of the Secretary General to the General Assembly (2005) On international cooperation on humanitarian assistance in the field of natural disasters, from relief to development. Document A/60/227 [https://www.un.org/press/en/2005/ga10420.doc.htm]

UN—United Nations (2019) Summary for policymakers. In Pörtner HO, Roberts DC, Masson-Delmotte V et al (eds) IPCC Intergovernmental Panel on Climate Change, Special Report on the Ocean and Cryosphere in a Changing Climate. United Nations, WMO & UNEP, Washington, DC [https://www.ipcc.ch/srocc/download-report/]

Acknowledgements

Our appreciation is extended to the authors and keynote speakers of the NATHAZ'19 for their hardworking in producing high-quality contributions. We would like to thank the international scientific committee for their in-depth reviews and great efforts in improving the overall quality of the contributions. Also, thanks are extended to Letícia Moniz and Filipe Marques who carefully handled the submission and peer-review system for the special volume, as well as to Helena Brasil, Roberto Dutra and Paulo Amaral for the continuous support and to João Carlos Nunes for leading the scientific field trip in Pico Island. In addition, our thanks are to Whaling Museum, Lajes do Pico Municipality and São Roque do Pico Municipality for the full support to the event. Special thanks to Diogo Macedo (LabGeo-Azores) for kindly shared the impressive image used in front cover of the book and in Fig. 1 of the preface. Many thanks also to José V. Cruz and Nabil Khélifi for the insightful foreword. Finally, to all the participants for the resilience and will in coming to Pico Island.

The guest editors are grateful for the continuous and the enthusiastic support of publishing staff of Springer headed by Nabil Khélifi, Senior Editor, Reyhaneh Majidi, Editorial Assistant, and Springer production team, for their efforts in completing this proceedings volume. All the above-mentioned efforts were very significant in making this book a success. The volume is included in the celebration of the 40 years old of LREC—Regional Civil Engineering Laboratory, Azores.

Contents

Hazard Assessment, Spatial Planning and Climate Change

Natural Hazards, Hydrodynamics and Engineering Design

Around of Pico Island Geology: Meeting Natural Hazards

About the Editors

Francisco Fernandes is a skilled Civil Engineer, with over 22 years of experience in practice and research in civil engineering. After an internship at the "Sociedade de Empreitadas e Trabalhos Hidráulicos", he began his career in 1997 in budgeting and cost control at Engil. He was also a controller and responsible for several management areas at Eng. Luís Gomes company. He worked as a project manager in the Marques, Engil and Eng. Luís Gomes companies, and as director of the production.

Senior Member of the Institution of Portuguese Engineers has a master's degree in Business Administration and Management from the University of the Azores (2003), is Qualified Expert in the Energy Certification System (SCE–RCCTE), ADENE (Agency for Energy and Technical Superior of Safety and Hygiene of Work—Level 5). Other activities include participation in various courses, seminars and vocational training actions in the fields of engineering, safety and hygiene at work, energy certification and management, as a designer and qualified expert of the energy certification system.

Presently, since 2012 he is Head of the Regional Laboratory of Civil Engineering in Azores. Currently, he is a board member of the Portuguese Seismic Engineer Society (SPES) and a member of the Fiscal Council of the Fund for the Development of Construction Sciences.

As President of the Organizing Committee, he was responsible for several scientific events, namely the 10th Portuguese Congress of Seismology and Earthquake Engineering, the 16th Portuguese Geotechnical Congress, the 6th Luso-Spanish Geotechnical Conference and the 1st and 2nd International Workshop on Natural Hazards. Currently, is the chair of the organizing committee of the 3rd International Workshop on Natural Hazards (Volcanological Risks)—NATHAZ'21 (Terceira Island, Azores, May 2021).

Ana Malheiro is a skilled Geologist with 34 years of experience in practice and research in engineering geology, geotechnics and geological hazards. She studied Geology at the Faculty Sciences of the University of Lisbon in 1983 and received her master's degree at the University of Azores in Volcanology and Geological Risks in 2002.

She works since 1984 in the Regional Laboratory of Civil Engineering where she responsible for the areas of engineering geology, in situ geological and geotechnical investigations, geotechnics and geological risks. She was Head of the Division of Ports (1995/1998), and since 1998 until now, she is Director of Services of the Department of Geotechnics and Exploration. She participated in several Coordination Commissions for Management Plans (namely Regional Plans and Municipal Plans). In addition, was also invited assistant lecturer (1985/1990; 1996/1999) at Azores University.

She participated in several organizing committees for National and International Congresses and Workshops, as well as Scientific Commissions (e.g. 15th World Conference on Seismic Engineering, Lisbon, 2012; 5th, 9th and 10th National Congress of Seismology and Seismic Engineering, respectively in 2001, 2014 and 2016; 16th National Congress of Geotechnics, 2018 2nd International Workshop "Natural Hazards—Hydrological Risks"). Currently, is the chair of the scientific committee of the 3rd International Workshop on Natural Hazards (Volcanological Risks)—NATHAZ'21 (Terceira Island, Azores, May 2021).

She has been co-authored several publications in indexed journals, conference proceedings/full papers, book chapters, technical, professional papers and technical reports. She co-edited the ISRM special volume *Volcanic Rocks* (Taylor & Francis). In general, technical reports they focus on the field of in situ geological-geotechnical investigations to support the larger constructions built in the Azores region, coastal erosion, slope stability; as well as, the evaluation of resources in quarries, evaluation of geological and/or geotechnical problems and the assessment of geological hazards. She has integrated some research projects in the fields of geotechnical characterization of volcanic materials, stability of slopes, monitoring of unstable areas, etc.

Helder I. Chaminé is a skilled Geologist and Professor of Engineering Geosciences at School of Engineering (ISEP) of the Polytechnic of Porto, with over 29 years' experience in multidisciplinary geosciences research, consultancy and practice. He studied geological engineering and geology (B.Sc., 1990) at the Universities of Aveiro and Porto (Portugal), respectively. He received his Ph.D. in geology at the University of Porto in 2000 and spent his postdoctoral research in applied geosciences at the University of Aveiro (2001–2003). In 2011, he received his Habilitation (D.Sc.) in geosciences from Aveiro University.

Before joining the academy, he worked over a decade in international projects for mining, geotechnics and groundwater industry and/or academia related to geodynamics and regional geology, hard-rock hydrogeology and water resources, engineering geosciences and applied geomorphology, rock engineering and georesources. His research interests span over fundamental to applied fields: GIS mapping techniques for applied geology, structural geology and regional geology, engineering geosciences and rock engineering, slope geotechnics, mining geology and hydrogeomechanics, hard-rock hydrogeology, exploration hydrogeology, urban groundwater and hydromineral resources. He has interests in mining geoheritage, history of cartography, military geosciences and higher-education dissemination, skills and core values.

Presently, he is Head of the Laboratory of Cartography and Applied Geology (LABCARGA | ISEP), Senior Researcher at Centre GeoBioTec | U.Aveiro and Centre IDL | U.Lisbon, as well as belongs to the executive board of the M.Sc./B.Sc. Geotechnical and Geoenvironmental Engineering (OE +EUR-ACE Label) and the Department of Geotechnical Engineering (ISEP). Currently, he belongs to the board of the Portuguese Geotechnical Society (SPG) and IAH—Portuguese Chapter. He was a board member of the APGeom—Portuguese Association of Geomorphologists (2009–2013). He was a consultant and or responsible over 70 projects of rock engineering, applied geology, hydrogeomechanics, slope geotechnics, mining geology, exploration hydrogeology, hard-rock hydrogeology, water resources, urban groundwater and applied mapping (Mozambique, Portugal and Spain).

He has been co-authored over 200 publications in indexed journals, conference proceedings/full papers, book chapters, technical and professional papers. He co-edited 13 special volumes, as well as is presently evolved in editing themed issues for three international journals (Geotechnical Research ICE, Springer Nature Applied Sciences, Water MDPI). He has a wide activity as a referee for several international journals. He served as invited Expert Evaluator of Bologna Geoscience programme for DGES (Portugal) and Scientific Projects Evaluation for NCST (Kazakhstan) and NRF|RISA (South Africa), as well as Coordinator of "Geology on Summer/Ciência Viva" programme at ISEP for geosciences dissemination. He has been also active with teaching and supervising of many Ph.D., M.Sc. and undergraduate students.

He has been on the editorial board, among others, of *Arabian Journal of Geosciences* (SSG+Springer), *Hydrogeology Journal* (IAH+Springer), *Geotechnical Research ICE* (UK), *Springer Nature Applied Sciences* (Springer), *Mediterranean*

Geoscience Reviews (Springer), *Geosciences MDPI*, *Revista Geotecnia* (Portugal/Brazil/Spain) and *Geología Aplicada a la Ingeniería y al Ambiente* (Argentina). Currently, is co-chair of the International Conference "Geoethics and Groundwater Management" (IAH+IAPG, Porto, May 2020) and in scientific committee of the 3rd International Workshop on Natural Hazards—NATHAZ'21 (Terceira Island, Azores, May 2021).

Hydrological Hazards, Hydrogeomorphology, Groundwater and Disasters

A Century and Half of Hydrogeomorphological Disasters in Mainland Portugal

José L. Zêzere, Susana S. Pereira, and Pedro P. Santos

Abstract

The Disaster database lists disastrous floods and landslides registered in mainland Portugal in the period 1865–2015, based on data collected from newspapers. The database includes 1950 hydrogeomorphological disaster cases that caused 1256 deaths, 14,884 evacuated people and 41,977 displaced people. Disastrous floods are clustered in the Lisbon region and the Tagus valley, in the Oporto region and the Douro valley, in the Coimbra region and the Mondego valley and along the Vouga valley. Disastrous landslides are clustered in the Lisbon region and the Douro valley. The period 1935–1969 registered the highest number of disastrous floods and landslides, whereas the last 45 years of the series (1970–2015) do not show any clear temporal trend. However, the 10-year moving average increased at the beginning of the twenty-first century, for floods and landslides, and it is apparent the increasing number of disaster events consisting of several cases, which may result from the increasing occurrence of rainfall extreme events related with climate change. The spatial and temporal trends observed on disastrous floods and landslides reflect the distribution of conditioning factors, the temporal incidence of triggering factors, but also the evolution of the exposure and the vulnerability of people, structures and infrastructures.

Keywords

Floods • Landslides • Disasters • Database • Mainland Portugal

J. L. Zêzere (✉) · S. S. Pereira · P. P. Santos
CEG—Institute of Geography and Spatial Planning, University of Lisbon, Lisbon, Portugal
e-mail: zezere@igot.ulisboa.pt

© Springer Nature Switzerland AG 2020
F. Fernandes et al. (eds.), *Advances in Natural Hazards and Hydrological Risks: Meeting the Challenge*, Advances in Science, Technology & Innovation,
https://doi.org/10.1007/978-3-030-34397-2_1

1 Introduction

Hydrogeomorphologic disasters are natural processes of hydrologic (flood, flash flood, urban flood) or geomorphologic (landslides and other slope mass movements) origin that produced harmful consequences like loss of life or injury, property damage, economic disruption or environmental degradation (Zêzere et al. 2014).

According to several sources (e.g. EM-DAT 2019), hydrogemorphological disasters registered an increasing trend during last decades worldwide, which has been related to the increasing frequency of meteorological extreme events associated to climate changes (IPCC 2012), but also to the population growth and the increasing exposure of people and assets (Zêzere 2007).

In this work, we present the Disaster database that lists disastrous floods and landslides registered in mainland Portugal for a long-lasting period (151 years).

2 Materials and Methods

The Disaster database contains both disaster cases and disaster events. A Disaster case is a unique location that was affected by a flood or landslide in a specific period of time, which generated fatalities, injuries or missing, evacuated or displaced people, independently of the number of affected people. A Disaster event is a set of Disaster cases sharing the same trigger, which can have a widespread spatial extension and a certain magnitude (Zêzere et al. 2014).

The information on hydrogeomorphological disasters was collected systematically by analyzing newspapers. We assume that, as a rule, the newspapers report those floods and landslides that generated human consequences in Portugal. A total of 17 newspaper titles were used to guarantee the largest temporal coverage, but also the best regional spatial distribution of the newspapers to cover the complete study area. For the period 1865–1907 only up to two newspapers were available, whereas the number of newspapers increases to four for the period 1907–1936. For the period 1936–2015 the number of surveyed newspaper ranges from nine to twelve.

Statistical tests have shown that the Disaster database can be considered complete for the period 1936–2015, whereas the incompleteness can be up to 42% for the period 1865–1935 (Zêzere et al. 2014).

3 Results and Discussion

The Disaster database includes 1950 hydrogeomorphological disaster cases (on average, 13 per year) that caused 1256 deaths, 14,884 evacuated people and 41,977 displaced people (Table 1). Most disasters were generated by floods (85%) that were responsible for 81% of total deaths, 94.5% of total evacuated people and 96.2% of total displaced people.

The geography of disastrous floods and landslides is shown in Fig. 1. Although widespread all over the country, disastrous flood cases are clustered in the Lisbon region and the Tagus valley, in the Oporto region and the Douro valley, in the Coimbra region and the Mondego valley and along the Vouga valley. On the contrary, landslides are typically constrained to the region located northward the Tagus valley, in association with the steepest slopes. The Lisbon region and the Douro valley are the most important clusters of landslide cases.

The yearly distribution of disastrous floods and landslides registered in mainland Portugal in the period 1865–2015 is shown in Fig. 2. The 10-year moving average is represented by the blue and red lines, for floods and landslides, respectively.

The first 70 years of the series (1865–1934) were characterized by a low incidence of floods and landslides generating adverse human consequences. During this period, the year 1909 was an outlier, associated to a single extreme

Table 1 Hydrogeomorphological disaster cases and their human consequences in mainland Portugal in the period 1865–2015

	Floods	Landslides	Total
Number of cases	1658	292	1950
Number of deaths	1015	241	1256
Number of missing people	71	23	94
Number of injured people	479	433	912
Number of evacuated people	14,061	823	14,884
Number of displaced people	40,365	1612	41,977

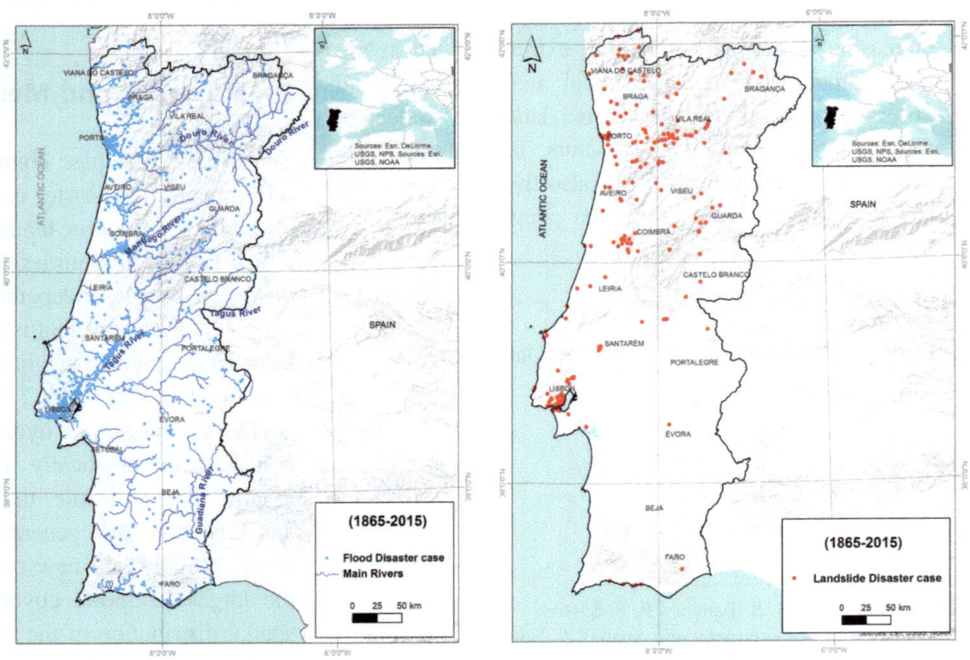

Fig. 1 Spatial distribution of disastrous floods (left) and landslides (right) occurred in mainland Portugal in the period 1865–2015

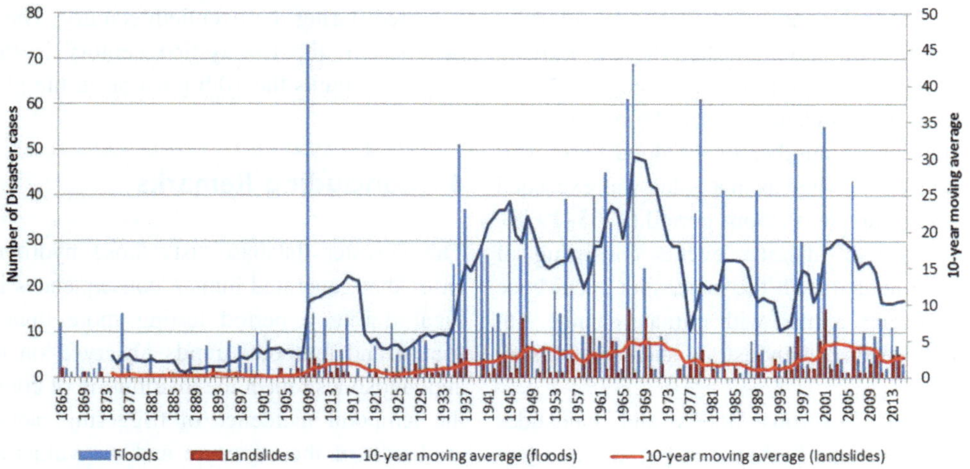

Fig. 2 Temporal distribution of disastrous floods and landslides occurred in mainland Portugal in the period 1865–2015

event occurred in December that affected the north and centre of Portugal. This event ranks the 2nd place in the top hydrogeomorphological disasters (Table 2) because it contains the highest number of flood and landslide cases in the Disaster database and generated 34 death people (Pereira et al. 2016).

The 35-year period lasting from 1935 to 1969 registered the highest number of disastrous floods (781, 47% of total flood cases) and landslides (133, 46% of total landslide cases). As a consequence, the 10-year moving average raised up to 30 and 5 at the end of the 1960 decade, for floods and landslides, respectively. This period includes the most catastrophic hydrogeomorphological disaster ever registered in mainland Portugal (Table 1), the flash flood affecting the Lisbon region in 25–26 November 1967 that killed more than 500 people (Trigo et al. 2016).

Table 2 Top 10 hydrogeomorphological disaster events occurred in mainland Portugal in the period 1865–2015

Rank	Event type	Date	Affected districts	Event duration (days)	#Disaster cases	Fatalities	Injured people	Evacuated people	Displaced people
1	FF; UF	25–26 Nov 1967	11, 15	2	67	522	330	304	885
2	F; FF; UF; L	20–28 Dec 1909	1, 3, 4, 5,6, 9,10,11,13, 14, 15,16,17,18	9	83	37	4	679	478
3	F; FF	15–17 Feb 1941	11, 15	3	6	33	0	109	0
4	F; L	9–12 Feb 1904	1, 3, 13, 17	4	4	27	1	1	3
5	F; FF	25–26 Nov 1865	11	2	9	21	0	0	0
6	FF; UF; L	18–19 Nov 1983	11, 14	2	37	18	0	255	3239
7	FF; UF	2–9 Nov 1997	2, 8, 14, 15	8	16	11	22	141	134
8	F; FF; L	5–16 Feb 1979	5, 6, 11, 13, 14, 17	12	67	8	3	4244	14,322
9	F; UF; L	2–6 Jan 1940	4, 7, 11, 14, 15, 17	5	26	7	3	35	1043
10	F; L	26–27 Jan 2001	1, 3, 6, 9, 10,13, 17, 18	2	28	6	5	402	570

Event type: F (flood); FF (flash flood); UF (urban flood); L (landslide)
Districts: 1—Aveiro; 2—Beja; 3—Braga; 4—Bragança; 5—Castelo Branco; 6—Coimbra; 7—Évora; 8—Faro; 9—Guarda; 10—Leiria; 11—Lisboa; 12—Portalegre; 13—Porto; 14—Santarém; 15—Setúbal; 16—Viana do Castelo; 17—Vila Real; 18—Viseu

The last 45 years of the series (1970–2015) do not show any clear temporal trend, although the 10-year moving average increased at the beginning of the twenty-first century, for both flood and landslide cases. Moreover, it is remarkable that the annual number of 40 flood cases was reached 6 times after 1978, whereas that value was exceeded in just 5 years in the complete previous period (1865–1978). The increasing number of disaster events consisting of several cases may be related with the increasing occurrence of rainfall extreme events related with climate change, but additional work is needed to demonstrate such relationship in the study area.

The seasonality of disastrous floods and landslides occurred in mainland Portugal is shown in Fig. 3. As it was expected, hydrogeomorphological events are uncommon during the driest period, from April to September. Floods are more frequent from November to February (75% of total flood cases), whereas landslides commonly occur from December to March (73% of total landslide cases). Flash floods and urban floods occur usually during the autumn and the early winter, in response to very intense rainfall events typically concentrated in just a few hours (Zêzere et al. 2014; Trigo et al. 2016). On the contrary, landslides are more common latter on the hydrologic year, being triggered by the rise of groundwater table, which is usually related to persistent rainfall periods lasting from several weeks to a few months (Zêzere et al. 2014). Such rainfall conditions are usually also responsible for floods along the major rivers (e.g. Tagus, Douro, Mondego).

The top 10 hydrogeomorphological disaster events occurred in mainland Portugal in the period 1865–2015 are summarized in Table 2 (Pereira et al. 2018). Besides the already mentioned 1967 and 1909 events that rank first and second, other relevant events were identified, most of them occurred during the twentieth century. The single top event occurred in the twenty-first century is the January 2001 event that ranks the 10th position in the global disaster list.

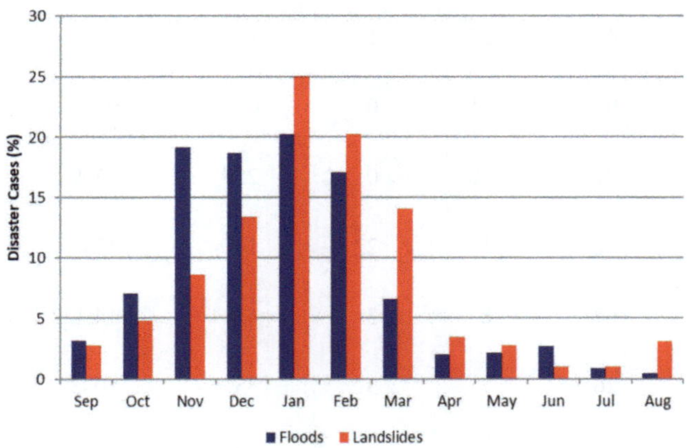

Fig. 3 Monthly distribution of disastrous floods and landslides occurred in mainland Portugal in the period 1865–2015

4 Concluding Remarks

The Disaster database lists those hydrogeomorphological cases that generated human consequences in mainland Portugal during a period lasting more than 150 years. The spatial and temporal trends observed on disastrous floods and landslides reflect the distribution of conditioning factors, the temporal incidence of triggering factors, but also the evolution of the exposure and the vulnerability of people, structures and infrastructures. Therefore, this database allows for the knowledge of the disaster drivers and their distinct incidence both in time and in space, which should be considered by stakeholders responsible for civil protection and spatial planning in order to manage and reduce disaster risk.

Acknowledgements This work was financed by national funds through FCT–Portuguese Foundation for Science and Technology, I.P., under the framework of the project FORLAND—Hydrogeomorphologic risk in Portugal: driving forces and application for land use planning [PTDC/ATPGEO/1660/2014].

References

EM-DAT (2019) The OFDA/CRED International disaster database—www.emdat.be—Université Catholique de Louvain, Brussels, Belgium. Last accessed 23 Mar 2019

IPCC (2012) Managing the risks of extreme events and disasters to advance climate change adaptation. A special report of working groups I and II of the intergovernmental panel on climate change. Cambridge University Press, Cambridge, UK, and New York, NY, USA

Pereira S, Ramos AM, Zêzere JL, Trigo RM, Vaquero JM (2016) Spatial impact and triggering conditions of the exceptional hydro-geomorphological event of December 1909 in Iberia. Nat Hazards Earth Syst Sci 16:371–390

Pereira S, Ramos AM, Rebelo L, Trigo RM, Zêzere JL (2018) A centennial catalogue of hydro-geomorphological events and their atmospheric forcing. Adv Water Resour 122:98–112

Trigo R, Ramos C, Pereira S, Ramos A, Zêzere JL (2016) The deadliest storm of the 20th century striking Portugal: flood impacts and atmospheric circulation. J Hydrol 541(A):597–610

Zêzere JL (2007) Riscos e Ordenamento do Território. Inforgeo 20 (21):59–63

Zêzere JL, Pereira S, Tavares AO, Bateira C, Trigo RM, Quaresma I, Santos PP, Santos M, Verde J (2014) DISASTER: a GIS database on hydro-geomorphologic disasters in Portugal. Nat Hazards 72:503–532

Hazard Analysis of Hydrometeorological Concatenated Processes in the Colombian Andes

Edier Aristizábal, María Isabel Arango Carmona, Federico José Gómez, Sandra Milena López Castro, Alfredo De Villeros Severiche, and Andrés Felipe Riaño Quintanilla

Abstract

The Colombian Andean region is characterized for its tropical climate and mountainous topography, where common short-duration and high-intensity rainfall events trigger progressive hydrometeorological phenomena that cause rapid concentration of streamflow as flash floods which, as their erosion capacity increases, may remobilize sediments and trigger slides creating debris floods torrents. When rainfall events exceed critical slope stability thresholds, clusters of landslides are triggered, especially on coarse residual soils increasing the sediment concentration and altering the fluid properties into a viscous mass surge of water and sediments. The final deposition of these materials is often located in basin fans that are usually highly populated resulting in huge disasters with high economic losses and fatalities. The most recent reports in Colombia include the events of Salgar in May of 2015 with 104 fatalities, Mocoa in March of 2017 with 400 people died. In this study, a regional hazard assessment methodology is proposed for flash floods, debris floods and channelized debris flows, considering the spatiotemporal variables, basin morphometry, sediment availability and rainfall data using statistical and physical methodologies to assess the susceptibility and hazard at a basin scale, resulting in a key tool for decision-making authorities.

Keywords

Colombian andes • Landslides • Debris flows • Flash floods • Debris floods • Hazard assessment

1 Introduction

Over the last few decades, human and economic losses from hydrometeorological hazards have grown dramatically and have occurred more frequently (Balbi et al. 2013; Changnon et al. 2000; Guha-Sapir et al. 2016; IPCC 2012). This risk scenario will increase in terms of human and economic losses, as a consequence of the more common occurrence of higher intensity of extreme weather events associated to global warming (Easterling 2000; Huntington 2006; Morss et al. 2011), and the increasing of population and property at risk from hydrometeorological hazards (Changnon et al. 2000; Hoppe and Pielke 2006; Morss et al. 2011).

Rainstorms in small and mountainous catchments generate a variety of flows involving a large volume of water and sediments in different concentrations along torrential streams (Ancey 2001). Stiny (1910) originally defined debris flow in a very general way as a flood in a mountain torrent that changed as the amount of sediment carried by the flow increased. In this sense, some authors (Coussot and Meunier 1996; Hungr et al. 2001; Hungr et al. 2014; Jakob and Hungr 2005) prefer to use the term debris flow to cover all this wide range of flows, where several types of flow-like landslides are included. According to these definitions, debris flow represents the entire process from the slope failure, to the transformation of shallow debris slides into debris avalanches, the transport of the rapid flow along steep confined channels, which correspond to channelized debris flows, and finally the deposition zone as debris floods.

In Colombia, according to the publicly accessible local disaster database DesInventar (http://www.desinventar.org/), 1.139 torrential flows have been registered from 1914 to 2015, which represent only 4% of the entire reports, but 2.195 deaths, which correspond to 52% of the flood-related casualties, showing its highly hazardous and destructive nature. Hazard from torrential flows is associated to high velocities, sediment content and size, and long-run out. The occurrence of these phenomena is associated to intense

E. Aristizábal (✉) · M. I. Arango Carmona · F. J. Gómez · S. M. López Castro · A. De Villeros Severiche · A. F. Riaño Quintanilla
Departamento de Geociencias y Medio Ambiente, Universidad Nacional de Colombia, Medellín, Colombia
e-mail: evaristizabalg@unal.edu.co

© Springer Nature Switzerland AG 2020

F. Fernandes et al. (eds.), *Advances in Natural Hazards and Hydrological Risks: Meeting the Challenge*, Advances in Science, Technology & Innovation,
https://doi.org/10.1007/978-3-030-34397-2_2

short-duration rainstorms in small and steep catchments typically from 0,1 to 100 km^2, with an optimum size range from 0,1 to 10 km^2 for Mizuyama (1982) and 25 km^2 for Rickenmann and Hunzinger (2008); and channel gradient ranges from 1 to 6% for Ancey (2001) and Aulitzky (1980), and larger than 5–10% for Rickenmann and Hunzinger (2008), which produces reduced evacuation time and prevent to take actions to reduce human and infrastructure losses (Liu and Lei 2003).

In this study the torrential flow hazard assessment in the northern Colombian Andes was carried out considering (i) susceptibility, which points out the spatial and morphometric catchment analysis, (ii) the temporal analysis of rainfall as triggering factor, and (iii) the magnitude analysis of the phenomenon according to the sediment supply. The results show the high hazard conditions of most of the catchment in the Colombian Andes due to the mountainous terrains and tropical conditions.

2 Materials and Methods

Hazard assessment was carried out by its three main components: (i) susceptibility analysis, (ii) temporal probability estimation of the triggering factor, and (iii) magnitude evaluation of the event.

Catchment morphometric control hydrologic and hydraulic torrential response. In this study, a discriminant multivariate statistical analysis, the Fisher analysis (Fisher 1940), was carried out to determine which morphometric parameters are critical to torrential flow occurrence and discriminate between alluvial and torrential catchments in the Colombian Andean. For this purpose, morphometric data was collected from 31 catchments where torrential flows have taken place in the Andean Colombian region between 1950 and 2017. Additionally, 42 catchments were chosen where there is no history of torrential flows but have similar topography and environmental conditions. More than 30 morphometric indexes were calculated for both group of catchments based on ALOS PALSAR 12.5 m DEM. In order to define the torrential flow type for each catchment, the geological map elaborated by the Colombian Geological Survey (SGC) with scales from 1:100,000 to 1:400,000 was used. Lithologic units from this cartography were categorized into coarse-grained or fine-grained soils according to the most extended mineralogy.

Temporal probability (FT) is computed from the occurrence probability of a rainfall threshold (P_t) and the conditional probability of occurring a torrential flow given the occurrence of the rainfall for a certain time $\left(P_{tf}|P_t\right)$ (Düzgün 2008; Düzgün and Grimstad 2007; Erener and Düzgün 2012) as it follows:

$$\mathrm{FT} = P_t \times \left(P_{tf}|P_t\right) \times \left(P_{\mathrm{landslide}}|P_t\right) \tag{1}$$

Then, in order to find the spatial distribution of the torrential flow probability, first the rainfall probability P_t is calculated and then multiplied with the conditional probability of having a torrential flow with the given rainfall $\left(P_{tf}|P_t\right)$ and the conditional probability of having a landslide with the given rainfall ($P_{\mathrm{landslide}} \mid P_t$). The first one is obtained from CHIRPS database as explained later, and the torrential flow and landslide conditional probabilities are obtained from torrential flows and landslide inventories from the area correlated with rainfall database through logistic regression. Then, debris flood and channelized debris flows probabilities are calculated from Eq. (1). For debris floods only comes into the calculation the rainfall and torrential flow conditional probabilities and for channelized debris flows the landslide conditional probability is also considered.

The CHIRPS (Climate Hazard Group Infrared Precipitation with Stations) database (Funk et al. 2015) was implemented to estimate the occurrence probability of three rainfall thresholds P_t 50,100 and 150 mm/h. The occurrence probability of a rainfall threshold (P_t) is then calculated with the free statistical software *R* using raster and rgdal libraries. In order to estimate the conditional probability of a torrential flow given the rainstorm thresholds, $\left(P_{tf}|P_t\right)$ a rainfall-event correlation analysis is accomplished using logistic regression (Glade et al. 2000; Guo et al. 2013). These methodologies establish a relationship between inventory events and historical daily rainfall data records.

The torrential flow magnitude was estimated according to the potential volume of sediments that could be included in the channels using well known physically-based model SHALSTAB. Geotechnical parameters for all lithological units in the study area were estimated from an international database called GeoTech Data (http://www.geotechdata.info/). This database compiles geotechnical studies and proposes estimations of these parameters according to its USCS classification. Soil depth distribution was estimated using a model proposed by Catani et al. (2010).

3 Results

The result of Fisher analysis leaves five morphometric parameters that determine torrential flow susceptibility. Using these parameters, the entire Antioquia region was divided into catchments between 8 and 40 km^2 and according to the discriminant equations, each catchment was discriminated between torrential or alluvial regime.

Then, according to the granulometric distribution of soils, torrential catchments are further sub-classified by its sort of susceptibility. Catchments, where only fine-grained soils are found, are classified as flash or debris flood, and catchments were coarse-grained soils are found are classified as susceptible to channelized debris flows.

From rainfall analysis, it is found the spatial distribution of the probability that, for each pixel, a 50 mm/day-rainstorm is exceeded. The fitting models obtained from logistic regression analysis link occurrence probability from different precipitation values from which the intercept and coefficient are found and the probability equations are made for torrential flows and landslides being function of the triggering rainfall (x).

Later, catchments are assigned the maximum probability value found within and the reclassified according to natural breaks of the histogram of all catchments into three categories. Probability values range from 0 to 4.9×10^{-4} and as seen in the probability of the rainfall, the highest values are found on west, northeast and northwest and southeast the study area, meanwhile the lowest probabilities are found in the centre.

Magnitude values are referred to volume susceptible to failure with a 50-mm/day rainfall per area. Thus, volume values can be normalized per catchment area and can be compared. These values range from 0 to 2.15 m^3/m^2. Higher magnitude is related to coarse-grained soils in the centre and southeast.

Finally, temporal probability and magnitude in their respective categories are combined through the hazard matrix to obtain the final hazard for torrential flows.

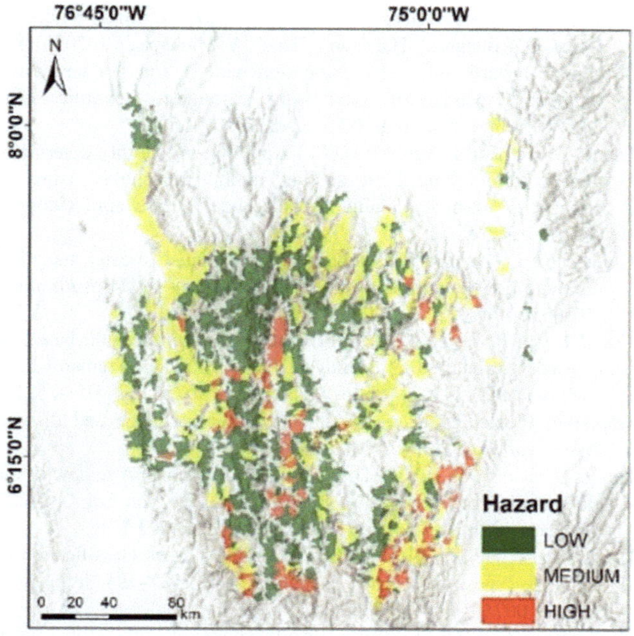

Fig. 1 Hazard assessment of the northern Colombian Andes

4 Concluding Remarks

The multi-hazard scenario of the Colombian Andean terrains permits a complex spatial and temporal hazard-triggering interaction. These complex processes generate different flow-like landslides triggering mechanisms, and in many cases increases the magnitude or impact of torrential flows. The most common and catastrophic triggering mechanism corresponds to the cascade effect by heavy rainfall. Short-duration and high-intensity rainfall events often trigger progressive hydrometeorological hazards caused by concatenated process.

A torrential flow hazard assessment in the northern Colombian Andes was carried out considering a spatial and morphometric catchment analysis, the temporal analysis of rainfall as triggering factor, and the magnitude analysis of the phenomenon according to the sediment supply. The results indicate that the considerable hazard scenarios of most of the catchment in the Colombian Andes. Figure 1 shows the hazard assessment of the northern Colombian Andes, where 53% of them in low levels; 32% in moderate and the final 15% in high scenario of hazard.

References

Ancey C (2001) Debris flows and related phenomena. Rapport Interne 1–21

Aulitzky H (1980) Preliminary two-fold classification of torrents

Balbi S, Giupponi C, Olschewski R, Mojtahed V (2013) The economics of hydro-meteorological disasters: approaching the estimation of the total costs, (August). http://dx.doi.org/10.2139/ssrn.2317437

Catani F, Segoni S, Falorni G (2010) An empirical geomorphology-based approach to the spatial prediction of soil thickness at catchment scale. Water Resour Res 46(5):1–15. https://doi.org/10.1029/2008WR007450

Changnon SA, Jr RAP, Changnon D, Sylves RT, Pulwarty R (2000) Human factors explain the increased losses from weather and climate extremes. Bull Am Meteorol Soc 81(3):437–442. https://doi.org/10.1175/1520-0477(2000)081%3c0437:HFETIL%3e2.3.CO;2

Coussot P, Meunier M (1996) Recognition, classification and mechanical description of debris flows. Earth Sci Rev 40(3–4):209–227. https://doi.org/10.1016/0012-8252(95)00065-8

Düzgün HSB (2008) A quantitative risk assessment framework for rock slides. In: 42nd US Rock Mechanics Symposium and 2nd U.S. Canada Rock Mechanics Symposium, (January 2008), 9

Düzgün HSB, Grimstad S (2007) Reliability-based stability analysis and risk assessment for rock slides in Ramnefjell. In: Proceeding of Applications and Statistics and Probability in Civil Engineering, (ICASP10), 189–198

Easterling DR (2000) Climate extremes: observations, modeling, and impacts. Science 289(5487):2068–2074. https://doi.org/10.1126/science.289.5487.2068

Erener A, Düzgün HSB (2012) Landslide susceptibility assessment: what are the effects of mapping unit and mapping method? Environ Earth Sci 66(3):859–877. https://doi.org/10.1007/s12665-011-1297-0

Fisher RA (1940) The precision of discriminant functions. Ann Eugen, Lond 10:422–429. https://doi.org/10.1111/j.1469-1809.1940.tb02264.x

Funk C, Peterson P, Landsfeld M, Pedreros D, Verdin J, Shukla S, Husak G, Rowland J, Harrison L, Hoell A, Michaelsen J (2015) The climate hazards infrared precipitation with stations—a new environmental record for monitoring extremes. Scientific Data 2:150066. https://doi.org/10.1038/sdata.2015.66

Glade T, Crozier M, Smith P (2000) Applying probability determination to refine landslide-triggering rainfall thresholds using an empirical "antecedent daily rainfall model". Pure appl Geophys 157(6–8):1059

Guha-Sapir D, Hoyois P, Below R (2016) Annual disaster statistical review 2015: the numbers and trends. CRED, 1–59. https://doi.org/10.1093/rof/rfs003

Guo XJ, Cui P, Li Y (2013) Debris flow warning threshold based on antecedent rainfall: a case study in Jiangjia Ravine, Yunnan China. J Mt Sci 10(2):305–314. https://doi.org/10.1007/s11629-013-2521-z

Hoppe P, Pielke RA Jr, (eds) (2006). Climate change and disaster losses workshop, (25–26 May), 234

Hungr O, Evans SG, Bovis MJ, Hutchinson JN (2001) A review of the classification of landslides of the flow type. Environ Eng Geosci 7 (3):221–238. https://doi.org/10.2113/gseegeosci.7.3.221

Hungr O, Leroueil S, Picarelli L (2014) The Varnes classification of landslide types, an update. Landslides 11(2):167–194. https://doi.org/10.1007/s10346-013-0436-y

Huntington TG (2006) Evidence for intensification of the global water cycle: review and synthesis. J Hydrol 319(1–4):83–95. https://doi.org/10.1016/j.jhydrol.2005.07.003

IPCC (2012) Managing the risks of extreme events and disasters to advance climate change adaptation: special report of the intergovernmental panel on climate change

Jakob M, Hungr O (2005) Debris-flow hazards and related phenomena. Springer (vol. 1)

Liu X, Lei J (2003) A method for assessing regional debris flow risk: an application in Zhaotong of Yunnan province (SW China). Geomorphology 52(3–4):181–191. https://doi.org/10.1016/S0169-555X(02)00242-8

Mizuyama T (1982) Analysis of sediment yield and transport data for erosion control works. Recent Dev Explan Predict Eros Sediment Yield, IAHS Publ No 137:177–182

Morss RE, Wilhelmi OV, Meehl GA, Dilling L (2011) Improving societal outcomes of extreme weather in a changing climate: an integrated perspective. Annu Rev Environ Resour 36(1):1–25. https://doi.org/10.1146/annurev-environ-060809-100145

Rickenmann D, Hunzinger LKA (2008) Hochwasser und Sedimenttransport w¨ahrend des Unwetters vom August 2005 in der Schweiz. In Conference proceedings, vol 1. Internat. Research Society Interpraevent (pp 465–476)

Stiny J (1910) Die Muren : Versuch einer Monographie mit bes. Berücksichtigung der Verhältnisse in den Tiroler Alpen

Statistical and Physically-Based Rainfall Triggered Landslides Susceptibility Assessment in the Tagus River Left Margin Basins, Almada County (Portugal)

Fernando Marques, Sónia Queiroz, Luís Gouveia, and Manuel Vasconcelos

Abstract

The Tagus River left margin slopes (Almada County, Portugal) are rainfall triggered landslide-prone areas, which are a serious threat to life and property. To improve knowledge on this topic, a study on landslide susceptibility assessment and mapping was carried out using simple statistical and physically based methods but using improved databases on geology and land use mapping. The statistically-based information value was used to relate an inventory of landslides with the landslides predisposing factors slope angle, lithology, aspect, curvatures, Topographic Wetness Index and land use. The physically-based methods were the SHALSTAB hydrological module coupled with an infinite slope stability model complete solution. The model results were validated against the landslides inventory using ROC curves and provided values of the Area Under the Curve for the Information value (0.90) higher than those of the infinite slope solution (0.80). These differences raise questions on the influence of database and model selection on the reliability of results. The used approaches, which require a careful analysis of the input data, while providing different results, had a satisfactory global performance, being adequate for the landslide susceptibility assessment at regional scale and relevant tools for hazard prevention.

Keywords

Landslides • Rainfall triggered • Susceptibility • Physical models • Statistical models • Database improvement

F. Marques (✉)
Departamento de Geologia and Instituto D. Luiz, Faculdade de Ciências, Universidade de Lisboa, Lisboa, Portugal
e-mail: fsmarques@fc.ul.pt

S. Queiroz · L. Gouveia
FCiências.ID, Universidade de Lisboa, Lisboa, Portugal

M. Vasconcelos
ESRI, Lisboa, Portugal

1 Introduction

Landslides are a source of major natural hazards and a serious constraint for the human activities in areas prone to the occurrence of these phenomena. In this study, carried out in a landslide-prone area located in the northern part of the Almada County (Portugal) (Marques et al. 2017), a preliminary shallow landslide susceptibility study was carried out using simple methods and focusing mainly on database improvement effects in results. The susceptibility assessment was made using the proven bivariate statistically-based Information Value method (Yin and Yan 1988) and a physically-based model which includes the hydrological module of SHALSTAB (Montgomery and Dietrich 1994) and a complete solution of the infinite slope stability method (Sharma 2002).

In the paper are presented the results obtained, validated using ROC curves and the corresponding area under the curve (AUC), and the questions raised by the influence of the database quality, including landslide inventory completeness and the susceptibility model selection.

2 Setting

The study area includes several small basins of the northern part of Almada County, located southwards of Lisbon, being limited at the north by the Tagus river estuary. The geological structure is dominated by the northern limb of the Albufeira syncline, which affected the Miocene formations that dip 4°–6° to SSE. The geological formations in the study area (DEGAS/CMA 2005) correspond to alternating layers of sands, weak limestones and clays (Fig. 1) deposited in fluvio-marine environment.

The morphology of the study area, with maximum elevation exceeding 120 m, is composed of successions of scarps cut in the weak limestone, separated by gentle slopes that correspond mainly to the outcrop of the weaker clayey

© Springer Nature Switzerland AG 2020

F. Fernandes et al. (eds.), *Advances in Natural Hazards and Hydrological Risks: Meeting the Challenge*, Advances in Science, Technology & Innovation,
https://doi.org/10.1007/978-3-030-34397-2_3

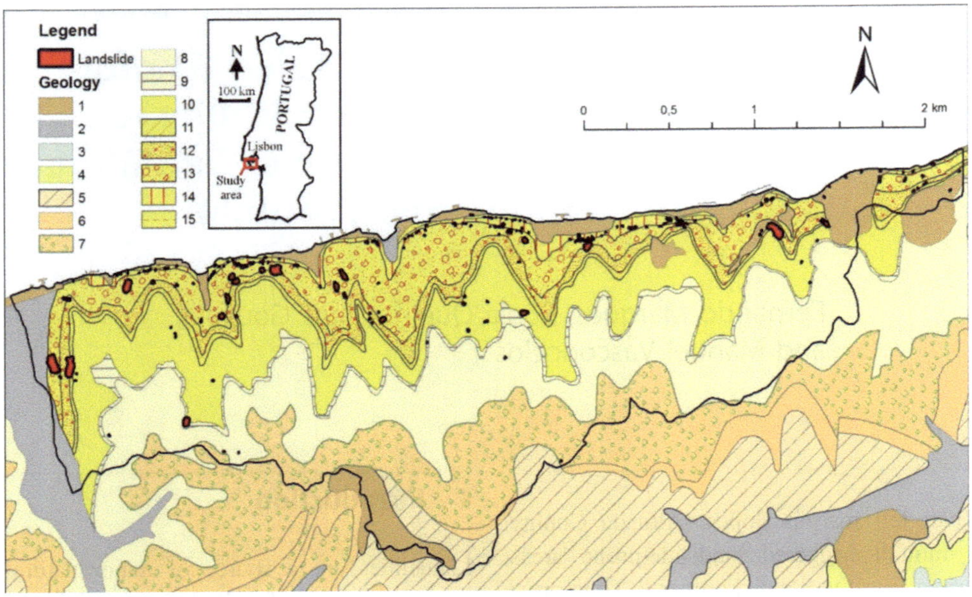

Fig. 1 Localization and geological map of the study area (geological unit local designations): 1—landfills; 2—alluvium; 3—Pleistocene gravel, sands, and clays; 4—Pliocene silty sands (3 and 4 do not occur in the study area). Miocene units: 5—fine silty sands (MVIIa + b); 6—sands and weak limestones (MVIb + c); 7—clays (MVia); 8—weak limestones and sands (MVb + c); 9—weak limestones (MVa3) 10—sands (MVa2); 11—weak limestones (MVa1); 12—sands (MIVb); 13—clays (MIVa); 14—weak limestones (MIII); 15—fine silty sands (MII). Landslide inventory (red dots) and study area limits (black line). Adapted from DEGAS/CMA (2005) with modifications

and sandy units. These slopes are deeply incised by short creeks with steep bed and flanks.

The study area was affected by 198 shallow landslides between 1942 and 2015 (Fig. 1), which were mapped by (a) Digital photogrammetric comparison of aerial photos of 1942 (1:17,000 scale) and 2010 (digital, field size pixel of 0.3 m); (b) 2011 aerial photos interpretation; (c) Interpretation of internet-based aerial imagery (Google Earth, Bing Maps); (d) interpretation of terrestrial long-range photos (e) Field surveys.

3 Methods

3.1 Statistically Based Landslide Susceptibility

The statistically-based slope instability assessment and mapping was performed using the bivariate Information Value (IV) method (Yin and Yan 1988), using the inventory of shallow landslides, a set of landslides predisposing factors (geology, land use, slope angle, exposure, plan, and profile curvature, Topographic Wetness Index) and a 5 m pixel terrain unit basis. In this method each factor is divided into classes, and each class corresponds to a variable. The information value Ii of each variable Xi is:

$$Ii = \log[(Si/Ni)/(S/N)] \quad (1)$$

where Si is the number of pixels with landslides in the pixels with the variable Xi, Ni is the number of pixels with the variable Xi, S is the total number of pixels with landslides, and N is the total number of pixels in the study area. The positives values of Ii indicate that the variable is prone to the occurrence of landslides, with the negative ones indicating low susceptibility. The total information value Ij for a given pixel j is:

$$Ij = \sum_{i=1}^{m} Xji.Ii \quad (2)$$

where m is the number of variables, Xji is 0 if the variable is absent in the pixel j, or 1 if the variable is present.

The predisposing factors used were: (a) Geology information extracted from the geological map of Almada (DEGAS/CMA 2005), with formation boundary lines corrections based on field surveys and aerial photo interpretation; the map includes 13 units (Fig. 1; units 3 and 4 do not occur in the study area). (b) Land use information which was derived from the vegetation map provided by the municipality, completed with the topographic information on structures and roads; the map includes the following units: roads; empty spaces in urban areas; buildings; tree areas; agricultural areas; shrub and herbaceous vegetation areas; shrub areas; herbaceous vegetation areas; beach sands; main infrastructures. (c) Morphometric information, which was derived by processing a detailed (1:1000 scale) and recent (2011) digital topographic map with contour lines separated by 1 m, also provided by the municipality. The maps included slope angle, classified in 6 classes with the following slope limits (in degrees): 0–10; 10–20; 20–30; 30–45; 45–60; and 60–75. Nine classes were considered in the slope exposure map: Flat; N; NE; E; SE; S; SW; W; NW. The plan and profile curvatures of the slopes were separated in five quantile-based classes: very concave; moderately concave; rectilinear; moderately convex; very convex. The Topographic Wetness Index was computed using the expression TWI $= \ln(a/\tan b)$, in which a is the local upslope area draining through a certain point and β is the slope angle (Beven and Kirkby 1979); they were considered the following quantile-based classes: 2.4–6.5; 6.5–7.3; 7.3–8; 8–9.1; and 9.1–20.7.

3.2 Physically Based Landslide Susceptibility

The physically-based model was based on Shalstab (Montgomery and Dietrich 1994) and was used a DTM derived from the topographic map, with a pixel of 5 m. Data on soil unit weights and permeability were obtained from literature, and soil strength parameters were obtained by back analysis of the landslides inventory in each geological unit, made using the infinite slope Sharma (2002) equation.

As calculations with the full Shalstab model provided unreliable results, the hydrological part (*a*/*b* ratio) of this model was retained, and the slope Safety Factor (SF) computing was made using the Sharma (2002) equation. Soil properties used for susceptibility assessment are summarized in Table 1.

4 Results and Discussion

The results obtained with the statistical and physical models are summarized in the corresponding receiver operating characteristics (ROC) curves and areas under the curve (AUC) (Figs. 2 and 3), which provided satisfactory values. The statistically-based method, despite being a simple one, provided a high adjustment to the input data, suggesting that the geological and land use base maps

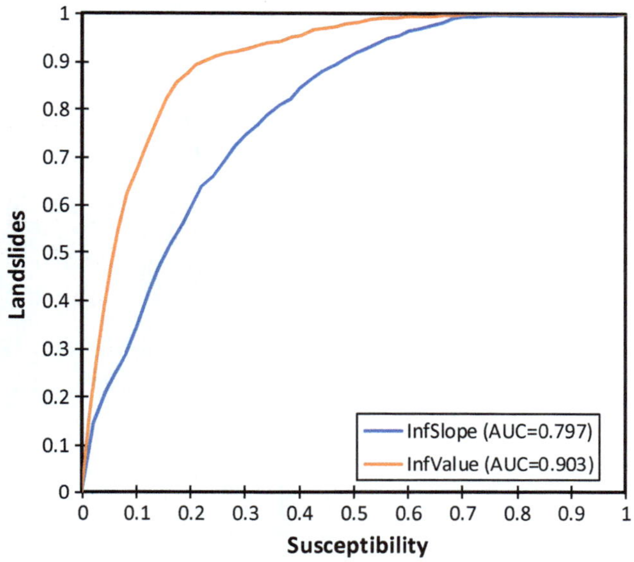

Fig. 2 ROC curves and corresponding AUCs for the information value and the infinite slope landslide susceptibility models

Table 1 Soil properties used for physically based landslide susceptibility assessment. (c'—effective cohesion; φ'—effective friction angle, γ—soil unit weights; k—permeability)

Geological unit	c' (kPa)	ϕ' (°)	γ (kN/m^3)	γ_{sat} (kN/m^3)	γ_{sub} (kN/m^3)	k (m/day)
Alluvium	11.0	24	16.00	18.81	9.00	0.864
Landfill	1.2	30	16.00	18.81	9.00	0.346
MVIIa + b	2.0	34	16.00	18.81	9.00	0.864
MVIb + c	2.0	34	16.00	18.81	9.00	0.864
MVIa	3.0	30	19.00	21.00	11.19	0.086
MVb + Vc	2.0	34	18.70	20.70	10.89	0.864
MVa3	2.0	34	23.00	25.00	15.19	0.605
MVa2	1.0	30	17.00	19.00	9.19	0.864
MVa1	2.0	30	21.00	23.00	13.19	0.605
MIVb	2.0	30	17.00	19.00	9.19	0.864
MIVa	1.0	26	19.00	21.00	11.19	0.086
MIII	1.5	34	21.00	23.00	13.19	0.605
MII	0.8	30	16.00	18.81	9.00	0.605

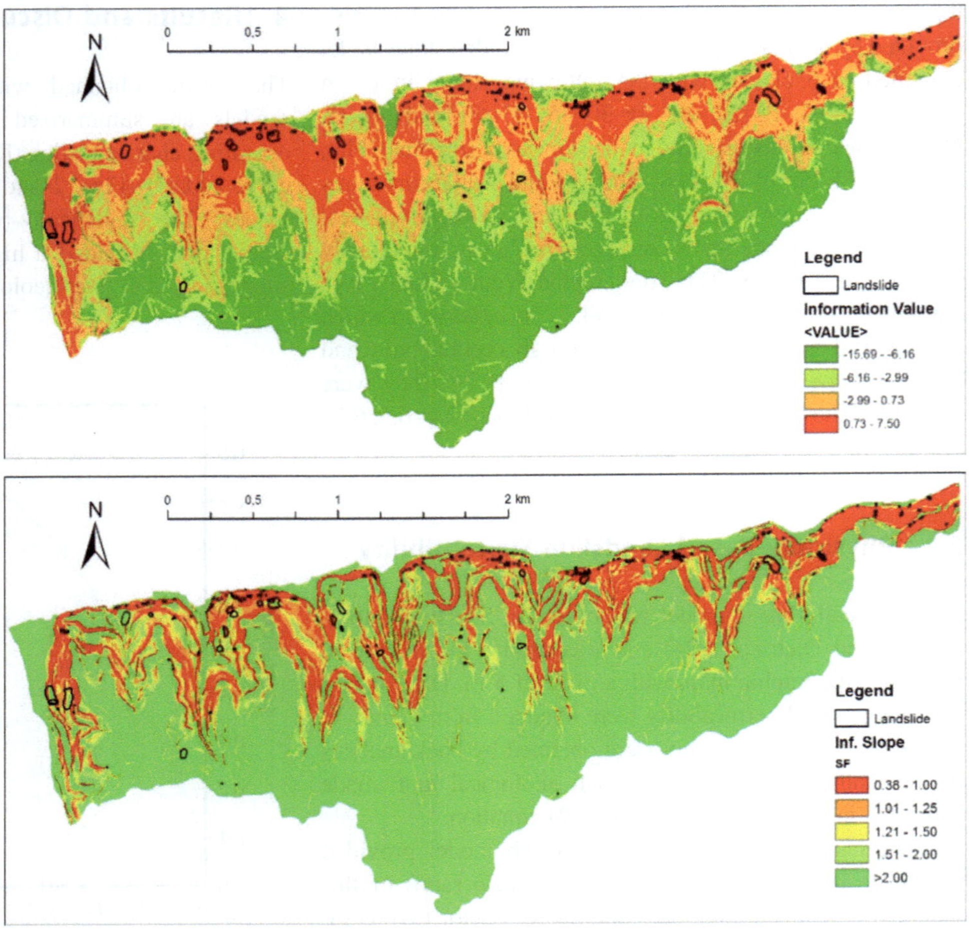

Fig. 3 Landslide susceptibility maps produced. Above, using information value method. Below, infinite slope safety factor map

improvement played a relevant role on the model reliability. The physically-based method provided a less good agreement with the landslide inventory, which may be due to a combination of factors: (1) limitations inherent to the hydrological module of Shalstab, which only considers lateral flow of water and all the rainwater infiltrating in the terrain and flowing freely across the basin, without taking into account the lateral permeability of the soils; (2) insufficient calibration of the strength parameters based on back analysis, which seem to require further adjustments; (3) partially inaccurate identification of lithological units for which upper soil properties may exhibit significant variations, which could not be separated due to the complex geological setting mainly composed by beds with strong lateral facies variations. It also must be noticed that the AUC values for the statistically-based method correspond to a success rate, while in the physically-based method, the AUC corresponds mainly to a prediction rate since the landslide inventory only contributed to the model for strength parameters in the back analysis.

5 Conclusions

The simple statistical and physically based landslide susceptibility methods used in this study provided useful results for the prevention of these hazardous phenomena in the northern part of Almada County. The Information Value method produced a high adjustment to the input landslide inventory, suggesting that the geological and land use base maps improvement may have played a significant role in these results. The physically-based model produced less successful results, probably due to a combination of factors mainly related to input soil strength data and the very crude hydrological model of Shalstab. The results obtained indicate the need to calibrate the model and apply more robust methods, especially in the physically-based approach.

Acknowledgements Publication supported by FCT- project UID/GEO/50019/2019—Instituto Dom Luiz and by FCT-project BeSafeSlide, PTDC/GES-AMB/30052/2017. Acknowledgements are due to the two anonymous reviewers which helped to improve the manuscript.

References

Beven KJ, Kirkby MJ (1979) A physically based, variable contributing area model of basin hydrology. Hydrol Sci Bull 24:43–69

DEGAS/CMA (2005) Geological map of almada county, 1:20,000 scale. Accessible at Almada county, department of energy, climate, environment and mobility, Almada, Portugal

Marques F, Queiroz S, Gouveia L, Vasconcelos M (2017) Slope, scarp and sea cliff instability susceptibility mapping for planning regulations in Almada County, Portugal. IOP Conference Series: Earth and Environmental Science 95:022022

Montgomery D, Dietrich WE (1994) A physically based model for the topographic control on shallow landsliding. Water Resour Res 30 (4):1153–1171

Sharma S (2002) Slope stability concepts. In Abramson LW, Lee TS, Sharma S, Boyce GM (eds) Slope stability and stabilization methods. John Wiley & Sons, New York, pp 329–461

Yin KL, Yan TZ (1988) Statistical prediction models for slope instability of metamorphosed rocks. In: Bonnard C (ed) Landslides, Proceeding of 5th International Symposium Landslides Lausanne, Switzerland, Balkema, Rotterdam, vol 2, pp 1269–1272

Aquifer Contamination by Coastal Floods in the Plain of Costa Da Caparica, Almada (Portugal)

Carolina A. Marques, Maria Rosário Carvalho, and Rui Taborda

Abstract

Climate change might lead to sea level rise, changes in the frequency, intensity and duration of wave storms, which will lead to an increase of coastal flooding events. One of the most important long-term effects of coastal floods is saltwater intrusion induced by the vertical infiltration of the salt water behind the overtopped and/or breached coastal barriers. A single overflow event may contaminate the freshwater aquifer for several years until it is remediated naturally under the effect of precipitation and subsequent seaward directed flow. The main objective of this study is to understand the effects of a maritime storm and the induced hinterland inundation on the water quality of a coastal aquifer. The study targets the municipality of Almada (Portugal) which has an extensive coastline, with coastal aquifers with high susceptibility to contamination, namely the unconfined aquifer of the coastal plain of Costa da Caparica. Groundwater flow modelling and mass transport in the aquifer were modelled using the MODFLOW and MT3DMS software. The aquifer contamination by an overflow was modelled considering wave overtopping and flooded area on the coast was estimated from previous works that coupled a wave transformation model (SWAN) with a high-resolution swash model (XBEACH). The extent of the subsurface contamination is a function of the flood extent. Results showed that at an extreme storm event, waves can overtop the coastal dune causing a coastal flood that extends approximately 160 meters inland with a significant increase in chlorine concentration in the aquifer. Recover of the aquifer to previous concentrations was found to take several years.

Keywords

Coastal flood • Coastal aquifer • Salinization • Numerical modelling • Mass transport

1 Introduction

Coastal aquifers are highly susceptible to several natural and anthropic related environmental changes such as sea-level rise, reduction of recharge, pollution or coastal floods by wave overwash. These latter processes can induce significant saltwater intrusion in freshwater aquifers (Elsayed and Oumeraci 2017). Contamination of groundwater by infiltration of seawater is a consequence that deserves attention.

In the coastal region of the municipality of Almada, there is an unconfined aquifer developed in sandy sediments that are exploited for irrigation purposes. This aquifer presents high susceptibility to contamination, especially in relation to the advance of the salt wedge, caused by sea-level rise and overexploitation. Coastal flooding events can also have a significative impact because they are fast and cover vast areas of infiltration. The salinization of the aquifer can undermine local economic activity linked to agriculture and the exploitation of groundwater in shallow wells.

The main objective of this study is to understand the effects of an extreme maritime storm and the associated hinterland inundation on the water quality of a coastal aquifer. The study targets the municipality of Almada (Portugal).

The municipality of Almada is located on the central coast of mainland Portugal, on the left margin of the Tagus River (NW region of the Setúbal Peninsula), Setúbal district. The plain of Costa da Caparica is located on the NW extreme of the municipality. It is limited by the Tagus River (and its estuary) at the north and at west by the Atlantic Ocean, and by a cliff at the east (Fig. 1).

The regional climate is marked by an oceanic regime with warm winters and hot summers, by proximity to the Atlantic

C. A. Marques (✉) · M. R. Carvalho · R. Taborda
Departamento de Geologia, Faculdade de Ciências, Universidade de Lisboa, Lisbon, Portugal
e-mail: carol.mia24@gmail.com

M. R. Carvalho · R. Taborda
IDL—Instituto D. Luiz, Universidade de Lisboa, Lisbon, Portugal

© Springer Nature Switzerland AG 2020

F. Fernandes et al. (eds.), *Advances in Natural Hazards and Hydrological Risks: Meeting the Challenge*, Advances in Science, Technology & Innovation,
https://doi.org/10.1007/978-3-030-34397-2_4

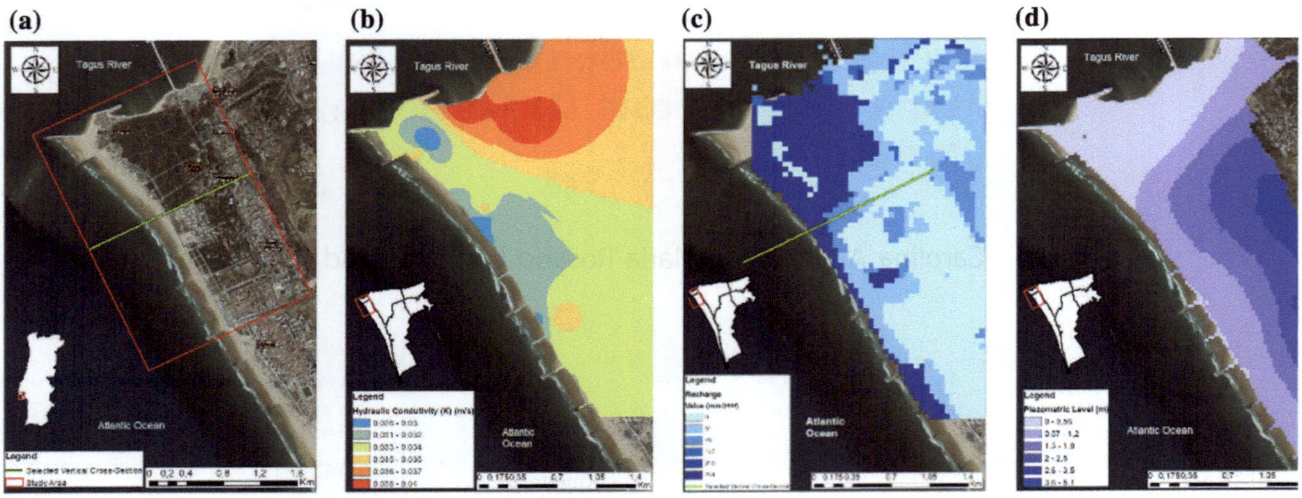

Fig. 1 Location and limits of the study area; spatial distribution of the hydraulic conductivity; variation of the direct recharge of the aquifer; piezometric map of the study area; (from left to right)

Ocean (and whose weak annual thermal amplitude allows regularization of temperature throughout the year). The annual average temperature is between 16 and 17.5 °C, and the average annual precipitation between 500 and 700 mm (in Caria et al. 2013). During the winter period, the plain of Costa da Caparica is recurrently affected by wave overtopping, with consequent flooding of inland areas.

The plain of Costa da Caparica is formed by deposits composed of dune sands, slope deposits and alluviums and/or landfills of Pliocene–Holocene (Pais et al. 2006). Those deposits cover sedimentary formations of the Miocene also represented on the eastern cliff. The Miocene formations correspond to continental deposits alternated by other marine ones, characteristic of a broad alluvial plain in the form of an estuary, subject to transgressions.

The aquifer developed in Pliocene–Holocene deposits is unconfined, with a thickness between 20–25 m. The hydraulic conductivity of the formations varies from 3.4 to 4.5 m/day, and Ferreira (2012) estimated average value of 704 m^2/day for the transmissivity. The aquifer receives direct recharge from precipitation, around 250 mm per year, and lateral recharge from the Miocene formations at east (Ferreira 2012) but the amount of the water transfer is not well known.

2 Materials and Methods

To achieve the objectives, the groundwater level and flow in the coastal aquifer were modelled using the MODFLOW software. The mass (chloride concentration) transport using the MT3D allowed to understand the aquifer contamination associated with floods during storm events. Both software are compiled by the Processing Modflow (PMWIN) software (Processing Modflow-version 5.3.1. Copyright©1991–2001 W.H.Chiang and W. Kinzelbach).

The aquifer was simulated in 3D and at steady-state regime, using a different element mesh and one unconfined layer with 20 m of thickness. The mesh is composed of 440 columns and 430 lines with variable cell width (10, 5 and 2.5 m) with higher resolution in the area affected by floods. The boundary conditions define the groundwater influx from the east, from the Miocene aquifers, and the constant head at seaside. The hydraulic conductivity of the layer was obtained by interpolation (inverse distance method) of values obtained through granulometric analysis of selected samples (equivalent diameter) and varies between 0.035 and 0.028 m/s (Fig. 1). Considering the sandy composition of the plain aquifer a porosity value of 0.25 was assumed. The recharge by precipitation was obtained from Caria (2012), Caria et al. (2013) and ranges from 0 to 254 mm per year (Fig. 1). The zero represents the impermeable zones (anthropic uses and actions), while the higher permeable zones are associated to agricultural land. The initial hydraulic heads, interpolated by kriging, are represented in Fig. 1; the water level varies from 4.43 m at east to close to 0 m (mean sea level).

The mass transport model was simulated in the transient regime for a particular section of the model (location in Fig. 1), representing the area flooded during the storm on 8–10 of February of 2014. The highest volume of flood was estimated in 0.2 L/m/s, coupling a wave transformation model (SWAN) with a high-resolution swash model (XBEACH), (Pires 2017). The extension of the flood was estimated in approximately 162 m inland, considering a wave overtopping of 600 s. The initial conditions for the chloride concentration were assumed as 60 mg/L for the regional groundwater, 5 mg/L for rainwater and 20,000 mg/L for seawater. The simulation was carried out in two steps: the first consider the infiltration of the seawater on the top of the layer, representing a flood of 1 day; the second is equivalent to the period of natural remediation, and the chloride concentration was simulated for different periods (stress periods), 1, 2 and 5 years, to understand how long the aquifer will take to recover.

3 Results and Discussion

3.1 Groundwater Flow Modelling

The aquifer numerical modeling shows that the groundwater flows from east to west into the sea, corroborating the results of Ferreira (2012) and Storm (2014). Figure 2 shows a cross-section of the simulated groundwater level in the aquifer, at steady-state conditions. It is possible to see that the infiltration of the seawater in the inundated area causes a rise in the groundwater level, which can be higher than the beach profile. In this simulation the piezometric level varies between 5 m at east and 0.16 m at the beach. The second simulation shows (Fig. 3) that the heads recover when the storm ends and return to the normal values, i.e., 0 m close to the sea.

3.2 Mass Transport Modelling

The mass transport modelling results, in the transient regime and two steps, are represented in Figs. 4 and 5. The first simulation corresponds to the flood period, with an inundation period of 1 day; the second concerns to the evaluation of the time the aquifer needs to recover from seawater contamination during the flood.

At an observation point in the aquifer(⊕), located at the limit of the flooded area, it is possible to see that the concentration simulated in the aquifer after 1 day of flood is 19,177 mg/L (stress period 1; Fig. 4).

The evolution of the chloride concentration in the aquifer over time, at the observation point, can be seen in Figs. 5 and 6 where is clear that the aquifer needs several years to recover the groundwater quality. At the time steps length of 1–5 years, the simulated chloride concentration at the observation point will be 318, 84 and 67 mg/L, respectively.

4 Concluding Remarks

The numerical simulation of the flow model and mass transport of the unconfined aquifer of the Costa da Caparica Plain shows that the coastal flood associated with extreme events can lead to the contamination of the aquifer over a long period of time.

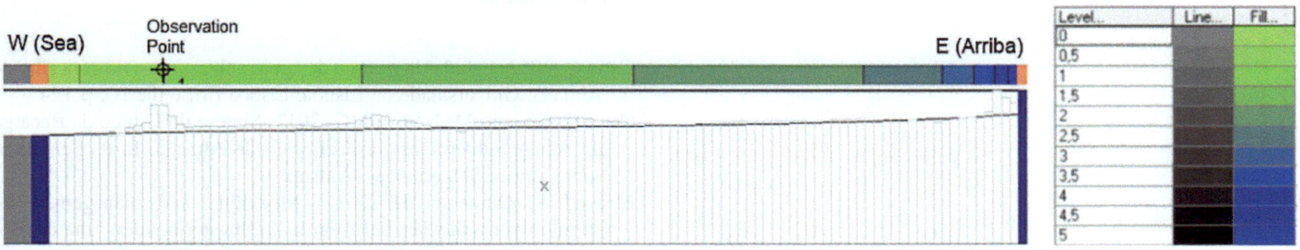

Fig. 2 Vertical section of the aquifer with the simulated piezometric level (dark line), when wave overtopping occurs on the study area (stress period 1)

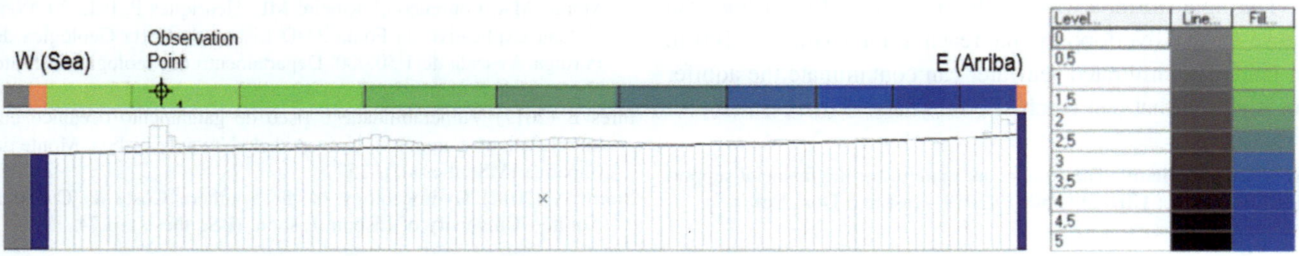

Fig. 3 Vertical section of the aquifer with the simulated piezometric level (dark line), after the storm (stress period 2)

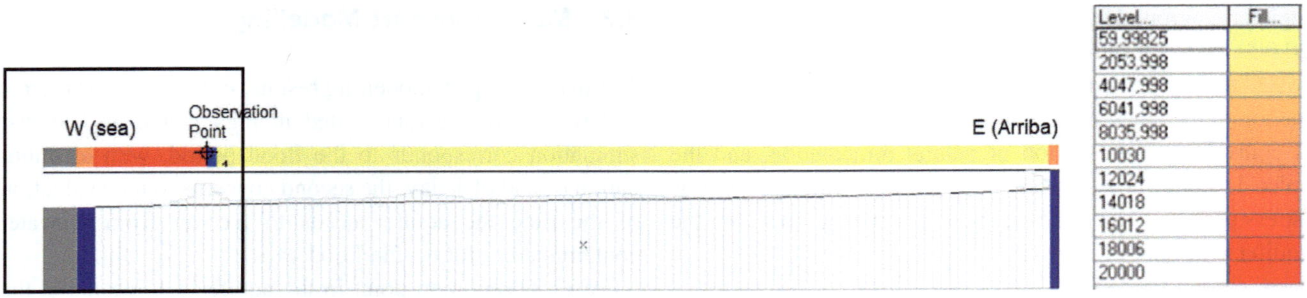

Fig. 4 Vertical section of the aquifer with the simulated chloride concentration in observation point after 1 day of flood; square represents the aquifer zone in Fig. 5; dark line is the groundwater level

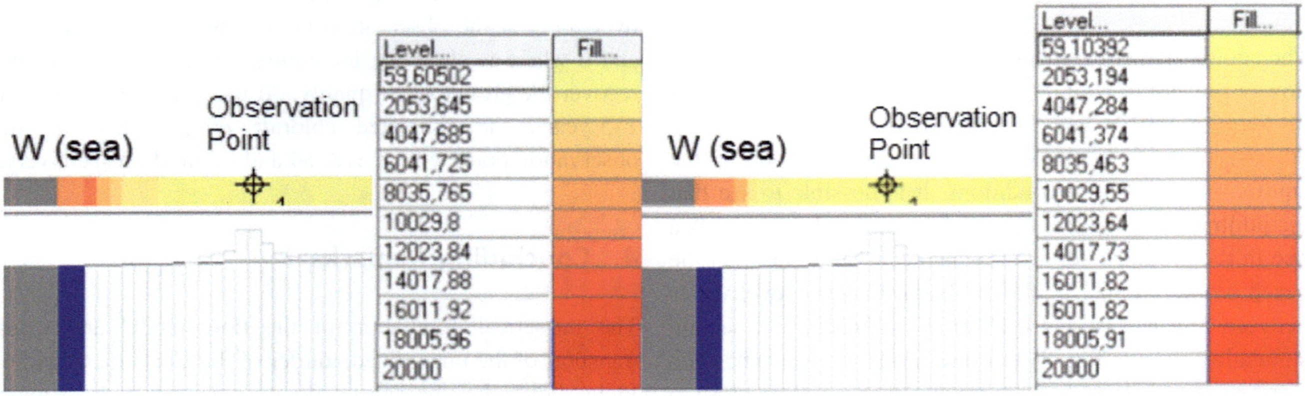

Fig. 5 Vertical section of the aquifer zone delimited in Fig. 4, with the simulated chloride concentration after 1, 2 and 5 years of the storm; dark line is the groundwater level

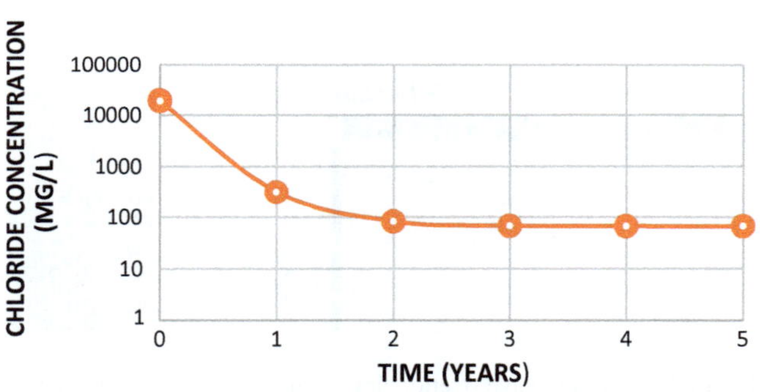

Fig. 6 Chloride concentration calculated on the vertical cross-section, after the wave overtopping occurs in the study area, versus time (years)

During a storm, a wave with a flow rate of 0.2 L/m/s can cause a wave overtopping that reaches approximately 160 m in land. The infiltrated seawater can contaminate the aquifer for a period that can reach several years.

Acknowledgements The authors are grateful for the financial support of FCT—project UID/GEO/50019/2019—Instituto Dom Luiz.

References

Caria C (2012) Estudo da recarga de águas subterrâneas do concelho de Almada. Universidade de Lisboa, Lisboa (MSc thesis), p 124

Caria C, Oliveira MM, Silva MC (2013) Síntese do Estudo da Recarga de Águas Subterrâneas do Concelho de Almada. In 9° Seminário sobre Águas Subterrâneas. Almada

Elsayed S, Oumeraci H (2017) Modelling and management of storm-driven saltwater intrusion in freshwater aquifers: the case of near Bremerhaven, Germany. p 20. In: Silva R, Chavez V (eds) Integrating ecosystems in coastal engineering practice. Puerto Morelos, Mexico, p 19

Ferreira F (2012) Avaliação dos Impactes das Alterações Climáticas nos Aquíferos Costeiros do Concelho de Almada. Universidade de Lisboa, Lisboa (MSc thesis), p 100

Pais J, Moniz C, Cabral J, Cardoso JL, Legoinha P, Machado S, Morais MA, Lourenço C, Ribeiro ML, Henriques P, Falé P (2006) Notícia Explicativa da Folha 34-D Lisboa da Carta Geológica de Portugal à escala de 1:50.000. Departamento de Geologia, Instituto Nacional de Engenharia, Tecnologia e Inovação, p 75

Pires B (2017) Vulnerabilidade e risco de galgamento oceânico em litorais baixos e arenosos. Universidade Nova de Lisboa, Monte da Caparica (MSc thesis), p 257

Storm A (2014) Hydrogeology of the Southern Costa da Caparica Aquifer. University of Lisbon, Lisbon, (BSc thesis), p 74

Groundwater Monitoring in Regional Discharge Areas Selected as "Hydrosensitive" to Seismic Activity in Central Italy

Marco Petitta, Domenico Marino Barberio, Maurizio Barbieri, Andrea Billi, Carlo Doglioni, Stefania Passaretti, and Stefania Franchini

Abstract

The aim of this study was to identify potential patterns of hydrogeological and geochemical changes in response to seismic activity, including possible variations of ion concentrations, gas compositions, and isotopic ratios in groundwater. Different monitoring areas in Central Italy have been selected, located in regional discharge areas with abundant groundwater resources. They are located near active normal faults, along the southeast-ward prolongation of the faults that nucleated the L'Aquila 2009 Mw 6.3 event and more recently the seismic sequence of Amatrice–Norcia 2016–2017. The test-site selection has been based on both seismic and hydrogeological criteria. The areas are characterized by low strain rate, long frequencies of earthquake occurrence and high depth of the brittle-ductile transition, which are indicators of potential future earthquakes. In this framework, the hydrogeological setting is considered essential for obtaining reliable results applied to seismic purposes. Springs with long and deep flow paths in the regional aquifers are under monitoring. Indeed, the regional groundwater flows are also clearly influenced by deep fluids contribution. Consequently, the hydrogeological monitoring has been enhanced by continuous active radon and carbon dioxide probes, installed in some of the monitored springs. The results show significant inferences between hydrogeology and seismicity, having different characteristics in relationship with Amatrice–Norcia 2016–2017 seismic sequence and with global minor seismic activity. Based on these preliminary results, the second phase of this study is aimed to reinforce monitoring in selected springs and wells.

Keywords

Earthquake hydrology • Central Italy • Geochemistry • Radon • Isotopes

1 Introduction

Among all-natural disasters, earthquakes are the most feared. Their destructive power is unmatched. To make matters worse, many seismologists believe that earthquakes are inherently unpredictable. Despite the different points of view, earthquake forecasts represent the holy grail for most geologists. The main advances in the field have been achieved by observing variations in the sequences of foreshocks (Reasenberg 1999), variations in the ratios of the velocities of the seismic waves V_p/V_s (Lucente et al. 2010), changes in electromagnetic fields (Fraser-Smith et al. 1990), gas emissions (Toutain and Baubron 1999), superficial deformations (Riguzzi et al. 2013), variations in circulation and chemistry of groundwater (Barberio et al. 2017 and reference therein) and others (Cicerone et al. 2009). One of the main problems encountered in the analysis of the potential pre-seismic anomalies is that the signals tend to be modest and often subject to processes of overlapping and/or interference with others of a non-seismic type and often of a local nature; they also tend to be site-specific and of a very different nature (Cicerone et al. 2009). Another major problem for those studying the seismic precursors is the lack of clarity in the correlations between the anomaly found and the triggering deformation process underway at tens of kilometres of depth (Scoville et al. 2015). A practical solution to such complex problems could be sought in systematic measures developed with high-performance multi-parametric networks working on large regions and for long periods of time (Cicerone et al. 2009; Wang and Manga

M. Petitta (✉) · D. M. Barberio · M. Barbieri · C. Doglioni · S. Passaretti · S. Franchini
Earth Sciences Department, University of Rome "La Sapienza", P.le Aldo Moro 5, 00185 Rome, Italy
e-mail: marco.petitta@uniroma1.it

A. Billi
IGAG-CNR, Institute for Environmental Geology and Geoengineering, National Research Council, Rome, Italy

© Springer Nature Switzerland AG 2020
F. Fernandes et al. (eds.), *Advances in Natural Hazards and Hydrological Risks: Meeting the Challenge*, Advances in Science, Technology & Innovation,
https://doi.org/10.1007/978-3-030-34397-2_5

2015). The aim of this study was to identify potential patterns of level changes in response to several earthquakes, and possible variations of ion concentrations, gas compositions and isotopic ratios in groundwater. A multiparametric observatory has been established in Central Italy, near the active normal faults of the Sulmona basin.

2 Materials and Methods

The choice of the site for hydrogeological and hydrogeochemical monitoring is the first and most important step to be taken. Starting from a conceptual physical model that describes the behaviour of the most superficial layers of the earth's crust (Doglioni et al. 2014; 2015), the effort was made to identify potential structural-geological and hydrogeological criteria able to define the best conditions to achieve the stated objectives.

Geological and seismological criteria:

- Identification of areas with high seismic hazard from the seismic hazard map (http://zonesismiche.mi.ingv.it/mappa_ps_apr04/italia.html);
- Estimate the maximum magnitude expected (Doglioni et al. 2015; Riguzzi et al. 2013);
- Paleo-seismological analysis (Galli et al. 2008).

Hydrogeological and hydrogeochemical criteria:

- Availability of water resource and monitoring points;
- Knowledge of underground hydrodynamics;
- Regional representativeness of monitoring points;
- Identification of points potentially influenced by deep circulation.

These criteria led to the identification of the first pilot site to be submitted to hydrogeological and hydrogeochemical monitoring in Central Italy, the area of the Sulmona plain and the Gole di Popoli.

Hydrological and hydrogeological monitoring was carried out from July 2014 until February 2019. During this period, the rainfall and temperature data of the Bussi Officine thermo-pluviometric station (near Sulmona basin) were collected. Two piezometers have been equipped, through the installation of multiparameter probes for continuous monitoring of temperature, electrical conductivity, and piezometric level.

From the hydrogeochemical point of view, 8 springs were identified and monitored in the area of the Sulmona plain and the Gole di Popoli. Sampling started in November 2014. For each spring temperature, electrical conductivity and pH were measured, and samples were taken for the analysis of the major, minor elements and for isotopic analysis of H_2O ($\partial^{18}O$-D), SO_4 ($\partial^{34}S$-$\partial^{18}O$), B ($\partial^{11}B$) and Sr ($^{87}Sr/^{86}Sr$). In addition, in an attempt to identify variations in deep degassing, it was decided to also monitor Rn and CO_2 concentrations, two of the gases that showed higher sensitivity to seismic activity. The isotopic approach will improve the geochemical model related to the circulation path.

3 Results

3.1 Hydrogeological Results

Hydrogeological monitoring has allowed us to shed light on the peculiar hydrosensitive characteristics of this site. The Sulmona site has shown a high sensitivity to the events of both the seismic sequence of Amatrice–Norcia 2016–2017 (Barberio et al. 2017) and to events of high magnitude occurrence thousands of kilometers away (Fig. 1). The latter is an effect of the propagation of superficial seismic waves.

3.2 Hydrogeochemical Results

Figures 2 and 3 show the time series of the minor and trace elements determined by mass spectrometry for the sources of the Sulmona area (Barberio et al, 2017).

The concentrations are stable over time for most of the characterizing elements (Li, B, Co, Cu, Pb, U, Cs and Mn). They appear to have a strong potential correlation with the seismic activity (As, V, Cr, Sr and Fe). In particular, a strong increase in As and V can be observed starting from April 2016 until the sampling on 29 August 2016. In the following samples and until December 2016 there is a progressive decrease until the return to the pre-April conditions. 2016. The post-seismic phase is accompanied by a significant increase in Cr while during the entire period in which anomalies are recorded (from April 2016 to December 2016) Sr concentrations in the higher mineralization sources have more than doubled (Figs. 2 and 3).

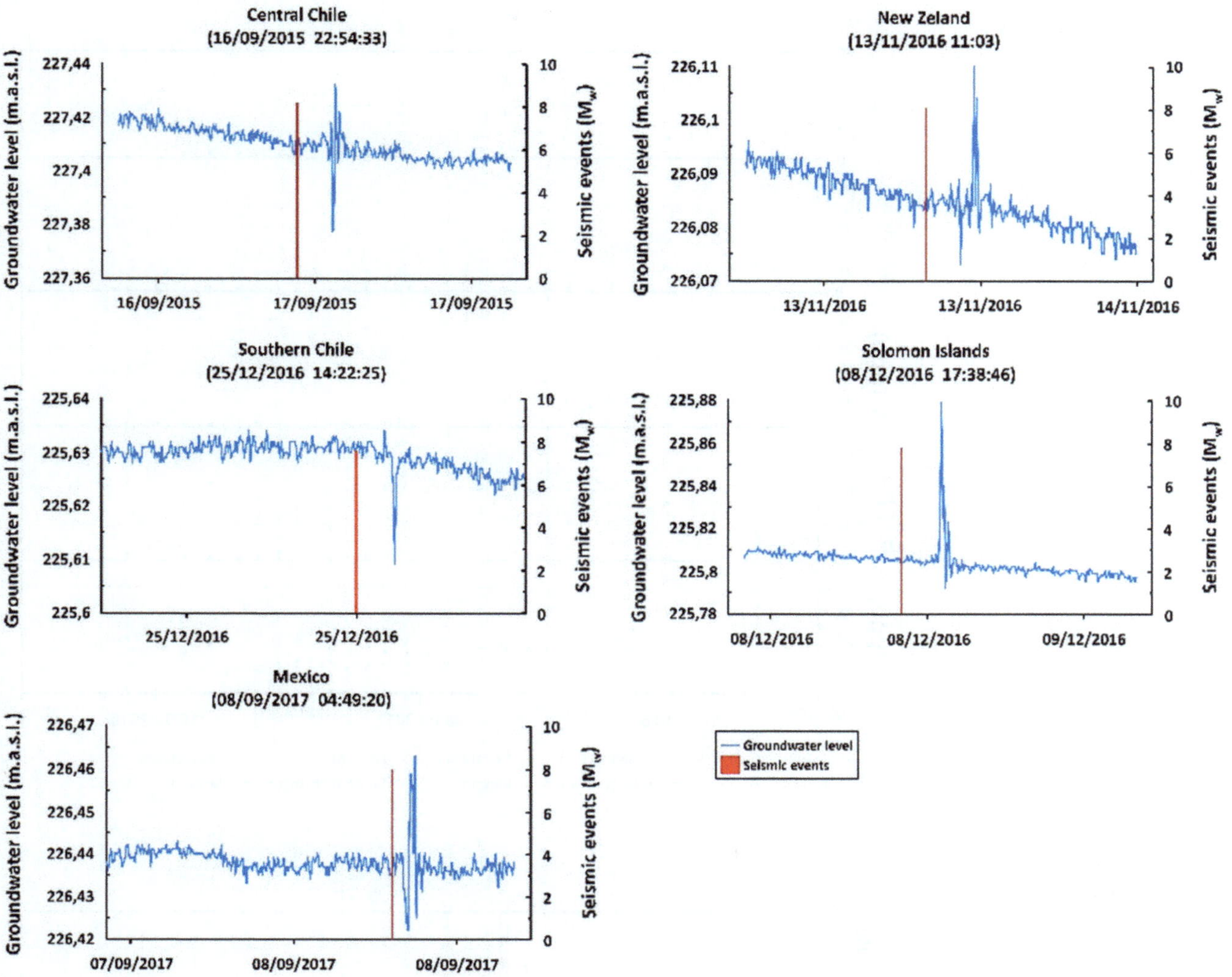

Fig. 1 Groundwater-level response to seismicity occurring far from the measurement location

4 Discussion

In the first analysis, the hydrogeological monitoring allowed to identify the peculiar hydro-sensitive characteristics of the chosen site. There is a clear response of groundwater levels to strong earthquakes occurring thousands of kilometers away and to the main earthquakes of Amatrice–Norcia 2016–2017 seismic sequence.

Rayleigh waves (surface waves) produce strong fluctuations in piezometric levels even at considerable epicentral distances. On the basis of the distances from the PF60.3 piezometer and the arrival times, it can be said that the seismic waves that caused oscillations in the piezometric trend are the superficial ones (Fig. 1) characterized by a longer travel time and a lower speed compared to the volume waves (waves P and S). In addition, geochemical monitoring has made it possible to identify possible seismic precursors in the increases of As and V (Barberio et al. 2017). The systematic geochemical monitoring of the selected springs has allowed to identify characteristic trends in the time series of the various geochemical parameters allowing to identify the basic values typical of the analyzed waters. The imminent release of the seismic sequence of Amatrice–Norcia 2016–2017 about 70 km from the Sulmona site meant that some of these concentration trends diverged from the typical values identified (As, V, Cr, Fe Sr).

Fig. 2 Time series of trace elements Li, B, V, Mn and Fe

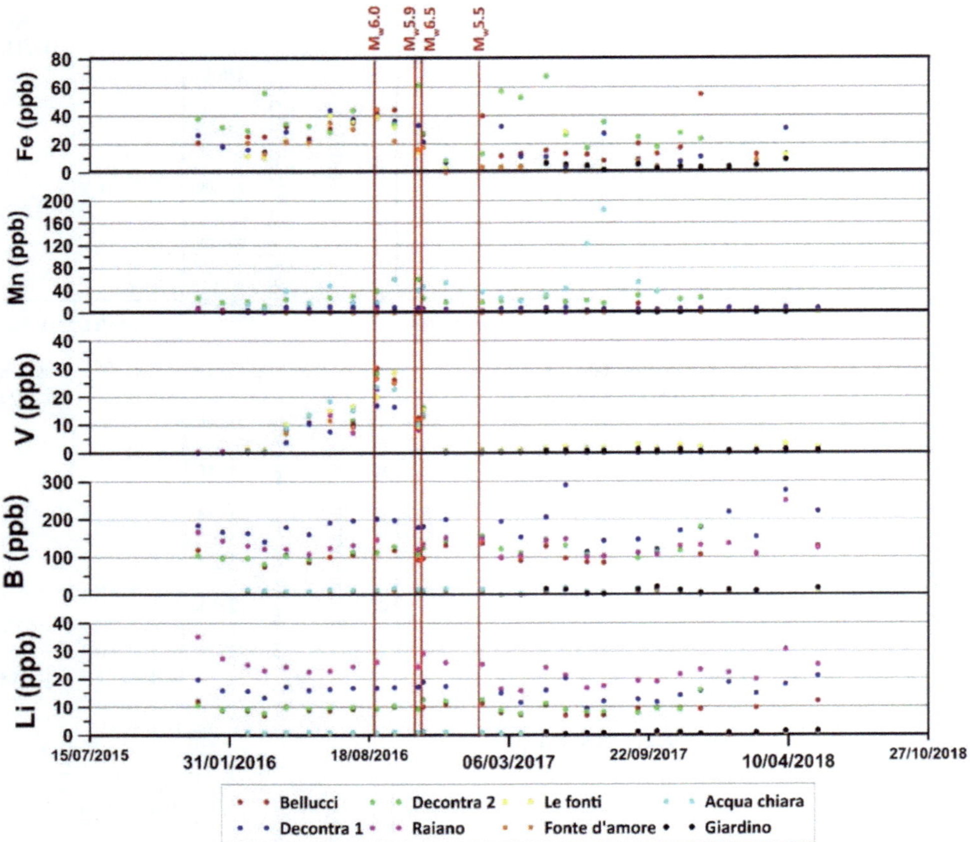

Fig. 3 Time series of trace elements Cs, As, Sr, Ni and Cr

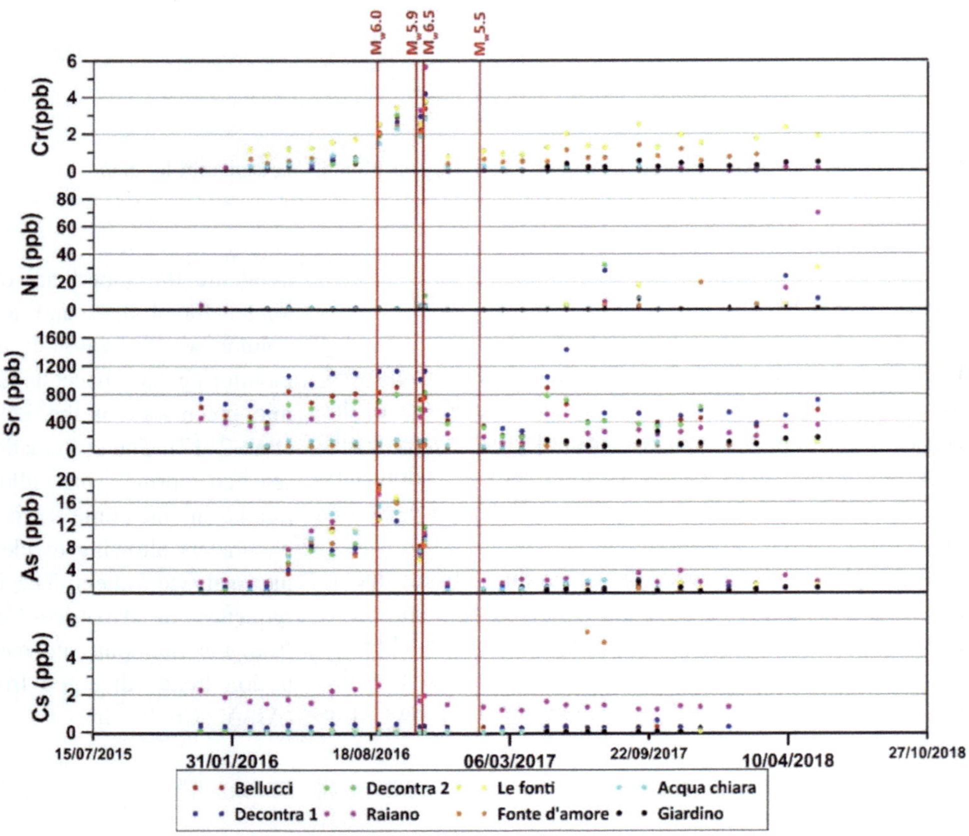

5 Concluding Remarks

- Hydrogeological and hydrogeochemical monitoring in the central Apennines carried out before and during the seismic sequence of Amatrice–Norcia 2016–2017 has allowed to identify as potential seismic precursors the concentration increases of As and V in the groundwater.
- The regional response of the carbonate aquifers to the major events of the seismic sequence (Mw > 5.5) highlights for the umpteenth time the sensitivity of the fractured aquifers to the triggering of these events. However, further studies are necessary for a complete understanding of the phenomenon.
- Identification of oscillations in the piezometric levels of the regional water table in the Gole di Popoli as a result of the propagation of surface seismic waves induced by strong events (Mw > 7.5) occurring thousands of kilometres away.

References

Barberio MD, Barbieri M, Billi A, Doglioni C, Petitta M (2017) Hydrogeochemical change before and during the 2016 Amatrice-Norcia seismic sequence. Sci Rep 7:11735. https://doi.org/10.1038/s41598-017-11990-8

Cicerone RD, Ebel JE, Britton J (2009) A systematic compilation of earthquake precursors. Tectonophysics 476:371–396

Doglioni C, Barba S, Carminati E, Riguzzi F (2014) Fault on–off versus coseismic fluids reaction. Geosci Front 5:767–780

Doglioni C, Carminati E, Petricca P, Riguzzi F (2015) Normal fault earthquakes or graviquakes. Sci Rep 5:12110

Fraser-Smith AC et al (1990) Low-frequency magnetic field measurements near the epicenter of the Ms 7.1 Loma Prieta Earthquake. Geophys Res Lett 17:1465–1468

Galli P, Galadini F, Pantosti D (2008) Twenty years of paleoseismology in Italy. Earth Sci Rev 88:89–117

Lucente FP et al (2010) Temporal variation of seismic velocity and anisotropy before the 2009 MW 6.3 L'Aquila earthquake, Italy. Geology 38:1015–1018 (2010)

Reasenberg PA (1999) Foreshock occurrence before large earthquakes. J Geophys Res Solid Earth 104:4755–4768

Riguzzi F et al (2013) Strain rate relaxation of normal and thrust faults in Italy. Geophys J Int 195:815–820

Scoville J, Sornette J, Freund FT (2015) Paradox of peroxy defects and positive holes in rocks. Part II: outflow of electric currents from stressed rocks. J Asian Earth Sci 114:338–351. https://doi.org/10.1016/j.jseaes.2015.04.016

Toutain JP, Baubron JC (1999) Gas geochemistry and seismotectonics: a review. Tectonophysics 304:1–27

Wang CY, Manga M (2015) New streams and springs after the 2014 w6.0 South Napa earthquake. Nat Commun 6:7597. https://doi.org/10.1038/ncomms8597

Seismic Vulnerability Assessment of a Water Supply Network: City of Aveiro Case Study

Hélder Pires, Hugo Rodrigues, Armando Silva Afonso, Carla Pimentel-Rodrigues, and Fernanda Rodrigues

Abstract

Critical infrastructures have critical importance on societies as they rely on fundamental development and safety aspects. The inoperability of these networks can cause great damage to strategical state-related activities and may reduce the ability to address the results in the aftermath of an accident or catastrophe. Taking in mind that damages in a water supply network do not depend only on the intensity of the disaster, but also on the vulnerability of each component of the entire system, it is important to ensure the ability of these infrastructures to remain operational during and after any disruptive event to prevent the loss of availability of vital goods and services, but also due to the high level of interdependence between different networks, that can lead to a propagation of effects, increasing the level of losses. The seismic vulnerability assessment of water systems is one of the most effective mitigation strategies that will allow to estimate the serviceability of the system and evaluate the estimated direct losses in terms of replacement cost. This paper aims to revise the main seismic risk assessment methodologies available for seismic vulnerability assessment of water networks and based on the selected methodology to present an application to a real network.

Keywords

Critical infrastructures • Water supply network • Seismic vulnerability • ArcGIS • ArcMap

1 Introduction

In the recent years the awareness towards Critical Infrastructures and networks and the need for their protection is becoming more noticeable, as societies rely on them not only to maintain their normal functioning but also to increase their development levels.

This can be observed with the implementation, at European level, of the Directive 2008/114/CE of 8 December 2008, on "*Identifying and designating the European Critical Infrastructures and the assessment of the need to improve their protection*", followed by the revision SWD(2013) 318 Final—"*Commission Staff Working Document on a New Approach to the European Program for Critical Infrastructure Protection Making European Critical Infrastructures more secure*" (European Commission Homepage) answering to the request made by the European Council in June 2004, for planning a general strategy on Critical Infrastructure Protection (CIP).

A Critical Infrastructure is an asset, system or part thereof located in Member States which is essential for the maintenance of vital societal functions, health, safety, security, economic or social well-being of people, and which disruption or destruction would have a significant impact in a Member State as a result of the failure to maintain those functions (EUR-LEX Homepage).

In this way, were developed, over the years, a vast array of methodologies to assess all the different risks that an infrastructure can be exposed to. The present paper is focused on the seismic vulnerability assessment of the water supply network (WSN) located in the Aveiro municipality, midland Portugal, following the work developed by Rodrigues et al. (2017).

H. Pires
Department of Civil Engineering, University of Aveiro, Aveiro, Portugal

H. Rodrigues
RISCO, School of Technology and Management, Polytechnic Institute of Leiria, Leiria, Portugal

A. S. Afonso · C. Pimentel-Rodrigues
ANQIP—Portuguese Association for Quality and Efficiency in Building Services, Department of Civil Engineering, University of Aveiro, Aveiro, Portugal

A. S. Afonso · C. Pimentel-Rodrigues · F. Rodrigues (✉)
RISCO, Department of Civil Engineering, University of Aveiro, Aveiro, Portugal
e-mail: mfrodrigues@ua.pt

© Springer Nature Switzerland AG 2020

F. Fernandes et al. (eds.), *Advances in Natural Hazards and Hydrological Risks: Meeting the Challenge*, Advances in Science, Technology & Innovation,
https://doi.org/10.1007/978-3-030-34397-2_6

2 Seismic Risk Assessment Methodologies

As water is essential for Human life and is considered a resource that has to be preserved, the analysis, mitigation or elimination of the risks that can damage the water supply system is a priority (Halfaya et al. 2012).

Urban water supply systems (UWSS) are generally composed of water sources, transmission pipes, treatment plants, and distribution networks from source to tap and usually are exposed to a variety of uncertain threatening hazards that may affect its functioning (Roozbahani et al. 2013).

However, the inventory of the network, on which that analysis is based, comprises the distinctive features of the components (e.g., geometry, material, age, etc.), which is a time and cost consuming process with large uncertainties due to the lack or/and poverty of data, the spatial distribution of the networks and sometimes the reluctance of manufacturers, owners and managers of the lifelines to share information (Alexoudi et al. 2007).

Based on the bibliographic review, Table 1 depicts methodologies for seismic assessment of WN (water network).

3 Case Study

The selected case study consists on the application of the model proposed by Zohra et al. (2012) to the water supply network of Aveiro municipality that supplies almost 78,450 people (Census 2011) and is composed of eight significant points as represented in Fig. 1 (AdRA 2017).

For the integral understanding of the network the software GIS, ArcGIS desktop, of ESRI—Environmental Systems Research Institute, was used. Geographic Information System—GIS is an informatics system that capture, store, analyses and shows geospatial data (Chang 2007).

Zohra's et al. (2012) proposes pipe evaluation based on a Vulnerability Index (VI) that is obtained relating data as pipe diameter (C_{d}) and material (C_{p}), fault crossings (C_{f}), settlement and landslide risk (C_{s}), ground type (C_{g}), seismic intensity (C_{i}) and liquefaction potential (C_{l}), given by Eq. (1):

$$\mathrm{VI} = C_{\mathrm{d}} \cdot C_{\mathrm{p}} \cdot C_{\mathrm{f}} \cdot C_{\mathrm{s}} \cdot C_{\mathrm{g}} \cdot C_{\mathrm{i}} \cdot C_{\mathrm{l}} \tag{1}$$

This methodology was chosen amongst the others as it does not require specific software or complex mathematical models as well as the fact that all the needed information to perform the implementation was easily obtained.

For this evaluation, the used range for C_{d} is a value between 0.7 and 1.6, according to pipe diameter. C_{p} is between 0.1 (PEHD) and 2.5 (Asbestos cement), depending on pipe material. It was assumed one intersection for fault crossing factor, $C_{\mathrm{f}} = 2.0$. Regarding settlement and landslide, it is proposed an evaluation with average risk, $C_{\mathrm{s}} = 2.0$; $C_{\mathrm{g}} = 2.9$; $C_{\mathrm{l}} = 1.0$. The used seismic intensity factor was weighted regarding the zone's normal seismic activity, resulting in an assessment for an earthquake with an intensity MMI < 8 − $C_{\mathrm{i}} = 1.0$.

Table 1 Methodologies for seismic assessment of WN

Model	Authors	Characteristics
Vulnerability Assessment of Water Supply Networks	Zohra et al. (2012)	Estimate pipeline damages regarding several factors as pipe diameter, materials, crossings, ground type, settlement risk, seismic intensity, and liquefaction. The final results are presented as a Vulnerability Index
Seismic Reliability Assessment of Water Distribution Networks (WDN)	Laucelli et al. (2013)	Evaluates the probability of failure for each pipe, as a consequence of an assumed earthquake; Searches the most destructive scenarios for a WN, using multi-objective optimization to search the minimum number of damaged pipes that would result in the minimum supplied water demands
Estimating earthquake-induced failure probability and downtime of Critical Facilities	Porter and Ramer (2012)	With a fault tree approach, this methodology seeks to estimate both the probability that a main and reserve facility would be simultaneously rendered inoperative in a single earthquake and the required time to restore normal functionality
Seismic Vulnerability Analysis of Waste-Water System	Alexoudi et al. (2008)	Seismic risk assessment methodology that includes the estimation of the system's inventory, typology, seismic loads, fragility curves and damages in absolute numbers, as well as its spatial distribution

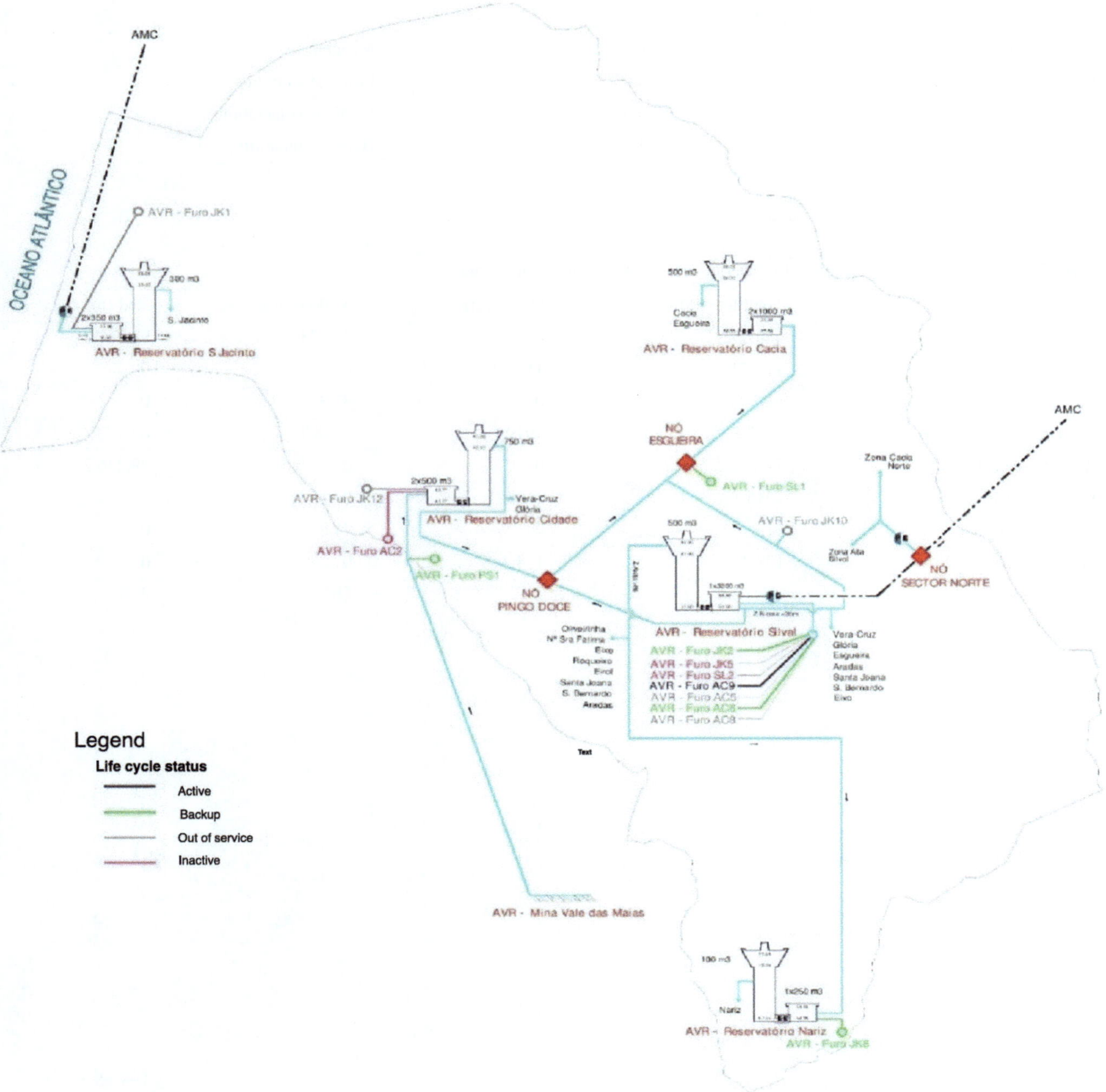

Fig. 1 Water supply network—Aveiro

With these values, vulnerability indexes for each pipe are calculated and can be divided into three intervals, low vulnerability, medium vulnerability, and high vulnerability (see Table 2).

The obtained results (see Table 3 and Fig. 2) show that the total extent of PEHD, Ductile Cast Iron and Polyethylene pipes present a lower vulnerability index, representing 3% of the total network, for the defined scenario. However, asbestos cement and smaller diameter PVC ($\phi < 75$ mm) pipes present higher vulnerability level, representing 29% of the network, with vulnerability indexes that vary from VI = 29 to VI = 46.4, values that represent more than the double of the limit that represents High Vulnerability (VI = 12). This is mainly due to the material's fragility and behavior towards external stress as well as the fact that asbestos cement pipes are known to have moderate to high vulnerability, especially in liquefaction areas (EPA 2018).

Table 2 Vulnerability index—spatial representation

Vulnerability index (VI)	Color	Meaning
0–5	Green	Low vulnerability
5–12	Yellow	Medium vulnerability
12>	Red	High vulnerability

Table 3 Vulnerability index according to materials and network's total pipe length

Material	Total length (m)	Green (m)	Yellow (m)	Red (m)
PVC	488,052.2	0	396,284	91,768.2
PEHD	11,542.5	11,542.5	0	0
Cast Iron	33.3	0	33.3	0
Fiber reinforced polyester	6167.8	4919.1	1248.7	0
Asbestos cement	79,246.2	0	0	79,246.2
Ductile cast iron	1615	1615	0	0
Galvanized iron	41.8	0	41.8	0
Polyethylene	134.4	134.4	0	0
Percentage		3%	68%	29%

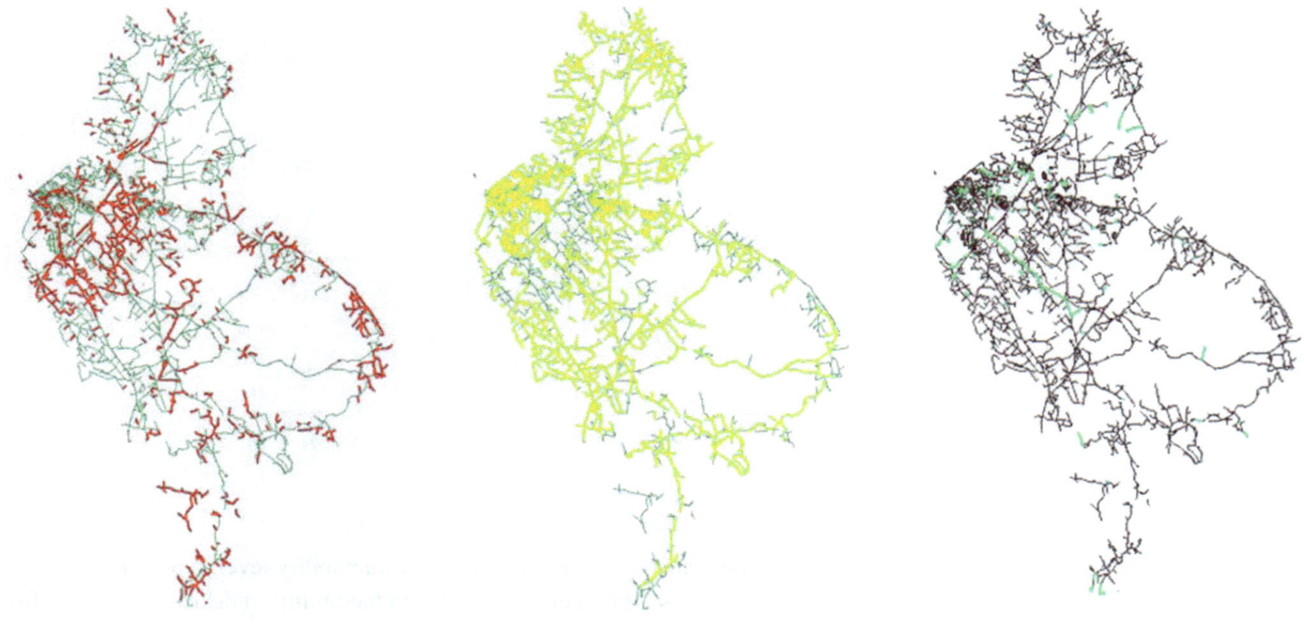

Fig. 2 Vulnerability indexes for the network under analysis

4 Final Remarks

With the current preliminary study, it was possible to observe that only approximately 3% of the network presents a low vulnerability index interval, being mainly composed of PEHD and ductile iron pipes. 29% of the network shows a high vulnerability index, with the totality of asbestos cement and the smaller diameter PVC pipes contributing to this fact. The remaining 68% represent medium vulnerability indexes for the high diameter PVC pipes ($\phi > 75$ mm).

To decrease the seismic vulnerability of this WDN it should be considered the replacement of the 29% high vulnerability index part with more ductile materials.

References

AdRA Homepage (2017). http://www.AdRA.pt/content/index.php?action=detailfo&rec=1800&t=Quem-somos. Last accessed 2017/06/12

Alexoudi M, Anastasiadis A, Pitilakis K (2007) Seismic risk assessment of the water system of Thessaloniki. In: 4th international conference on earthquake geotechnical engineering. Paper no. 1632

Alexoudi M, Manou D, Pitilakis K (2008) Seismic vulnerability analysis of waste-water system. In: Methodology and application for Duzce and Kocaeli earthquakes based on the microzonation study of Duzce. The 14th world conference on earthquake engineering

Census (2011) Population density. National Institute of Statistics, Lisbon. http://mapas.ine.pt/map.phtml Accessed 5 Nov 2017. (in Portuguese)

Chang K (2007) Geographic information system. In: The international encyclopedia of geography. http://onlinelibrary.wiley.com/doi/10.1002/9781118786352.wbieg0152/abstract;jsessionid=AF6C1BE20AFB7E1D304C290D656A8B74.f01t02?userIsAuthenticated=false&deniedAccessCustomisedMessage. Last accessed 2019/02/12

EPA Website (2018). https://www.epa.gov/sites/production/files/2018-02/documents/180112-earthquakeresilienceguide.pdf. Last accessed 11/02/19

EUR-LEX Homepage. https://eur-lex.europa.eu/legal-content/EN/TXT/PDF/?uri=CELEX:32008L0114&from=PT. Last accessed 2018/02/10

European Commission Homepage. http://ec.europa.eu/transparency/regdoc/rep/10102/2013/EN/10102-2013-318-EN-F1-1.PDF. Last accessed 2018/02/10

Halfaya FZ, Bensaibi M, Davenne L (2012) Vulnerability assessment of water supply network. Energy Procedia 18:772–783

Laucelli D, Berardi L, Politano G, Giustolisi O (2014) Seismic reliability assessment of water distribution networks. Procedia Eng 70:998–1007

Porter K, Ramer KN (2012) Estimating earthquake-induced failure probability and downtime of critical facilities. J Bus Contin Emer Plan 5(4):352–364

Rodrigues F, Borges M, Rodrigues H (2017) Risk management in water supply networks: a case study. In: CIB W062—44th international symposium—water supply and drainage for buildings, pp 310–320. 27–30 Agosto. Ponta Delgada, Açores, Portugal. ISBN: 978-989-97476-2-3

Roozbahani A, Zahraie B, Tabesh M (2013) Integrated risk assessment of urban water supply systems from source to tap. Stoch Environ Res Risk A 27(4):923–944. http://doi.org/10.1007/s00477-012-0614-9. Last accessed 2019/03/18

Zohra HF, Mahmoud B, Luc D (2012) Vulnerability assessment of water supply network. Energy Procedia 18:772–783. https://www.sciencedirect.com/science/article/pii/S1876610212008636. Last accessed 2019/03/18

Natural Hazards Coming from Trace Elements Natural Enrichment: The Bevera Valley Basin (Northern Italy) Case History

Giuseppe Sappa, Maurizio Barbieri, and Francesca Andrei

Abstract

Trace elements are natural constituents of soils and their concentration varies depending on parental materials. In order to investigate the inorganic pollution conditions of soil in Bevera Valley Basin, Northern Italy, two selected samples were taken from Rainer Quarry and Femar Quarry were analyzed. High levels of Arsenic (As) have been reported in groundwater. Exposure to As in the environment is hazardous to biota, this study aims to investigate mechanism involved in natural enrichment of As, considering interactions of underground waters with soils. The concentrations of trace elements were determined by using ICP-MS. Laboratory analysis shows a greater enrichment in copper for both quarries, maybe due to the geological nature of the soils. Then the potential pollution risks of trace elements in the soil were evaluated by method of geoaccumulation index (I_{geo}) and enrichment factor analysis (EF). Results show that the two sites are not contaminated, and the trace elements, with reference to Arsenic, are not linked to anthropic contributions.

Keywords

Trace elements • Soil pollution • Arsenic • Geoaccumulation index • Enrichment factor analysis

G. Sappa (✉) · F. Andrei
Department of Civil, Building and Environmental Engineering (DICEA), Sapienza University of Rome, Rome, Italy
e-mail: giuseppe.sappa@uniroma1.it

F. Andrei
e-mail: francesca.andrei@uniroma1.it

M. Barbieri
Department of Earth Sciences (DST), Sapienza University of Rome, Rome, Italy
e-mail: maurizio.barbieri@uniroma1.it

1 Introduction

Soil has a protective function for the environment, because of its "self-purifying" power able to mitigate the effects of any pollutants. The soil pollution is characterized by migration of trace elements, which can be conveyed from the soil to other components of the ecosystem, such as aquifers and plants, thus affecting human health through drinking water and the food chain.

Trace elements (for example, As, Mn, Ni, Pb and Cd) are not sensitive to any process of decomposition in soils and mobilizing within the environment, reducing its fertility conditions. In particular, Arsenic is naturally present as the main constituent of 200 minerals (Smedley and Kinniburgh 2002). Interaction processes of volcanic-sedimentary rock aquifers with the soils, further increased by precipitation, cause the occurrence of Arsenic in groundwater (Ergul et al. 2013; Sappa et al. 2014). Soil-groundwater interaction, characterized by the presence of organic matter or silty and clayey sediments, under reducing conditions, conveyed by precipitation, represents a phenomenon determining dissolution and release of Arsenic in environment (Smedley 2008; Tufano and Fendorf 2008).

The aim of this work was to present a suitable procedure for the assessment of the contamination status soils of trace elements, considered serious soil pollutants due to environmental persistence, toxicity and the ability to be incorporated into food chains (Liao et al. 2008). The study area is Bevera Valley Basin, characterized by exceeding the threshold for Arsenic in groundwater. This procedure involves sampling of soils, laboratory analysis of samples (ICP-MS), the definition of the Geoaccumulation Index (I_{geo}) and the Enrichment Factor (EF).

© Springer Nature Switzerland AG 2020
F. Fernandes et al. (eds.), *Advances in Natural Hazards and Hydrological Risks: Meeting the Challenge*, Advances in Science, Technology & Innovation,
https://doi.org/10.1007/978-3-030-34397-2_7

2 Materials and Methods

The study area is located in the Bevera Valley Basin, in the Municipality of Arcisate (VA), in the northeast part of Lombardy Region (Northwest Italy), and presents three big quarrying areas: Rainer Quarry on the right side of the Bevera Creek, in the southeast part of the Municipality of Arcisate (non-operative since the 90s), and on the left side of the Bevera Creek, Femar Quarry (non-operative since the 90s) and Valli Quarry (operative) (Fig. 1).

This area has hilly morphology, with mild slopes determined by the presence of morainic apparatuses and fluvioglacial terraces. The Bevera Aquifer System is characterized by silty clayey and arenaceous interlayers, separating the free upper aquifer from aquifers under conditions that vary from confined to semi-confined.

Laboratory analysis for the determination of concentrations of some trace elements present in soil samples from Rainer Quarry and Femar Quarry were performed at the Laboratory of the Department of Earth Sciences of the University of Rome "La Sapienza". Two selected samples of Rainer Quarry and Femar Quarry picked out from a mixture of 6 subsamples for 3 kg and submitted to the quarter method, with the aim to reduce a large sample of soil to yield a representative sample weighing about 100 g. The contents of trace elements are analyzed by Thermo Scientific Application Note 40,619 (US EPA SW-846 Method 6020A Using the XSERIES 2 ICP-MS).

In Rainer Quarry an excavator was used to remove the covering of topsoil at depth 50 cm, on the contrary, in Femar Quarry it was possible to pick out directly the 6 subsamples to depth 10 cm (Table 1).

The soils sampled are mainly silty clayey soils, with lenses of peat: the presence of organic matter in soil or sediments is an important factor for leaching As mainly under reducing conditions (Redman et al. 2002; Wang and Mulligan 2006).

3 Results

The concentrations of trace elements are shown in the line chart, and it is possible to verify the analogy of the trends in the two quarries (Fig. 2).

This trend is linked, in fact, to the common geological origin of soil constituting two deposits, with high values for Aluminium (Al) and Iron (Fe). About arsenic values, only the "public green" threshold, defined by Legislative Decree

Fig. 1 Location of study area

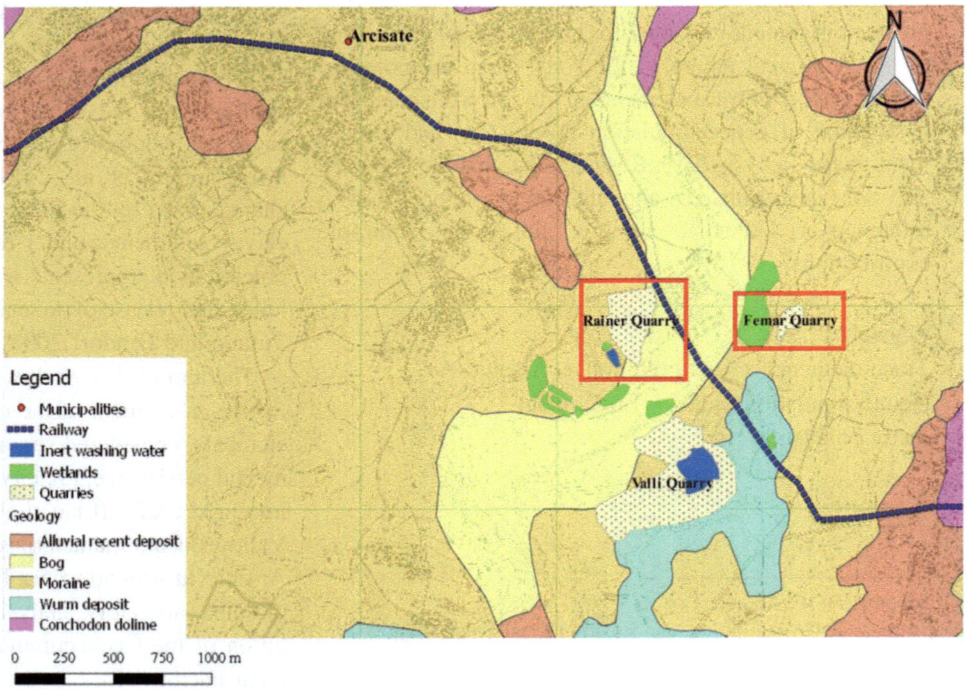

Table 1 Sampling location of the study area

	North (WGS84)	East (WGS84)	Depth (cm)
Rainer Quarry	45.845715	8.888601	50
Femar Quarry	45.847282	8.900802	10

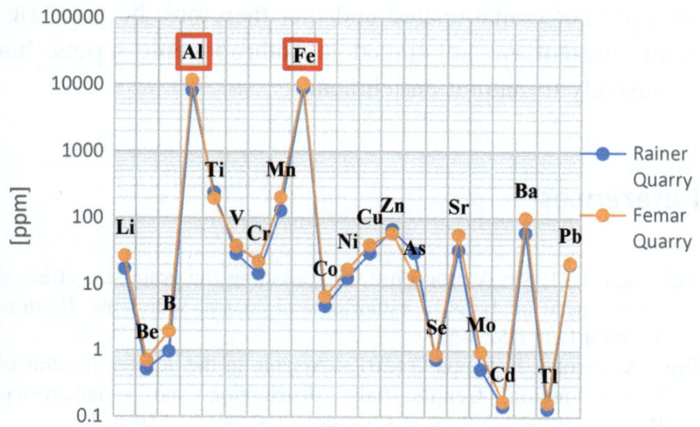

Fig. 2 Trend concentrations trace elements Rainer and Femar Quarry

152/06, for Rainer Quarry is exceeded, considering it, however, an acceptable risk.

4 Discussion

The pollution conditions of Rainer Quarry and Femar Quarry soil samples are analyzed through determination of the geoaccumulation index and the enrichment factor.

Since the 1970s, the geoaccumulation index (I_{geo}) has been used as a measure of contamination of bottom sediments (Muller 1969) and has been used evaluating the contamination of trace elements on soils (Micó et al. 2006). It is used to evaluate enrichment of the upper horizons respect to "local background". The I_{geo} was calculated (in the fraction $\leq$ 2 μm) by following way:

$$I_{geo} = \log_2\left(\frac{C_n}{1.5 * B_n}\right) \quad (1)$$

where

C_n is the measured concentration of the examined trace elements in the soil,

B_n is the geochemical background concentration of metal 'n', equal to the 95th percentile (Sena and Umlauf 2015). Factor 1.5 is the background matrix correction factor due to lithogenic effects.

Based on the classification of the soils proposed by Muller (1981), the following results were obtained (Table 2).

The results of the geoaccumulation index highlight the absence of contamination for the two sites. This leads to affirming the absence of soil contamination in the two quarries.

The enrichment factor (EF), instead, allows to evaluate the variation of potentially toxic element respect to a reference element. A reference element is an element particularly stable in the soil, which is characterized by absence of vertical mobility and/or degradation phenomena. In the following discussion, the Aluminum (Al) was chosen, because it is a highly conservative element and an important constituent of clay minerals (Ackerman 1980).

The Enrichment Factor is expressed as follow:

$$EF = \frac{(C_{nMetal}/C_{nAl})}{(B_{nMetal}/B_{nAl})} \quad (2)$$

where

C_{nMetal} is the measured concentration of the examined metal in the soil;
C_{nAl} is the measured concentration of Aluminum;
B_{nMetal} is the geochemical background concentration of metal (background value);
B_{nAl} is the geochemical background concentration of Aluminum.

It is possible to verify that Rainer Quarry is enriched more in Arsenic and Copper. The enrichment in Arsenic is justified by the greater concentration of metalloid, as shown in Fig. 3. The greater enrichment in Copper, also present in

Table 2 Geoaccumulation index (I_{geo})

	Rainer Quarry		Femar Quarry	
I_{geo} As	−0.56	Unpollution	−1.70	Unpollution
I_{geo} V	−1.62		−1.24	
I_{geo} Cr	−2.45		−1.86	
I_{geo} Ni	−2.50		−2.08	
I_{geo} Cu	−1.41		−1.00	
I_{geo} Zn	−1.68		−1.87	
I_{geo} Cd	−3.36		−3.13	
I_{geo} Pb	−2.79		−2.74	

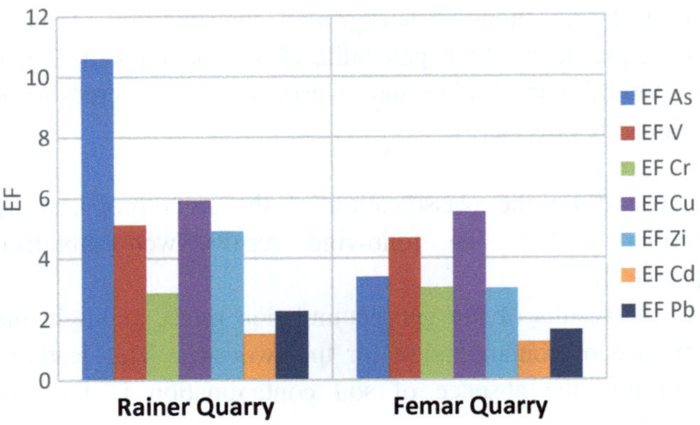

Fig. 3 Enrichment factor analysis

Femar Quarry sample, is linked to the clayey nature of the land (morainic and alluvial deposits).

5 Concluding Remarks

The area of the Bevera Valley Basin was characterized by high values of Arsenic concentrations in groundwater. The proposed procedure involves the definition of the geoaccumulation index and the determination of the enrichment factor. The results allow to state that trace elements are linked to the syngenetic phase of minerals, excluding an epigenetic nature, on the contrary combined to soil contamination phenomena. This condition is confirmed by the geoaccumulation index that is negative, thus allowing the affirmation of the absence of soil contamination. However, for Rainer Quarry there is a greater enrichment in Arsenic and Copper, linked to the clayey nature of soils in study area. In particular, the greater presence of Arsenic is linked to high concentrations of iron in soils, defining an effective absorption of Arsenic and therefore anomalous values. The proposed evaluations make it possible to state that the two sites are not contaminated and that therefore the trace elements present are not linked to anthropogenic inputs, but exclusively to natural enrichment.

References

Ackerman F (1980) A procedure for correcting the grain size effect in heavy metal analyses of estuarine and coastal sediments. Environ Technol Lett 1:527–528

Ergul S, Ferranti F, Sappa G (2013) Arsenic in the aquifer systems of Viterbo region, Central Italy: distribution and geochemistry. Rendiconti Online Società Geologica Italiana 24:116–118

Liao GL, Liao DX, Li QM (2008) Heavy metals contamination characteristics in soil of different mining activity zones. Trans Nonferrous Met Soc China 18(1):207–211

Micó C, Recatalá L, Peris M, Sánchez J (2006) Assessing heavy metal sources in agricultural soils of an European mediterranean area by multivariate analysis. Chemosphere 65(5):863–872

Muller G (1969) Index of geoaccumulation in sediments of the Rhine River. GeoJournal 2:108–118

Muller G (1981) Schwermetallbelstung der sedimente des neckars und seiner nebenflusse: eine estandsaufnahme, pp 157–164

Redman AD, Macalady DL, Ahmann D (2002) Natural organic matter affects Arsenic speciation and sorption onto hematite. Environ Sci Technol 36(13):2889–2896

Sappa G, Ergul S, Ferranti F (2014) Geochemical modeling and multivariate statistical evaluation of trace elements in Arsenic contaminated groundwater systems of Viterbo Area (Central Italy). Springer Plus 3(1):237

Sena F, Umlauf G (2015) Progetto di Monitoraggio Ambientale su tutto il Territorio della Regione Lombardia (Progetto Soil)

Smedley PL, Kinniburgh DG (2002) A review of the source, behaviour and distribution of arsenic in natural waters. Appl Geochem 17(5): 517–568

Smedley PL (2008) Sources and distribution of arsenic in groundwater and aquifers, pp 4–32

Tufano KJ, Fendorf S (2008) Confounding impacts of iron reduction on arsenic retention. Environ Sci Technol 42(13):4777–4783

Wang S, Mulligan CN (2006) Effect of natural organic matter on arsenic release from soils and sediments into groundwater. Environ Geochem Health 28(3):197–214

Radon Gas and Groundwater: Study of Risks in Water Galleries in Tenerife (Canary Islands, Spain)

Juan C. Santamarta, Luis E. Hernández-Gutiérrez, Jesica Rodríguez-Martín, Rafael J. Lario-Bascones, and Ángel Morales-González-Moro

Abstract

Water galleries are the main system for capturing groundwaters on the western islands of the Canarian archipelago (Spain). This research focuses on taking measurements of radon, a natural radioactive gas present in the galleries, and may cause health problems for workers in volcanic territories, such as in the Canary Islands. The aim of this research is to determine the values of radon gas, as a natural risk, in the water galleries in the exploitation of water resources, by means of officially approved measuring devices and the subsequent scientific analysis of the results showing how they are related to health at work.

Keywords

Radon gas • Groundwater • Volcanic islands • Safety at work • Mining works

1 Introduction

The island of Tenerife is one of the eight which make up the Canarian Archipelago and is volcanic in nature. The variety of formations, products, and volcanic lithotypes there are on the Canary Islands means they are one of the most interesting territories in the world from a volcanology point of view. The water resources on the island are specially managed due to the uniqueness of the volcanic insular systems. Factors such as origin, geology and formation process, have a considerable effect on the use of groundwaters, and in Tenerife the mines or "water galleries" are the main system for exploiting the aquifers (Santamarta 2016).

In general, the mines or water galleries are boreholes or tunnels with an average cross-section of 1.5 × 2 m. These collections have just one point of access to the interior but they may have one or more secondary divisions (branches) so they may reach lengths of over 6–7 km, although the most common ones are between 2 and 3 km. Normally, the most important extractions of groundwaters are carried out at the deepest point of the main gallery and/or the branches. However, there are usually other extractions of greater or lesser importance throughout the gallery route.

Radon is a radioactive element that is found in nature in different states (radioactive isotopes) and one of these, which is the most stable and plentiful, is ^{222}Rn, hereinafter, just radon. This radon isotope is what we are concerned within this research.

The IARC (International Agency for Research on Cancer) classifies radon as being "carcinogenic for humans". It must be considered that all radionuclides which give off alfa-particles and which have been studied in detail, including ^{222}Rn and its descendants, have been shown to cause cancer in humans (IARC 2001).

Spanish law in the sphere of health and safety at work makes it compulsory for the owners of certain places of work to declare the situation they are in as regards the ionizing radiation which may affect their workers. Volcanic territories are considered by the Nuclear Security Council (CSN) as "identified areas" since they require special attention as they create and are conducive to the emission of ionizing radiation caused by radon (CSN 2012).

The lithotypes on the Canary Islands cover a wide range of petrological classifications of volcanic rocks and can be

J. C. Santamarta (✉)
Departamento de Ingeniería Agraria, Náutica, Civil y Marítima, Universidad de La Laguna (ULL), La Laguna, Spain
e-mail: jcsanta@ull.es

L. E. Hernández-Gutiérrez
Servicio de Laboratorios y Calidad de la Construcción, Viceconsejería de Infraestructuras y Transportes Consejería de Obras Públicas y Transportes, Gobierno de Canarias, Las Palmas de Gran Canaria, Spain

J. Rodríguez-Martín · R. J. Lario-Bascones
Departamento Técnicas y Proyectos en Ingeniería y Arquitectura, Universidad de la Laguna (ULL), La Laguna, Spain

Á. Morales-González-Moro
Servicio de Minas, Dirección General de Industria y Energía, Gobierno de Canarias, Las Palmas de Gran Canaria, Spain

© Springer Nature Switzerland AG 2020

F. Fernandes et al. (eds.), *Advances in Natural Hazards and Hydrological Risks: Meeting the Challenge*, Advances in Science, Technology & Innovation,
https://doi.org/10.1007/978-3-030-34397-2_8

found rocks from the basic field, such as basalts, to felsics or salics, such as the rhyolites, through to intermediate ones such as trachytes and phonolites, among others. Recent studies have determined that many of these rocks, most of which are on the felsic and intermediate fields of the geochemical scale, surpass the standard criteria for natural radiation security (Hernández et al. 2012).

The main aim of this project is to promote the application of the regulations in force on health protection against ionizing radiation caused by radon gas in areas of work in the Canary Islands, especially the underground mining systems.

The following special objectives are covered:

- Making a report about incidences in the workplace of the measurements taken.
- An analysis of the relationship between the measurements taken and the type and hydrogeological framework of the galleries.
- Drawing up a report about the works made which is suitable for direct publication in the specialized press.
- Recommendations for future research.

To reach the general objective, some operating objectives were considered, which were:

- Assessing the air quality and how healthy it was, with respect to the concentration of radon gas in the water galleries. All of this was carried out in observance of the regulations in force and by applying the corresponding protocols, and as a pilot project which may be used in the other water galleries in the Canary Islands region.
- Drawing up the action procedures for evacuating the underground work areas, which covered the following items:
  a. Designing campaigns for measuring the concentration of radon gas in the interior of the underground systems.
  b. Managing the devices for measuring the concentration of radon gas.
  c. Declaring these activities before the authorized bodies for an industry in the region, in accordance with Royal Decree 1439/2010 of the 5th of November.
  d. Applying the provisions of European Directive 2013/59/EURATOM which by transposition to Spanish legislation makes it obligatory to comply with.

2 Methodology

The methodology used in this project is that developed by the Nuclear Security Council (CSN) in its Security Guide 11.4 Methodology for assessing exposure to radon in places of work (CSN 2012) and in compliance with Royal Decree 1439/2010, of 5th of November, which amends the Regulation on Protecting Health against Ionizing Radiation, passed by Royal Decree 783/2001, of the 6th of July (BOE num. 279 of 18.11.2010). This guide was conceived as one for applying the Regulation on Health Protection against Ionizing Radiation (RPSRI). The guide is the official document for studying radon in places of work.

When planning this study nine water galleries were selected, which were preferably ones with works in progress, and which were representative of the island, in the sense that they could show the possible conditions of most of the galleries, depth, water line, air, water temperature, etc. For this reason, due to the varied types of works, as this project was carried out on the island of Tenerife, a greater number of variables were covered which enables the state regulations on the sector for the exploitation of groundwaters to be adapted to in the future.

According to the Security Guide 11.4 the studies on the radiological risk linked to radon must be representative of the annual exposure the workers and, if applicable, the public, are subject to. Therefore, the results must be based on measurements with passive detectors which are exposed for a minimum of three months. For this reason, nuclear traces passive detectors were chosen, with the Radtrak2 model from the Radnova laboratory (Sweden) in order to meet all the CSN requirements. These cutting-edge detectors consist of a radiation-sensitive film located within a capsule made of a special antistatic plastic which enables radon gas to enter by diffusion. These detectors are harmless and do not affect people. They do not require a power source and carry out integrated measurement for a recommended period of 3 months. The specifications of these detectors are:

- Measuring range: From 15 Bq/m^3 up to 25,000 Bq/m^3.
- Measuring period: From 2 to 6 months.

Measurements are carried out in places representative of the real occupation or reasonably foreseeable place of work for the workers. A detector is installed at the deepest area of the gallery and the other halfway along the route. Measurements are taken for 3 months. At the end of this period, the detectors are replaced for new ones and so on until a full year has passed, as established by the CSN. The measurement devices are deposited by the engineers of the exploitations.

After the passive detectors are taken away they are immediately sent to the accredited laboratory for analysis. At this laboratory, the detectors are analyzed using a cutting-edge image scanner. For each place of work, the laboratory issues a certifying report with the average concentration of radon gas, and its uncertainty, during the measurement period, expressed in Bq/m^3.

Table 1 Results of the radon gas measurements in the different water galleries

Code	Municipality	Length (m)	Minimum (Bq/m^3)	Average (Bq/m^3)	Maximum (Bq/m^3)
01	Güimar	2000	5040	6150	7260
02	Santiago del Teide	1800	1860	2270	2680
03	La Guancha	1407	10,070	12,590	15,110
04	Guía de Isora	2500	1860	2270	2680
05	Los Realejos	1300		>25.000	
06	Fasnia	3525	2970	3630	4290
07	Guía de Isora	1600	10,280	12,860	15,440
08	Arico	3015	3200	3910	4620
09	La Orotava	3000	6100	7630	9160

3 Results

Below, the results are given for the front of the gallery in the first cycle for measuring radon gas, carried out between April'18 and June'18 (Table 1).

4 Discussion

In the absence of any conclusions for the measurements and the performance of a global analysis of them, for now, it can be said that the results of the concentration of radon gas in the nine galleries on Tenerife are well above the thresholds established by the Spanish regulations in force (300 Bq/m^3) and by that set by the European Union in its directive 2013/59/EURATOM.

5 Concluding Remarks

In all the measurements carried out to date, the limits of radon gas set by European legislation from safety at work point of view have been greatly surpassed.

Apart from the conclusions as regards safety at work other matters of a hydrogeological nature have been observed which are listed below:

- The presence of water extraction leads to an increase in the amount of radon gas measured.
- The greater the porosity and permeability of the materials which go through the galleries, the greater the amount of radon gas detected by the devices.
- The greater the concentration of endogenous CO_2 type gases in the galleries, the higher the value of radon gas.
- Galleries with less mechanical ventilation showed higher values of radon gas.
- The extracted waters may work as a vehicle for transporting radon gas in the gallery until it reaches the open-air.

Once the measurement results for the whole year have been obtained, the corresponding rectifying recommendations will be made so that workers in the galleries will not be exposed to levels of radon gas which are higher than those set by legislation.

5.1 Recommendations for Future Research

For future research, the following activities have been recommended:

Extending this research to other systems on different islands has been recommended.

One recommendation is to gain more knowledge about the variables which may affect or control the activity of radon gas in the atmosphere inside the galleries: location and flow of water extractions, water transport in the interior (open channel or closed conduction) atmospheric pressure and air temperature outside, etc.

Another suggestion is to make a technical recommendation by way of supplementary technical instructions in order to improve the working conditions in the water galleries as regards the radon gas.

Inform the pertinent industrial sector, by means of conferences or similar activities, of the project results in order to raise awareness about the problem of radon gas in underground areas in volcanic terrains.

Acknowledgements Research Project funding by Servicio de Minas. Dirección General de Industria y Energía, Gobierno de Canarias.

References

Consejo de Seguridad Nuclear (2012) Guía de Seguridad 11.04 del CSN: Metodología para la evaluación de la exposición al radón en los lugares de trabajo. Ed. Consejo de Seguridad Nuclear. Madrid

Hernández LE et al (2012) Radiology of Canarian volcanic rocks. In: Qian Q, Zhou Y (eds) Harmonising rock engineering and the environment. Taylor & Francis Group, London. ISBN 978-0-415-80444-8, pp 641–644

International Agency for Research on Cancer (2001) Ionizing radiation, Part 2: some internally deposited radionuclides. In: IARC monographs on the evaluation of carcinogenic risks to humans, vol 78. IARC Press and The World Health Organization Distribution and Sales, Lyon, France

Santamarta JC (2016) Tratado de Minería de Recursos Hídricos en Islas Volcánicas oceánicas. Ed. Colegio Oficial de Minas del Sur de España. Sevilla

Hazard Assessment, Spatial Planning and Climate Change

Hydrological Risks in Natural Hazards Focused on the Role of the Water: Studies on Landslides

Luis Sousa, Eurípedes Vargas Jr., Rita Sousa, and Helder I. Chaminé

Abstract

Landslides are one of the most important natural hazards on Earth. Landslide mechanisms related to hydrology are analyzed with a focus on the major risks, and formal risk assessment methodologies are presented. The management of accidents in slopes is discussed, with particular emphasis on Hong Kong and to Rio de Janeiro. Specific aspects of slope instability in the state of Rio de Janeiro are discussed, with a focus on cases that occurred in the mountainous region of the State. The megadisaster occurred in 2011, with high number of deaths and significant economic losses, is studied with particular incidence in individual landslides. A description of occurrence of large landslides in China is also presented. The modeling of catastrophic flow-type slide at Shenzhen landfill induced by intense rainfall is introduced and referred to large number of landslides happened during the Wenchuan earthquake in 2008 is mentioned.

Keywords

Natural hazard • Hydrological risk • Groundwater • Landslide • Earthquake

1 Introduction

Landslides are one of the most important geotechnical risks on Earth. Landslides are difficult to predict because their initiation depends on many factors and on the interaction between these factors. To satisfy the societal demand for protection against landslides, it is necessary to systematically assess and manage landslide hazard and risk. This can be done using principles of decision making under uncertainty.

The presence of water is one of the major triggers. The hydrologic response of a slope to rainfall is very complex and depends on a great number of factors, both on local and catchment basin scales. Figure 1 illustrates some of these factors. In this study landslide mechanisms related to hydrology are analyzed with a focus on the major risks, and formal risk assessment methodologies are presented. Concepts concerning the problem of risk management in slopes are also highlighted. The existing experience in Hong Kong is reported, as well in Rio de Janeiro city, with emphases the important activity of the Geo-Rio Foundation. Specific aspects of slope instability occurring in the State of Rio de Janeiro, with incidence in the mountainous region are discussed. Also, important events that occurred in China are illustrated.

2 Risk and Risk Assessment

The systematic risk assessment and management in slopes is essential to ensure the safety of such systems. The risk assessment begins with the site selection. During the design and planning stages, it is necessary to evaluate risks associated to the geology and to obtain information on-site that provides the input necessary for the risk assessment evaluation. The stage related to the identification and description of threats will be the next step, in order to evaluate risks and the formal incorporation of uncertainties in the slope issue. The problem of risk identification models, in particular with

L. Sousa (✉)
Department of Geotechnical Engineering, Tongji University, Shanghai, China
e-mail: sousa-scu@hotmail.com

E. Vargas Jr.
Department of Civil Engineering, Catholic University of Rio de Janeiro, Rio de Janeiro, Brazil

R. Sousa
Department of Civil, Environmental and Ocean Engineering, Stevens Institute of Technology, Hoboken, USA

H. I. Chaminé
Laboratory of Cartography and Applied Geology, Department of Geotechnical Engineering, School of Engineering (ISEP), P.Porto, Porto, Portugal

© Springer Nature Switzerland AG 2020
F. Fernandes et al. (eds.), *Advances in Natural Hazards and Hydrological Risks: Meeting the Challenge*, Advances in Science, Technology & Innovation,
https://doi.org/10.1007/978-3-030-34397-2_9

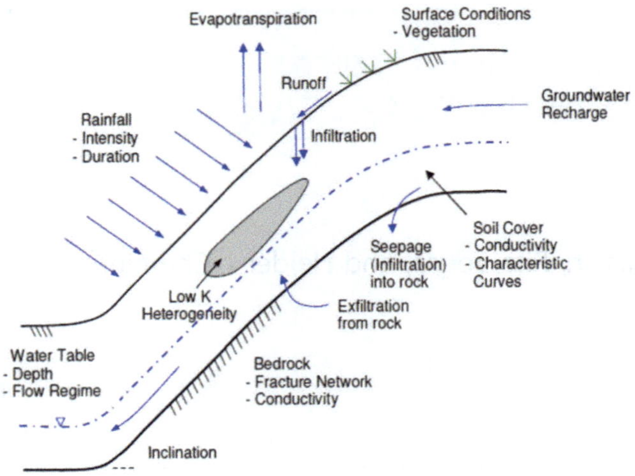

Fig. 1 Factors affecting slopes due to rainfall. Adapted from Karam (2005)

regard to instability mechanisms, deterministic and probabilistic models, geotechnical risk identification and risk management.

The identification and description of accidents in slopes are very important, with the collection of information on movements and geometry of accidents. The establishment of databases with accidents and their structuring aiming the application of Data Mining techniques (DM) and the direct description of the instability of slopes is processed in order to analyze the risk. Special focus is given to various DM techniques, with special emphasis on Neuronal and Bayesian Networks (BN). The use of the probabilistic techniques of BN is relevant in order to manage the risk and sensitivity analyses are recommended. The application of a BN to a decision problem in slope instability is well illustrated in Einstein and Sousa (2012).

Risk assessment and risk management for slopes requires an initial step the identification of the threats that are possible which can lead to detrimental consequences. The major threats are in general associated to precipitation, subsequent infiltration and groundwater flow (Sousa et al. 2017). The occurrence of earthquakes can be very relevant as a triggered factor (He et al. 2019).

3 Management of Accidents in Slopes

Natural or constructed slopes are defined as inclined surfaces in rocks or soils, or mixed masses originated from geological, geomorphological and human action processes. The project of a slope involves the collection of geological and geotechnical information, the evaluation of the resistance properties of the formations and discontinuities and the presence of the water with emphasis on the pluviometric regime.

For slope movements, there are several classifications (e.g., Bronnimann 2011). The risk management practice is described for Hong Kong region and the city of Rio de Janeiro.

Honk Kong is a region of China with a large population density, in a territory of only 1100 km^2, and very mountainous. About 75% of the territory has slopes greater than 15° and more than 30% has slopes greater than 30°. There is a history of tragic accidents in the slopes. The territory has a tradition of a high standard in the practice of slope engineering. The establishment of an efficient security system developed by Geotechnical Engineering Office (GEO) was established in the territory. In each year GEO reported hundreds of slope instabilities, whose data are stored on a database, being the large majority relatively small although larger slips occur, with more than 5000 m^3. It was also introduced as a long-term program designated *Landslide Preventive Measures* to systematically analyze the slopes of government works and perform a safety analysis of private works. Maintenance programs were established by the various governmental sectors. The definition of risk reduction strategies is one of the most relevant activities being developed in order to avoid or eliminate certain accidents and includes stabilization, mitigation; this is a reduction of vulnerability, and the establishment of warning systems (Sousa et al. 2013). Detailed risk analyses are made consistently in the territory, as well as the establishment of guidelines for the study of accidents. A flowchart of the slope stabilization actions is illustrated in Fig. 2.

The city of Rio de Janeiro is another important situation in terms of hydrological risks. The occupation of hills and slopes throughout its history has been promoted. The occurrence of serious geotechnical accidents, particularly of

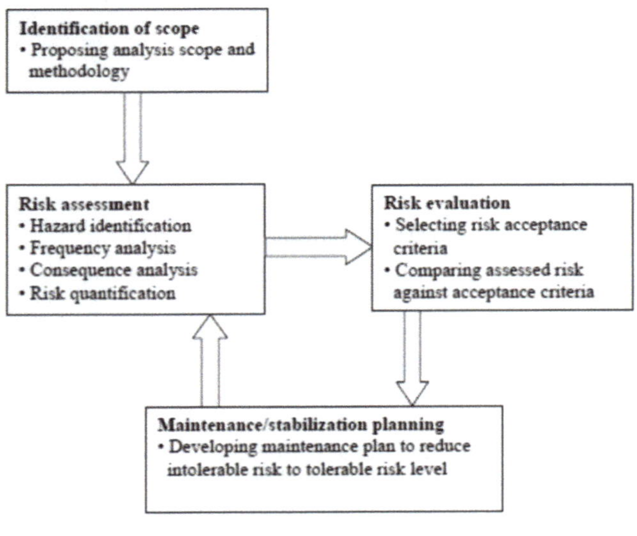

Fig. 2 A general flowchart of slope stabilization actions

falling blocks, derives from its unique and complex geographic and morphological characteristics. The studies conducted through surface mapping showed high susceptibility to rock falls. Sometimes there have been observed tears in the slopes during dry periods, and interpretative models originated in thermal actions were proposed. Several containment works have been carried out, as buttresses.

Almost 45 years ago, the *Instituto de Geotécnica*, a division of the Municipality of Rio de Janeiro, later renamed the Geo-Rio Foundation, was founded. Its creation occurred after severe rainy events that occurred in the city in 1966. Since then, the technical staff formed by engineers and geologists conducts surveys and defines the necessary works to ensure the safety of the population. Geo-Rio has implemented the Rio Alert system, which maintains pluviometric stations scattered throughout the city to monitor the rains (Fig. 3). Meteorological radar images are critical to the detection of storms. A radar is located in Sumaré, and its images are updated every two minutes and allow observing the location, displacement, and intensity of precipitation.

4 Instabilities of Slopes in Rio de Janeiro State

In 2010 and 2011, more than 1000 people died in natural disasters in the State of Rio de Janeiro, like in Angra dos Reis, in Niterói and in the mountainous region of the State. The megadisaster of the mountainous region of the State occurred between the 11th and 12th of January 2011, reaching seven cities mainly Nova Friburgo, Teresópolis, and Petrópolis and it was considered one of the larger events of generalized mass movements in Brazil (Sousa et al. 2013).

According to the documents of the time, doubts arose about what caused the disaster in the proportions that occurred and, in particular, on the factors that led to so many human losses. The region has always been characterized by a great natural vulnerability, located in the Serra do Mar, formed by rocks with a thin layer of land and covered by Atlantic forest, with high slope and heavy rainfall regime in summer, characteristics that generate soils more unstable and prone to landslides.

The urban slides that occurred affected slopes in the base of mid-slope elevations between 30° and 45°, which proved their associated high risk. In Fig. 4, aspects of the planar slipway occurred in *Três Irmãos* in 2012, responsible for the destruction of some buildings in the vicinity of the accident is evidenced.

5 Landslides in China

Considerable disasters due to landslides occurred in China. Flow-type landslides happened in several parts of China like in Gansu, China, in 2010 and recently a large flow-type occurred at a municipal solid waste in Guangming New District of Shenzhen, China. The landfill accident in Shenzhen that happened on December 20, 2015 caused the missing of 70 people and presumed dead and the destruction of 33 buildings (Li et al. 2018). The disaster involved an area with 1100 m in length and 630 m in maximum width. Figure 5 shows results of a numerical study of the accident (Li et al. 2018).

Earthquakes may trigger thousands of landslides in mountainous regions in the case of strong ground motion and associated not favorable geomorphic environment. This was what happens during the 2008 Wenchuan earthquake that occurred in Sichuan province, China. Most of catastrophic landslides were distributed along the central Longmen Shan fault system in a mountainous region, at the eastern margin of the Tibetan Plateau. The earthquake resulted in a large number of landslides and caused significant economic damage and it was the most important earthquake that had occurred in China. More than 60,000 landslides were triggered by the earthquake. The sudden damming of rivers by the generated landslides caused

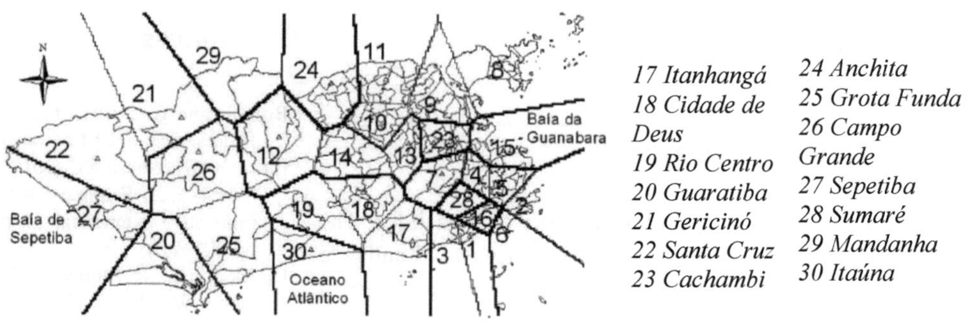

Fig. 3 Network of pluviometric stations at Rio de Janeiro Sousa et al. (2013)

Fig. 4 View of the landslide at *Três Irmãos* (Brazil)

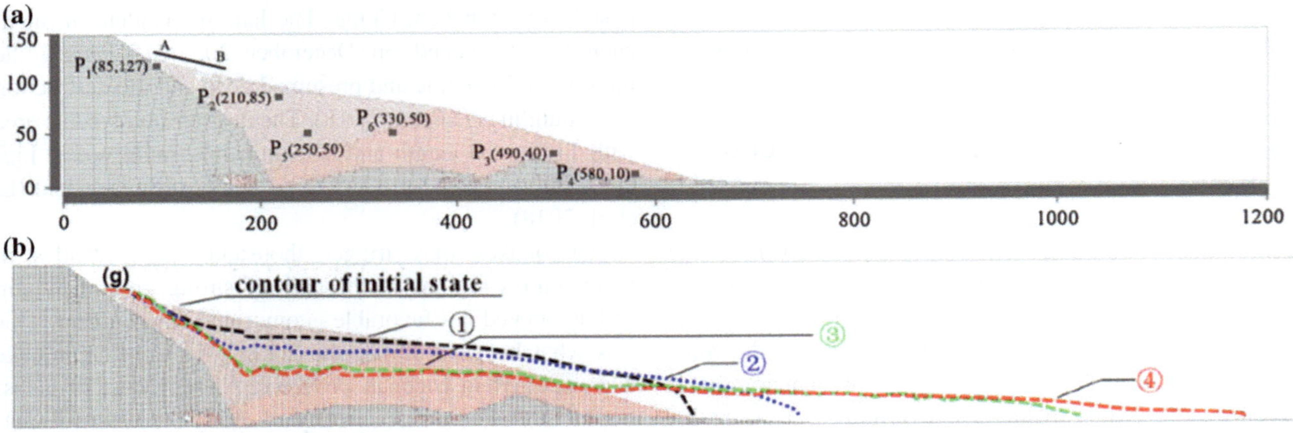

Fig. 5 Flowslide at Shenzhen: **a** numerical modeling; **b** flow configuration at different times Li et al. (2018)

flooding and generated a large number of landslide dams, more than 800. A description of more significant landslides is referred with focus on the Daguangbao landslide, involving a volume of 7.4×10^8 m^3. Numerical simulations were performed for understanding the kinetic mechanism that occurred at the Daguangbao landslide using MPM (Material Point Method) models using 2D and 3D approaches (He et al. 2019).

6 Concluding Remarks

This work focuses on hydrological risks of landslides. The presence of water is one of the major triggers. Landslide mechanisms related to hydrology are analyzed, and formal risk assessment methodologies are highlighted. The management of accidents in slopes is discussed, with emphasis on Hong Kong and to Rio de Janeiro sites. Specific aspects of slope instability in the state of Rio de Janeiro are discussed, with incidence to cases occurred in the mountainous region of the State. A description of occurrence of large landslides in China is also presented.

References

Bronnimann C (2011) Effect of groundwater on landslide triggering. Federal School of Lausanne. Swiss, 223. (Ph.D. thesis)

Einstein HH, Sousa RL (2012) Risk in slopes. In Ribeiro e Sousa L, Vargas E Jr, Fernandes MM, Azevedo R (eds) Proceedings of the Innovative Numerical Modelling in Geomechanics, CRC Press, pp 201–210

He M, Sousa LR, Muller A, Vargas Jr E, Sousa RL, Sousa Oliveira C, Gong W (2019) Numerical and numerical models and safety considerations for Daguangbao landslide triggered by the 2008 Wenchuan earthquake (submitted and in revision)

Karam K (2005) Landslide risk assessment and uncertainties. MIT, USA, 747. (Ph.D thesis)

Li ZH, Jiang YJ, Lv Q, Sousa LR, He MC (2018) Consistent modelling of catastrophic flowslide at the Shenzhen landfill using a hydro-elastoplastic model with solid-fluid transition. Acta Geotech 13(6):1451–1466

Sousa LR, Vargas Jr E, Sousa RL (2013) Analysis and risk management of slopes at Rio de Janeiro State. PUC-Rio, Report, 122 (in Portuguese)

Sousa LR, Sousa R, Vargas E Jr, Velloso R, Karam K (2017) Risk assessment on CO_2 injection processes and storage. Rock Mech Rock Eng 12:359–397

The Reliability of a Heuristic-Holistic Methodology for the Definition of Susceptibility to Destabilization—The Potential Sources of Uncertainty

Celeste Jorge

Abstract

The results of the approaches to assess the susceptibility to landslides are dependent on the type of approach process, the quantity and quality/resolution of data, and the experience of experts. In all processes, uncertainty is a certainty. This paper aims to identify and list the potential sources of uncertainty and to define the stages in which uncertainties can be inserted in this particular methodology. Using the Digital Elevation Model of the terrain and the geotechnical data, it is intended to establish the hazard related to the mass movements in the area and its spatial distribution. For that purpose, a heuristic-holistic approach is defined. That approach is done by phases. The first phase includes surveying and collecting existing data. The second phase corresponds to the organization, classification, and systematization of all information, using the Geographic Information System. Simultaneously is done a Multi-Criteria Analysis treatment of the data and there are selected the physical factors that more contribute to the instability. In a third phase, it is represented cartographically the variability of the selected factors in the area. The integration of all parameters, cartographically represented, allows to obtain maps of susceptibility. The reliability of the final susceptibility maps is dependent on the uncertainty, especially in a scenario of climate change.

Keywords

Reliability • Uncertainty • Heuristic-holistic approach • Multi-criteria analysis • Hydrological risks

1 Introduction

The occurrence of slope instabilities is an issue that affects many countries all over the world. The consequences of mass movements are not restricted to the loss of human lives. These events cause huge loss of assets and infrastructures, affecting entire communities. With the increasing demand for space in cities, more areas are being used for construction, including areas that are not suitable in terms of stability, or that will require human specialized intervention to improve their resistance to rupture. And as EEA (2016) stated, the design of the city, its buildings and its infrastructures are supposed to last for decades or even centuries.

In areas where there is already evidence or records of mass movements, it is necessary to do a stability analysis with the main aim of creating susceptibility, vulnerability and risk zonation maps. The spatial scale and the purpose of hazard assessment, as well as the data available, determine which methods or slope stability models can be applied (Almeida et al. 2017).

Considering science advice in a natural hazards context, many levels of uncertainty exist (Doyle et al. 2019). These range from the natural stochastic uncertainty (the variability of the system) to the epistemic uncertainty (lack of knowledge), to scientists being uncertain about their knowledge and data, through to disagreement amongst scientists due to (a) "incomplete information", (b) "inadequate understanding", and (c) "undifferentiated alternatives", as well as issues arising due to conflicting scientific advice from scientific advisory bodies and individuals (Doyle et al. 2019). Furthermore, uncertainty is a major issue and a constant in all geotechnical studies.

The uncertainty must be considered in all processes. In order to understand and trust in the final results of these studies, it is required a specialized critical sense to interpret them, and a huge knowledge about the mechanisms in analysis process. All of that is particularly important in a scenario of climate change, with the increase of extreme precipitation events that work as trigger of most of the landslides.

C. Jorge (✉)
Advisory Group of Board of Directors, Laboratório Nacional de Engenharia Civil, Lisbon, Portugal
e-mail: cjorge@lnec.pt

© Springer Nature Switzerland AG 2020

F. Fernandes et al. (eds.), *Advances in Natural Hazards and Hydrological Risks: Meeting the Challenge*, Advances in Science, Technology & Innovation,
https://doi.org/10.1007/978-3-030-34397-2_10

The aim of this communication is to identify and list the type of uncertainty presents in each stages of the established methodology, for the production of (in) stability susceptibility maps.

2 Methodology

It was defined and adopted an iterative methodology, based on indices. This method is essentially based on knowledge on the mechanisms and the processes that contribute for the instability in a given area, as defined by an expert, in a holistic frame—the land use, the geology/geotechnics, the geomorphology, etc.

The identification and the understanding of the contributing factors for the (in) stability, given their complexity and the number of factors involved in the processes of erosion, drainage and landslide slope, can only be achieved through intense office-based research. This research should include the analysis of previous studies in the area, and other relevant information, such as land use, type of vegetation, slope, drainage, roads system, among others; climatic data and appropriate treatment; interpretation of geological and geotechnical data; and systematic aerial elements (photos or satellite images) survey. Simultaneously or after, a field survey is required so that any suspicions can be confirmed and all the evidences of movements on the ground and in the built-up area taken into account.

The application of multi-criteria analysis was performed by Analytic Hierarchy Process (AHP) with a characterization of the area by the methods for the definition of geotechnical/geoenvironmental mapping. The assignment of weights for each attribute was made from a hierarchical parity analysis (larger weights given to greater potential for instability, and lower weights to less potential instability), as set forth in Jorge and Ramos (2012).

The performed mapping approach is analytical. The physical medium is divided into several components, which are represented and judged for subsequent integration in a holistic way. The developed and prepared documents for the geotechnical mapping phase include: the basic maps (e.g., topographical, geological, groundwater level, thickness of unconsolidated material), optional basic maps (e.g., geophysicist/partial climacteric, current or expected soil occupation), auxiliary maps (maps of documentation and data—both qualitative and quantitative) and derived or interpretative maps (obtained from the interpretation of the other maps already mentioned). In the present case, the derived maps include thickness of loose or soft material at different depths, depth of hard geologic formation, exposure and moisture content (wetness index—WI).

To carry out the work, several steps were established, as shown in Fig. 1. The work developed allows to obtain the resultant maps—susceptibility to (in) stabilization for different depths.

3 Results—Landslides Susceptibility Maps

According to Guzzetti (2005), susceptibility is the likelihood of the spatial distribution of the movements considering a set of geo-environmental conditions and physical factors. Following this definition, it could be assigned a given degree of susceptibility to each site of a shed, allowing to determine the locations most prone to the occurrence of landslides, not taking into account the probability or the magnitude of the occurrence.

For the production of landslides susceptibility maps, it is necessary to carry out a specific mathematical treatment, using the values defined for the main attributes/conditioning factors (lithology, slope, thickness of soil, land use, curvature, WI, exposure, distance to the watercourses, and distance to roads, etc.). The range of values of each of these attributes/factors has been divided into a given number of classes so that their contribution to the development of the instability can be characterized. The matrices obtained in the calculations and the respective sensitivity analysis were developed in Jorge and Ramos (2012).

After the weight vectors of the criteria and the scores vector of the alternatives were defined, relative to the criteria, these values were aggregated to obtain the final result. The integrated processing of data allowed producing the susceptibility maps to instability, with five potential classes for translational landslides and for rotational landslides at different depths, as presented in Fig. 2 (rotational landslides).

4 Discussion on the Uncertainty in the Maps Producing Process

The reliability of the susceptibility maps obtained depends on the treatment of the available data, during the process, and on the heuristic model design. The data are always considered not enough and the quality/resolution can be degraded during the classification. And agreeing with Almeida et al. (2017), it is impossible to remove researcher bias when interpreting and synthesizing findings.

In this specific methodology, the level of uncertainty can be associated with scenario uncertainty and to recognized ignorance. Relatively to the nature of the uncertainty, it can be considered the epistemic uncertainty and variability uncertainty. Table 1 shows all the uncertainty that can be defined in all stages of the process. Despite the uncertainty that appears in the process of landslide susceptibility maps construction, the results are the most reliable that can be

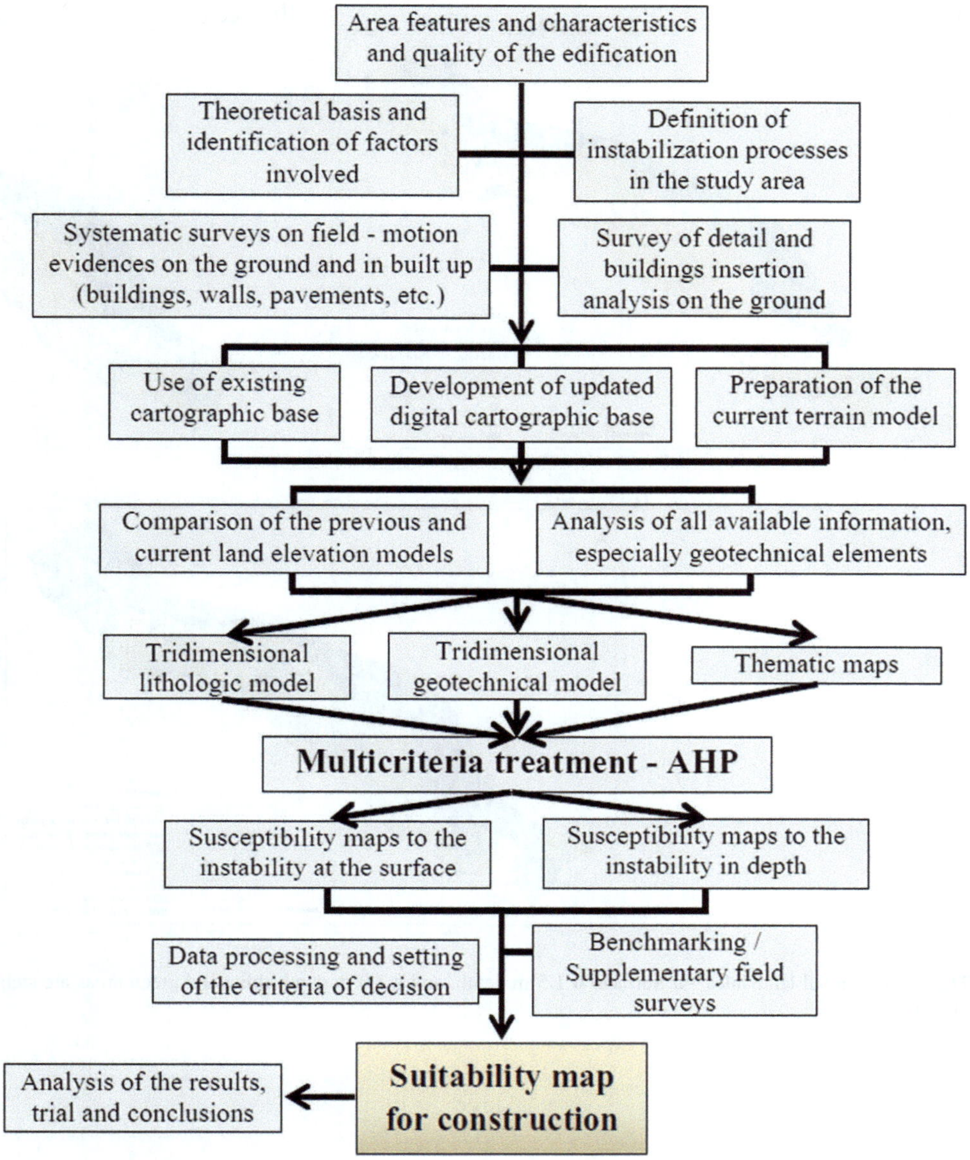

Fig. 1 Systematization of the procedure Jorge (2017)

obtained, with the data available, to respond to the request for zoning and classification of (in) stability in the study area.

5 Conclusion

The reliability of the susceptibility maps is determined by the precision of the inputs and the accuracy with which the model captures the relevant and physical processes contributing to the destabilization. The model is heuristics and in all the stages (understand the problem in question, establish a plan of approach, execute the plan and finally make a revision and interpretation of the result through the scientific method) the uncertainty is present. This kind of approach is very complicated and just expert knowledge can conduct to a good result, minimizing the uncertainty and increasing the reliability. The final susceptibility maps are the result of a progressive and accurate process. At the time of the study, with the knowledge and the information available, the situation outlined is the one presented for rotational landslides.

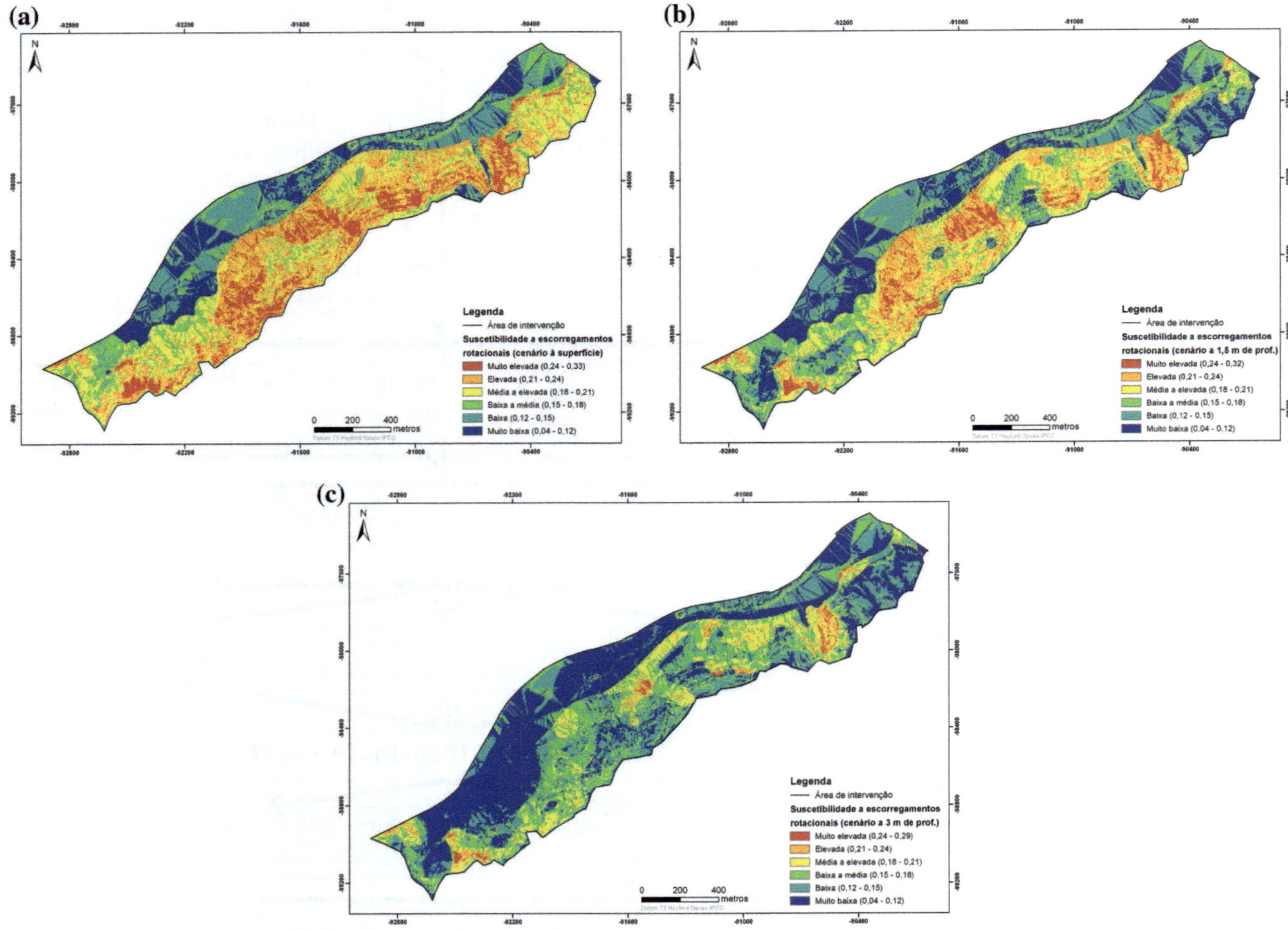

Fig. 2 Rotational landslides—**a** Surface, **b** 1.5 m depth, and **c** 3.0 m depth (blue and green areas are stable areas and reddish areas are unstable areas)

Table 1 Uncertainty classification—contribution to the reliability of the susceptibility maps, according to Walker et al. (2003)

Uncertainty type	Uncertainty describes the "doubt" about the measured value/model—qualitative analysis				
Location of uncertainty	Context of model	Model uncertainty	Inputs	Parameters	Model outcomes
	• Type, size, and magnitude of landslides • Physical processes • DEMs techniques and data treatment • Spatial uncertainty • System representation • Assumptions • Measurements	• Model uncertainty—action triggering movement • Specification of the problem • Formulation of conceptual and computational models • Estimation of input values • Calculation, interpretation, and documentation of the results • Physical processes • Measurement techniques • Classifications • Mathematical analyses (algorithms) • Data validation • Expert opinions • Attribute uncertainty • Sensitivity analysis (to reduce uncertainty)	• Numeric data • Qualitative data • Initial conditions • Boundary cond. • In situ test • Lab. tests • Databases construction • Data precision • Accuracy • No mean values • etc.	• Topography • Morphology • Lithology • Soil thickness • Groundwater • Curvature • Exposure • WI • Flow directions • Land use • Vegetation type • Dist. to roads • Dist. to streams	Cartographic representation - • Basic maps • Optional basic maps • Auxiliary maps • Interpretative maps • Model output
Level of the uncertainty	Scenario uncertainty (medium uncertainty)		Recognized ignorance (deep uncertainty)		
	• Insufficient data for the parameters		Trigger agent (frequency and magnitude) • Occurrence of extreme weather events (climate change) • Occurrence of seismic events		
Nature of the uncertainty	Epistemic uncertainty (knowledge imperfection)		Variability uncertainty (heterogeneity)		
	• Data missing • Extrapolation of data • Misclassification of a parameter • Parameter is capable (or incapable) of producing destabilization • Uncertainty about the appropriate model • Consistency of the method		• Depth and characteristics of loose material • Position of bedrock • Extrapolation of data • Variability in the parameters • Parameter is capable (or incapable) of producing destabilization		

References

Almeida S, Holcombe EA, Pianosil F, Wagener T (2017) Dealing with deep uncertainties in landslide modelling for disaster risk reduction under climate change. Nat Hazards Earth Syst Sci 17:225–241

Doyle EEH, Johnston DM, Smith R, Paton D (2019) Communicating model uncertainty for natural hazards: a qualitative systematic thematic review. Int J Dis Risk Red 28

EEA (2016) Urban adaptation to climate change in Europe. Transforming cities in a changing climate. European Environment Agency, Report N. 12/2016

Guzzetti F (2005) Landslide an hazard risk assessment—concepts, methods and tools for the detection and mapping of landslides, for landslide susceptibility zonation and hazard assessment, and for risk evaluation. Ph.D. thesis, University of Bonn, Italy

Jorge C (2017) An evaluation methodology of suitability for construction after a holistic and heuristic approach on a slope stability analysis. geo-risk 2017: impact of spatial variability, probabilistic site characterization, and geohazards, ASCE. Denver, Colorado

Jorge C, Ramos R (2012) Geological and geotechnical study for the slopes consolidation of the southern hillside of Odivelas. Vols. 1/2, Lisbon—03—0841—FEDER—00542. LNEC and CMO divelas, Lisbon (in Portuguese)

Walker W, Harremoes P, Rotmans J, van der Sluijs J, van Asselt M, Janssen P, Krayer von Krauss M (2003) Defining uncertainty: a conceptual basis for uncertainty management in model-based decision support. Integr Assess 4:5–17

Preliminary Analysis of Slope Instability Processes Triggered in the Guilherme Creek Watershed (Nordeste Municipality, S. Miguel Island, Azores)

Paulo Maciel Amaral, Rui Marques, Isabel Duarte, and António Pinho

Abstract

The Nordeste Volcanic Complex is located in the NE sector of S. Miguel Island (Azores) and it is composed of geological formations with about 4 million years old, extremely weathered. This fact, allied to the steep slopes on the area, promotes the occurrence of slope instability phenomena. In particular, the Guilherme Creek watershed has a high recurrence of this type of geological hazard, mainly triggered by precipitation, which promotes a high density of landslides in the area. Considering as main objective the landslide susceptibility analysis in the Guilherme Creek watershed, using statistical/probabilistic methods, it is ongoing an inventory process of these phenomena in a GIS, based on the aerial photography interpretation and fieldwork. Currently, the inventory is concluded based on aerial photography from 1982 and orthorectified aerial photography from 1995 to 2005. In order to validate and update the landslide inventory produced in the office, fieldwork has been carried out using, besides more traditional techniques, a drone (DJI Phatom 4 PRO). This tool has allowed studying steep slopes, which would otherwise be impossible to access in an area with a high forest density and very steep slopes. Until now, 517 landslides have been identified in the study area, mainly belonging to slides and falls typologies.

Keywords

Landslides • Geomorphological evolution • Geographic information systems • Nordeste volcanic complex • São miguel island • Azores

1 Introduction

Landsides are responsible for significant economic and social impacts in many areas of the world. They cause the death of people and animals, and destruction or damage of residential and industrial infrastructures and agricultural areas. The identification of areas that are susceptible to landslides is important for land-use and emergency planning, as stated by Silva et al. (2018). The Guilherme Creek watershed (Fig. 1) has a high recurrence of this type of geological hazard, mainly triggered by precipitation, which promotes a high density of landslides in the area. The Guilherme Creek watershed is located in the NE sector of S. Miguel Island (Azores) on the Nordeste Volcanic complex and its altimetry ranges from 0 to 1103 m (a.s.l.).

According to Fernandez (1980) and Duncan et al. (2015), this complex has 1300 m of exposed volcanic and hypabyssal intrusive rocks with a 4.01–0.95 M.y. In the higher peaks, road cuts expose highly weathered ash pumice profiles as much as 8 m thick, although in the lowlands thicknesses of 1–2 m are much more common. Furthermore, in the valleys, and in particular in the Guilherme Creek watershed, it is possible to trace alluviums and in some steeper cliffs the occurrence of aprons. All the geological formations are extremely weathered and allied to the steep slopes on the study area, promotes the occurrence of slope instability phenomena.

P. Maciel Amaral (✉)
Environment and Society, Department of Geosciences, University of the Azores, Ponta Delgada, Portugal
e-mail: amaral.paulomaciel@gmail.com

R. Marques
Centre for Information and Seismovolcanic Surveillance of the Azores, University of the Azores, Ponta Delgada, Portugal

Research Institute for Volcanology and Risk Assessment, University of the Azores, Ponta Delgada, Portugal

I. Duarte · A. Pinho
Centre GeoBiotec|UA and School of Science and Technology, Department of Geosciences, University of Évora, Évora, Portugal

© Springer Nature Switzerland AG 2020

F. Fernandes et al. (eds.), *Advances in Natural Hazards and Hydrological Risks: Meeting the Challenge*, Advances in Science, Technology & Innovation,
https://doi.org/10.1007/978-3-030-34397-2_11

Fig. 1 Guilherme Creek watershed partial view. Photo was taken by drone: Mário Nelson

2 Methods

The main objective of this ongoing work is the landslide susceptibility analysis in the Guilherme Creek watershed using the statistical/probabilistic methods proposed by Marques (2013). According to Silva et al. (2018), the assessment of susceptibility on the basis of statistical methods is based on the use of a landslides inventory, and their construction is generally a complex procedure that seeks the exact location and identification of specific characteristics and can be represented by cartographic points or polygons. First, with a GIS (QGIS 2.8) the landslide inventory of the area was carried out at the office, in a 1:1000

Fig. 2 Capturing the details of a landslide in the fieldwork, using a drone

scale, specifically by vectoring polygons from the monoscopic analysis of several orthophotos and the aerial photograph of the municipality of Nordeste. For this task were used orthophotos from 2005, at a scale of 1:5000 and resolution 0.4 m, orthophotos from 1995, at scale 1:15,000 and aerial photography from 1982, at scale 1:15,000. Additionally, were requested orthophotos of the year 2018.

Cartography fieldwork is being carried out, in particular, for validate, rectify and update the landslide inventory. In addition to the classic techniques, the fieldwork is supported by a drone (DJI Phatom 4 PRO), in steep slopes and high forest density areas, which would otherwise be impossible to access (Fig. 2).

3 Results

In total 571 landslides have been identified and mapped in the GIS (Fig. 3). From the preliminary results presented in Table 1, the density of landslides is 38/km^2 and the in the stabilized area is 78 m^2/km^2. The identified landslides correspond to slides and fall typologies.

4 Discussion

During the fieldwork it was possible to verify that for the massive majority of landslides identified, the deposits of the landslides were not in place, due to the high dynamics of the watercourses, erosion, and due to anthropic action.

In the south sector of the study area, the sector with higher altimetry, there are a large number of landslides reactivated, with landslides in progression, retrogression, reduction, enlargement. Often appear complex and composite landslides with multiple and successive phases of land movements. Thus, in this sector, landslides are the result of several reactivations, with advances and retreats and with a partial or total partition of rupture surfaces and displaced material, making it very difficult to identify and delimit each one of them. In this case, in the most complex areas, where the stability analysis is very complicated, it was decided to consider a single landslide, instead of a landslide with several types and phases of land movements.

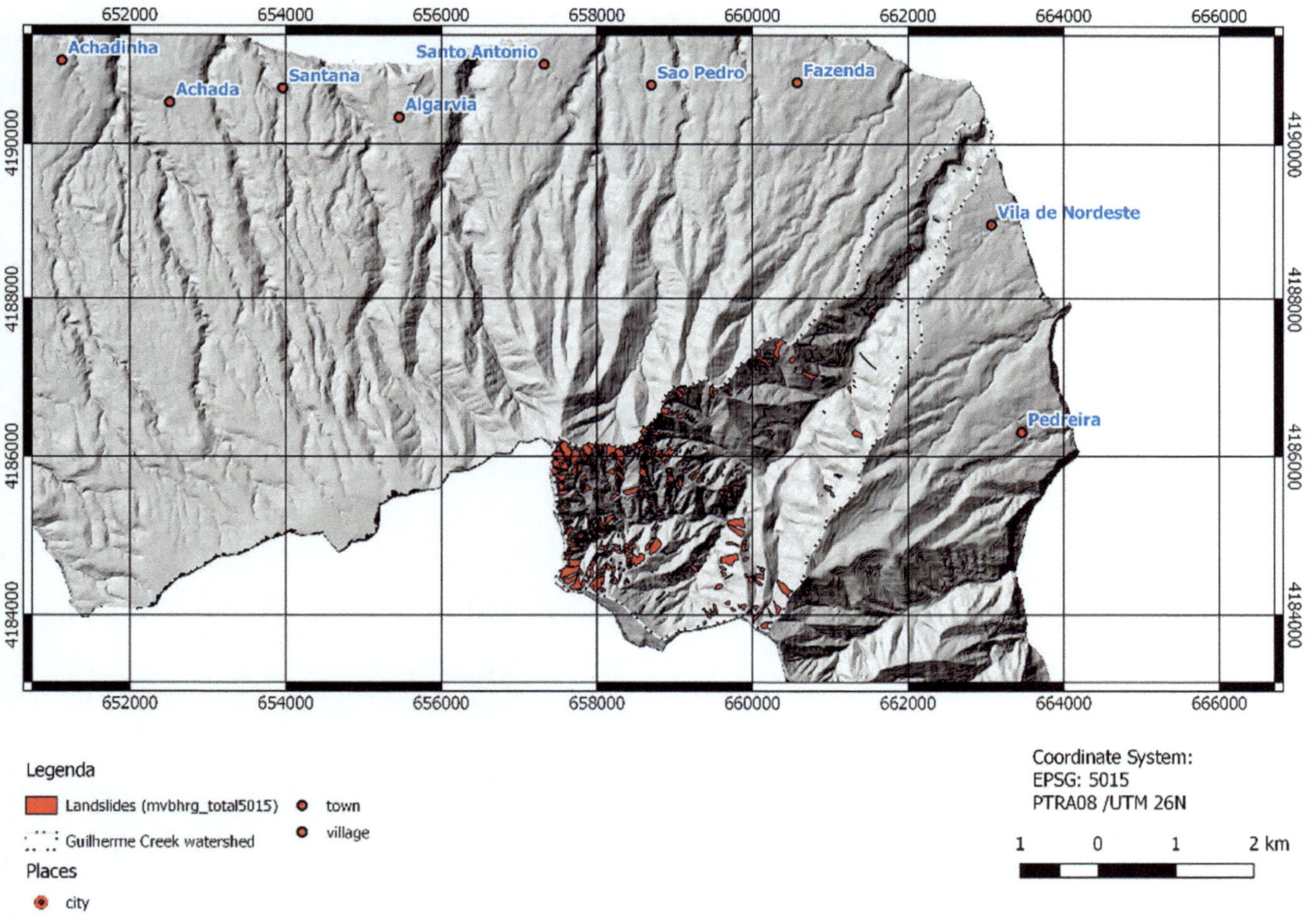

Fig. 3 Spatial distribution of the 571 landslides mapped in the Guilherme Creek watershed

Table 1 Main characteristics of the landslide inventory in the study area

Area (km^2)	Number of landslides	Planimetric unstabilized area (m^2)	Density	
			$\frac{\text{N.landslides}}{\text{Work area }(km^2)}$	$\frac{\text{Instabilized area }(m^2)}{\text{Work area }(km^2)}$
13.61	571	1,062,877.5	38	78,094

5 Concluding Remarks

This work represents the preliminary results of ongoing work. A lot of fieldwork has to be done in order to finish the landslide inventory and to study the morphometric characteristics of the landslides in the study area. After this task finished, statistical and probabilistic methods presented by Marques (2013) will be applied in order to study the spatial correlations between landslides and the predisposing factors (e.g., altitude, slope, insolation, and geology).

References

Duncan AM, Guest JE, Wallenstein N, Chester DK (2015) The older volcanic complexes of São Miguel, Azores: Nordeste and Povoação. In: Gaspar JL, Guest JE., Duncan AM, Barriga FJAS, Chester DK (eds) Volcanic geology of São Miguel island (Azores Archipelago). Geological Society, London, Memoirs, 44, pp 147–153

Fernandez AL (1980) Geology and petrology of the Nordeste volcanic complex, São Miguel, Azores. Geol Soc Am Bull, Part II 91:2457–2557

Marques R (2013) Estudo de movimentos de vertente no concelho da Povoação (Ilha de S. Miguel, Açores): Inventariação, caracterização e análise da susceptibilidade. Universidade dos Açores, Ponta Delgada (Tese de doutoramento)

Silva RF, Marques R, Gaspar JL (2018) Implications of landslides typology and predisposing factor combinations for probabilistic landside susceptibility models: a case study in Lajedo Parish (Flores island, Azores—Portugal). Geosciences 8(5):153

The Use of Total Station for Monitoring Mass Movements: Application to Fajãzinha Landslide at Flores Island (Azores Archipelago)

Paulo Amaral, Ana Malheiro, Filipe Marques, Letícia Moniz, Sílvia Furtado, and Nuno Loura

Abstract

The present work has as study area a deep-seated landslide, which affects a significant part of Fajãzinha parish, on Flores Island (Azores Archipelago). This study is part of a monitoring project composed of hydrological, geotechnical and topographic components used to monitor a large slow movement with a depth surface of rupture triggered by rainfall. With the main goal of detecting small soil movements that precede larger instabilities, a geodetic monitoring programme was established using an automatic total station. In total 34 benchmarks were reobserved by an automatic total station Leica TM50 of 0.5", with accuracy for distances of 0.6 + 1 mm. In this work, the total station data surveys are presented as well as the discussion of 27 campaigns carried out during the time span of 1 year (January 2018 until February 2019) to follow the evolution of the surface displacements. The maximum displacement rate observed in the area was about −57, 16 and −27 mm to E, N and Z component, respectively. The orientation of the displacement rate is compatible with the landslide that occurs in the area. Nowadays, this slope instability is confirmed by the slope geometry, ground cracks, and ground movement. This slow movement has caused, over the years, a number of disturbances to the inhabitants, due to damage to houses and structures, given the strong exposition of these vulnerable elements located on the body of the landslide.

Keywords

Landslide • Automatic total station • Rainfall • Fajãzinha

P. Amaral (✉) · A. Malheiro · F. Marques · L. Moniz
Laboratório Regional de Engenharia Civil (LREC), Ponta Delgada, Portugal
e-mail: Paulo.AP.Amaral@azores.gov.pt

S. Furtado · N. Loura
Direção Regional do Ambiente, Ponta Delgada, Portugal

1 Introduction

It is known that rainfall is one of the main triggering factors to cause landslides in Azores Archipelago, whenever their values promote the reduction of shear strength due to the decrease in apparent cohesion of the soil, or due to the increase of positive pore water pressure in the potential surface failure (Malet et al. 2005).

The detection and identification of possible precursors of landslides (initial failure slope, acceleration of the moving masses, fluidization) require high-frequency monitoring and very high precision of the displacements measurements (Malet et al. 2002). A wide variety of techniques is used for landslide displacements monitoring, from classical ground-based geodetic methods (tachometry, GNSS—Global Navigation Satellite System), to passive and active remote sensing methods (optics, radar or LiDAR) (Jaboyedoff et al. 2012). An overview of the accuracy, advantages, and disadvantages of methods and instruments for measuring displacements due to processes of geomorphological instability can be found, for example, in Gili and Corominas (1992) and Amaral (2005).

For surface displacement monitoring, classical methods allow to obtain 3D observations with very high accuracy (generally in the range of a few millimeters) and high frequency (minutes to hours). However, these methods provide only a local measurement (e.g., point) of the deformation and, hence, a low spatial resolution. In contrast, remote sensing methods provide a high spatial resolution because of their coverage but are characterized by a low temporal resolution and less accuracy (usually in the range of a few centimeters to a decimeter, with the exception of InSAR, reaching a few millimeters of accuracy for the altimetric component). In addition, these methods rarely provide 3D measurements, except for LiDAR observations (Travelletti et al. 2014). The great limitation of the remote sensing techniques with high spatial resolution is the temporal decorrelation derived from dense vegetation and high topographic variations, which is common in the Azores.

© Springer Nature Switzerland AG 2020

F. Fernandes et al. (eds.), *Advances in Natural Hazards and Hydrological Risks: Meeting the Challenge*, Advances in Science, Technology & Innovation,
https://doi.org/10.1007/978-3-030-34397-2_12

Observations using classical techniques with total stations (the integration of a theodolite and an EDM—Electronic Distance Meter), robotized and of high precision, are instruments considered as a reference in the evaluation of surface displacement of landslides and of structures such as dams and bridges (e.g., Afeni and Cawood 2013; Simeoni et al. 2015), due to the possibility of obtaining high precision in the 3D determination, of millimeter order, between the measurement point and the monitoring targets. These instruments enable the evaluation of unstable locations, helping in the definition of unstable mass geometry, determination of displacement directions and deformation velocities, as well as the understanding of the mechanisms involved in the instability process (Amaral 2005).

In December 2013, after a long period of rainfall, several signals were detected resulting from a process of active geomorphological instability (e.g, tensile cracks, cracking in dwellings), most of them in the Fajãzinha village, on Flores Island (Azores Archipelago).

In this sense, the Regional Directorate for the Environment of the Regional Government of Azores, which is responsible for the territorial planning and for the geological hazards impact under its control, requested LREC-Açores to develop and implement a multi-technical monitoring system in order to create tools to support the design of an alert and alarm system to mitigate risk on that area.

The present work constitutes the first step to understand the kinematic behavior of the landslide that occurs in Fajãzinha, through the use of a robotized total station with high precision. On what concerns the techniques/instruments to be implemented, the monitoring system is not yet closed. In the future, other techniques will be implemented (e.g., GNSS, inclinometers, vibrating wire piezometers, among others), in order to support decision-making in land use planning and emergency planning and consequently, mitigate the geomorphological risk.

2 Study Area

The present work has as study area a slope movement that affects a significant part of the urban center of Fajãzinha village, on Flores Island (Azores Archipelago) (Fig. 1).

Fajãzinha parish is located on a large coluvionar platform, with slopes ranging from 7° to 20°, and is bordered by imposing escarpments, with sub-vertical slopes, sometimes with heights exceeding 200 m. From the geological point of view, it includes a sequence of highly diversified volcanic products, from Base Complex deposits (lava flows, highly altered tufts) and Upper Complex deposits (lava flows and basaltic pyroclasts) to detrital products resulting of fluvial and geomorphological activity at platform level (Azevedo 1998).

Fajãzinha parish consists of a small set of houses (a few dozen), with its center located near the church and several dispersed houses. The access to this parish is made or possible through a road that leads to the main road connecting Lajes/Stª Cruz to Fajã Grande.

Several geomorphological indicators of slope movement are identified in the area, such as the presence of concave and convex forms, secondary scars, and hummocks, among others. As a result of the kinematic activity evidenced in the hydrological year of 2012–2013, several indicators of recent geomorphological activity were identified, expressed in the roads and buildings (e.g., ground cracks, echelon cracks, diagonal and longitudinal cracks in masonry walls). The geomorphological process corresponds to a complex slope movement, of slide—earthflow type, with a deep failure

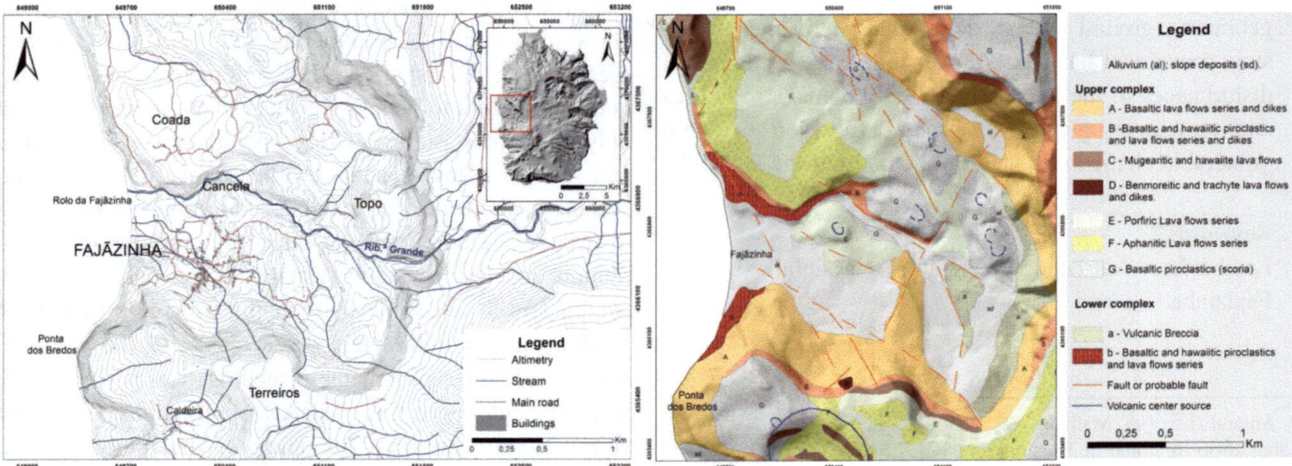

Fig. 1 Geographical location of the village of Fajãzinha and geological sketch of the place. Adapted from Azevedo (1998)

surface. The foot zone of the geomorphological instability corresponds to the sea cliff, where it is possible to observe the geological nature of the material and the presence of several springs.

3 Topographic Monitoring Network Architecture

In the landslide that affects the part of the urban center of Fajãzinha parish, the evaluation of surface deformation of the terrain is carried out with a total station Leica TM50 of 0.5" (0.15 mgon) with angular precision. Distance measurement in precise measuring mode is extremely accurate and has a precision of 0.6 + 1 ppm for circular prisms. The equipment used incorporates ATR (Automatic Target Recognition) technology for long reach, which allows the measurement of prisms up to 3000 m, being prepared to work continuously in intense sunlight conditions or total darkness conditions.

Figure 2 shows the identification and spatial distribution of the monitoring points. Current monitoring network consists of 2 base stations (PE1 and PE2), 30 control points (C1–C30) and 4 reference points (PR1, PR2, C15 and C16). All object and reference points are equipped with Leica GPR112 prisms, except PR1 reference point which is equipped with a GMP104 prism.

The periodicity of data acquisition is approximately 15 days, and these are assured by the Environmental Services of Flores Islands. In each data acquisition campaign, a minimum of 7 different scans (turns) are made for each observation mark. The atmospheric parameters (pressure, temperature, and humidity) were included in the instrument to correct the atmospheric refraction. The distance error measured, a priori, for a 95% confidence interval, was calculated from the constant part of the accuracy of the equipment (0.6 mm) and from the variable part of the accuracy based on the measured distance (1 ppm). The distance error varied between 1 and 3 mm, taking into account the distances between the station point and the observation marks, which reveals an excellent quality of the observed values when optimal conditions of observation exist. Considering the tolerance of the distances and the associated error in each measurement, that for 95% confidence interval the maximum obtained was 2 mm, the use of ±5 mm tolerance was considered for this variable, admitting system changes whenever this value is exceeded.

4 Results and Interpretation

The geodetic monitoring network was installed in November 2017. The distances between the base station and the observation points vary between 48 and 1137 m.

The data presented in this document covers the period between January 2018 and February 2019. During this period, 27 data acquisition campaigns were carried out. Figure 2 shows the variation of the displacements related to the slope distance of some representative benchmarks. In order to understand the movements in planimetry, the absolute deformation vectors obtained in the campaigns of 25-02-2019 (in purple) and of 22-01-2019 (in black) are projected in the same figure.

It is possible to see that the marks that are outside the unstable zone (PR1/C6, C15, and C16) do not present significant variation for the whole time series, being this variation associated with seasonal and daily fluctuations, caused by atmospheric variations and tidal effects. In contrast, the

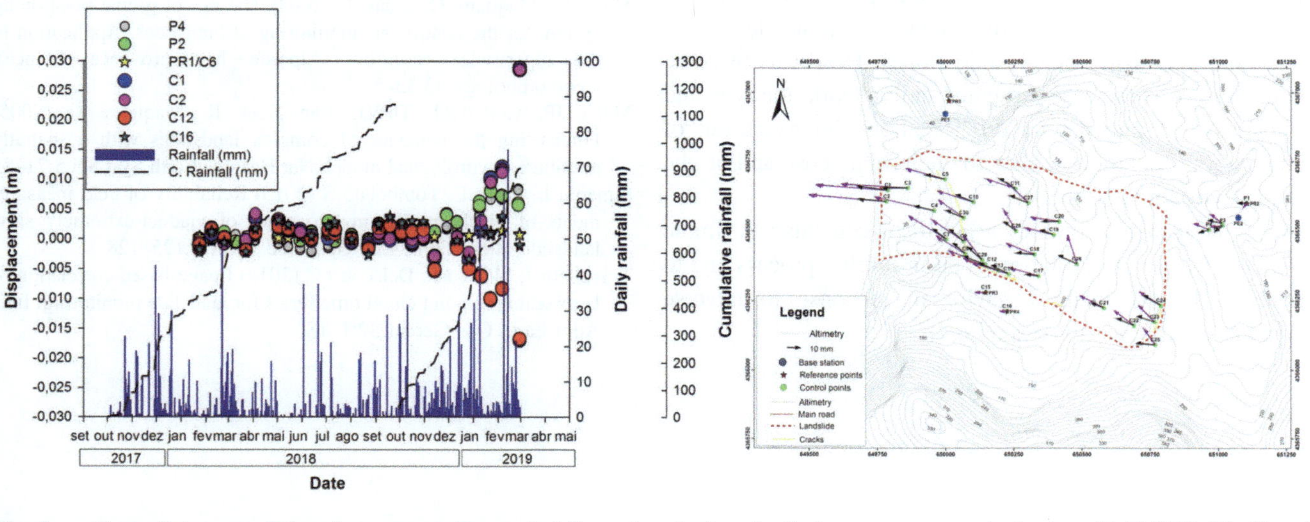

Fig. 2 a Slope distance variation for some points, and daily and accumulated precipitation; and **b** geometry of the monitoring network, with the identification of the observation marks and projection of the planimetric displacement vectors obtained on 01-22-2019 (black) and 27-02-2019 (purple)

other points showed a behavior within tolerance variation, with slight oscillations until December 2018, with a significant increase or decrease in displacement values, suggesting reactivation of the monitored system due to the precipitation quantitative.

In Fig. 2a, daily and accumulated rainfall obtained in the meteorological station of Terreiros, located in the area near the parish, is presented. It is possible to see that the months of December 2018, and January and February 2019 were particularly rainy. In the hydrological year of 2017/2018, the accumulated annual precipitation was around 1150 mm. In the hydrological year of 2018/2019, between December 2018 and February 2019, it already rained more than half (620 mm) than in the previous hydrological year, which is why the landslide that affects Fajãzinha has been active. Between December 2018 and February 2019, in 3 different days rainfall values were registered between 50 and 75 mm.

The maximum displacement observed in the monitoring period showed values of about −57, 16 and −27 mm for the components E, N, and Z, respectively. The orientation of the planimetric displacement is compatible with the preferential direction of the movement of the terrain, essentially towards west. The displacement data define an evolutionary history from the beginning of 2018, as an average annual planimetric velocity of 0.08 m/year. Indicators of movement in relation to references outside the unstable mass in a period of 40 years, obtained by reports of ancient people, allowed to estimate that the unstable terrain presents rates of annual movements in the order of 0.1 m/year, compatible with the registered ones during the monitoring period.

5 Concluding Remarks

The movement of landslides is a complex process that depends on many factors. In landslides prone zones, hazard management has been mostly oriented to avoid dangerous sites and stabilizing of unstable slopes. However, in some situations, as is the case presented in this work, the type and dimension of phenomena do not allow the execution of engineering techniques, due to their high cost and/or the high environmental impact.

So, the measurement of surface displacements is a simple way to observe the evolution of an instability process and to analyze the kinematic behavior in response to adverse weather conditions, for example.

This paper presents the initial phase of the monitoring network designed for the landslide phenomenon that affects the Fajãzinha, enabling an overall view of the kinematic behavior that occurs, in response to precipitation. The periods of reactivation of the system correspond to moments where rainfall exceeded 50 mm/day, indicating preliminarily a rainfall threshold responsible for the reactivation of the system.

Other techniques for evaluating surface deformation in real-time, such as GNSS, and deformation in-depth, such as inclinometers, as well as the continuous monitoring of the piezometric level through piezometers, will help to understand better the process that occurs in Fajãzinha, and prove to be extremely important in the global context of minimizing existing risks.

Acknowledgements The authors would like to thank the Environmental Services of Flores Islands, responsible for the monitoring campaigns.

References

Afeni TB, Cawood FT (2013) Slope monitoring using total station: what are the challenges and how should these be mitigated? S Afr J Geomat 2:41–53

Amaral P (2005) Monitorização de vertentes instáveis no concelho da Povoação, ilha de S. Miguel (Açores): ensaios com base na utilização de uma estação total. Tese de Mestrado em Vulcanologia e Riscos Geológicos. Universidade dos Açores, pp 147

Azevedo J (1998) Geologia e Hidrologia da Ilha das Flores (Açores—Portugal). Universidade de Coimbra, Tese de doutoramento. Departamento de Ciências da Terra

Gili R, Corominas J (1992) Aplicación de técnicas fotogrametricas y topograficas en la auscultation de algunos deslizamentos. III Simpósio Nacional sobre Taludes y Laderas Inestables, La Coruña 3:941–952

Jaboyedoff M, Oppikofer T, Abellan A, Derron MH, Loye A, Metzger R, Pedrazzini A (2012) Use of LiDAR in landslide investigation/a review. Nat Hazards 61(1):5–28

Malet JP, Maquaire O, Calais E (2002) The use of global positioning system for the continuous monitoring of landslides. Application to the super-sauze earthflow (Alpesde—haute-provence, France). Geomorphology 43:33–54

Malet JP, van Asch ThWJ, Van Beek R, Maquaire O (2005) Forecasting the behavior of complex landslides with a spatially distributed hydrological model. Nat Hazards Earth Syst Sci 5:71–85

Simeoni L, Ferro E, Tombolato S (2015) Reliability of field measurements of displacements in two cases of viaduct-extremely slow landslide interactions. Eng Geol Soc Territ 2:125–128

Travelletti J, Malet J-P, Delacourt C (2014) Image-based correlation of laser scanning point cloud time series for landslide monitoring. Int J Appl Earth Obs Geoinf 32:1–18

Subsidence Hazard in Limestone Cavities: The Case of "Grutas da Moeda" (Fátima, Central Portugal)

Isabel Duarte, António Pinho, Luís Lopes, Ricardo Sábio, and Micael Jorge

Abstract

"*Grutas da Moeda*" are natural touristic caves, located in the plateau of São Mamede, near Sanctuary of Fátima, in Portugal. They have been opened to the public since 1974, and receives about 75,000 visitors per year, from 76 nationalities. They are located in the "*Maciço Calcário Estremenho*", composed by limestone deposits of the Middle Jurassic (Bathonian), corresponding to the Serra de Aire formation. The geological risk assessment is fundamental to guarantee the safety of its visitors and staff; therefore, it is intended to develop geotechnical monitoring methodologies in order to collect data to understand the risks that may be associated with this natural cavity and to identify critical areas of collapse. There are four main alignments (faults/joints), which assume parallelism with the development of the caves. The trend of the layers in the cave is approximately N30ºE; 17ºSE. In the year 2015, a geophysical study was carried out, using the 3D electrical resistivity method and, for one of the areas in particular, the georadar method. The georadar method allowed to map the anomalies corresponding to empty spaces that have been identified with the 3D electrical resistivity technique. This paper presents the results obtained, as well as proposals for geotechnical measures to mitigate the risk of collapse subsidence.

Keywords

Subsidence risk • Limestone cavities • Geological hazards • Geophysical methods • Geotechnical monitoring

1 Introduction

The "Grutas da Moeda" are natural caves, located in Central Portugal (São Mamede plateau 39°37′26.83"N; 8° 42′18.49"W). This karstic region (*Maciço Calcário Estremenho*) is dominated mainly by limestone deposits from Middle Jurassic (Bathonian), more specifically the deposits of Serra de Aire Formation. According to (Azerêdo et al. 2015) this formation exhibits nice examples of pedogenic limestones (massive, laminar, nodular and brecciated calcretes, always with black-clasts included) interbedded with, or grading laterally or vertically into, a range of other deposits, including organic-rich sediments that are not known anywhere else in the Lusitanian Basin for the same stratigraphical interval. Besides the calcretes and the organic-rich marly/clayey seams and lenses, carbonates with evaporite traces, microbial laminites, black-clast, and fenestral limestones occur. This pedogenic-dominated facies association grades upwards into peritidal cyclothems and lagoonal limestones, with dinosaur footprints above. This formation has approximately 400 m thickness. Above the Serra de Aire deposits also occur a post - Jurassic argillaceous and sandy cover resulting both from the dissolution of the limestones and chemical and physical weathering of granites, schists, quartzites, and gneisses of the Lusitanian Basin basement rocks.

The existing fractures in the rock mass are responsible for the continuous infiltration of the water in the subsoil and its scarcity to the surface. For millions of years, water has shaped the rock through various weathering processes, physical and chemical, resulted in several characteristic calcareous forms. The limestone rock mass presents a variety

I. Duarte (✉) · A. Pinho
Centre GeoBioTec|UA and Department of Geosciences, School of Sciences and Technology, University of Évora, Évora, Portugal
e-mail: iduarte@uevora.pt

L. Lopes
ICT and Department of Geosciences, School of Sciences and Technology, University of Évora, Évora, Portugal

R. Sábio · M. Jorge
Grutas da Moeda, São Mamede, Fátima, Portugal

© Springer Nature Switzerland AG 2020
F. Fernandes et al. (eds.), *Advances in Natural Hazards and Hydrological Risks: Meeting the Challenge*, Advances in Science, Technology & Innovation,
https://doi.org/10.1007/978-3-030-34397-2_13

of karst structures, such as lapiés, mega-lapiés, dolines, poljes, uvalas, caves, sinkholes, and subterranean galleries.

Knowing the karstic nature of the rock, as well it is intensely fractured due to the previous action of tectonic processes, it is natural to have some geological risks associated.

According to the collected data, it was observed that the two principal family faults were crossing over one of the galleries turning it in one area on subsidence risk. Those are the characteristics that turns this natural cave in a study case.

In the visible cracks inside the cave is possible to find clay filling the empty spaces, but is possible that much more hidden cracks occur under calcitic deposits making them hard to identify and study turning it more difficult to analyze the associated risks.

It is important to understand that there is not a lot of information about this subject, having been some studies carried out but never applied in touristic caves like this one.

(McNeill et al. 2010) in a case study detailing the site investigation, hazard assessment, and development of a hazard mitigation strategy for development in karstic terrain, divided their study in several phases. The first was the desk study, which includes the review of aerial photographs, topographic and geological maps of the area, and analysis of technical articles. After the preliminary findings of the desk study, it was possible to identify potential presence of karst features, allowing to proceed to a microgravity and a 2D resistivity geophysical survey. The microgravity survey emphasized areas of decreased density, which could be related to karst features. The resistivity profiles also indicated potentially anomalous features.

According to (Potherat and Duranthon 2010) the elaboration of a monitoring system allows us to collect data and to predict the movements on this type of formations. Sudden movements (cliff collapse) are hard to predict, but in subsidence risk cases, what usually is done is to develop a monitoring system with several instruments like, wire extensometers, infrared mirrors, and trihedron for ultra-large band radar device. In conclusion of this study, with daily analysis it is possible to anticipate early enough any event that could happen by the changes in the rock mass behavior if the data are automatically collected and transferred to a monitoring center. A crisis management team is fundamental to estimate the maximum value of movements (for example 10 mm for this case study). If the value is above the maximum predicted is necessary to proceed to a measurement of four times a day instead of one. After being identified a possible case of subsidence, an exponential divergence model based on the last 24 h to predict the time to collapse is used. This analysis uses a Logm24 (log of the mean velocity/24 h) as function of the time and the intersection with the curve corresponds to the time left to the failure.

This work intends to understand the geological risks at the cave in the study, and identify critical areas to prevent. Furthermore, it is also important characterize the structural geology at this area, develop a methodology of monitoring to prevent possible subsidence cases and secure the safety of both workers and tourists.

2 Materials and Methods

GeoSurveys (2015) performed three studies about the geologic characterization of the “Grutas da Moeda”. The first study was a geophysical prospection with electrical resistivity, the works focused on the north zone of the space surrounding the Interpretative Center of the “Grutas da Moeda”, and the main objectives were the characterization of the prospected massif and the identification of possible karst structures susceptible of commercial exploitation under the prospected sites. Although 5 profiles of electrical resistivity were planned, 6 profiles were made, because it was considered useful regarding the addition of local context information in a karstic environment. The data acquisition was performed using a resistivimeter, the Syscal Junior 48 channel model from IRIS Instruments (Orleans, France). The elementary process of collecting data in tetra-electrode devices, i.e., in a four-point contact configuration, consists of the injection into the ground of electric current at two points called current electrodes. The reading of the electric potential is carried out in two other contact points–potential electrodes, due to the injection of electric current. The data were collected using the theoretical linear Dipole–Dipole device because it is sensitive to the lateral variations of electrical resistivity. Based on the tetra-electrode device, the current intensity and potential difference measured, it is possible to determine the electrical resistance to the passage of current in the prospected terrains. The electrode configuration used is affected by geometry effects, and in this way the apparent resistivity of the terrain under prospection is determined. By varying the spacing between the injection and potential electrodes, it is possible to obtain the first distribution of apparent resistivities (pseudo sections). The apparent resistivity distribution is used to calculate the electrical resistivity distribution models of the terrain (electrical resistivity tomography). In this study RES2DINV Software (Geotomo Software, Malaysia) was used, which uses the modern techniques of resolution of the inverse problem, namely the use of digital filtering to solve the direct problem with the technique of maximum descending gradient, in the resolution of the inversion of the parameters of the corresponding model. These techniques almost always require an approximate initial model, so that convergence to the final model occurs with a minimal error. In this study, electrical resistivity profiles were performed using the Dipole–Dipole tetra-electrode device, with a spacing of 5 m between electrodes and a maximum length of 235 m, a function of the multicomponent in use. The estimated theoretical maximum depth for the electrical resistivity distribution model was about 42.30 m.

In the second study, the 3D electrical resistivity was performed for the geophysical characterization of the "Grutas da Moeda", with the objective of locating geophysical anomalies correlated with the existence of cavities, as complement georadar tests were also carried out, with the purpose of verifying the suitability of the methodology in the local karstic environment. Two prospecting areas for 3D electrical resistivity were planned (Areas 1 and 2), on alignments, resulting from the interpretation of the results of the previous study. A positional adjustment of the electrodes was performed due to the obstacles encountered (concrete structures and external walls), having been given emphasis to the direction in which the karstic structure develops, which extends below the Interpretative Center of the "Grutas da Moeda". The acquisition of georadar data was executed in parallel lines to the largest dimension of the rectangle formed by Area 1. In this area, 7 profiles were made with 1 m spacing between profiles. The data were collected using the theoretical 3D device of the Pole–Dipole type, function of the availability of the areas of electric prospection. This device allows relatively high depths of investigation, depending on the reduced scanning spaces. Compared with the linear device of the Dipole–Dipole type, used in the previous study, it has less lateral sensitivity. The georadar method is based on detecting the response of the prospected medium to the propagation of high-frequency magnetic waves, commonly in the range of 25–2000 MHz and georadar frequencies outside this range may also be employed. The single-channel (bistatic) georadar system used has a pair of antennas (transmitter and receiver), which are arranged perpendicular to the data acquisition profile. Response events from the crossed medium (reflections, diffraction and others) are recorded continuously and are subsequently analyzed on digital signal processing platforms. The final interpretation is obtained by analyzing the various signatures observed in the processed signal.

In the third study, georadar trials were carried out in the SW zone of the Interpretive Center of the "Grutas da Moeda". A total of 44 georadar profiles were performed, parallel and spaced 1 m apart, and the acquisition of these data was performed with electromagnetically shielded antennas with nominal frequency 100 MHz.

3 Results

According to the results of the (GeoSurveys 2015) there are four main alignments (faults/joints) on the cave (see Table 1), which assume parallelism with the development of the caves, being the orientation of the stratification N30°E with 17°SE dip.

Through the electric resistivity and georadar methods, it was possible to identify anomalies that could result in potential cavities (see Fig. 1).

In situ observations shows that discontinuities are well visible in several galleries especially at the Pastor gallery ceiling (see Fig. 2). These discontinuities are associated to the joints sets observable at the surface.

4 Discussion

Several works allowed us to collect important data to understand better the structural geology and to identify the critical areas of the cave. The results of the three studies performed by GeoSurveys (2015) were essential and pointed out to the main family fractures set and also the new possible galleries that could be or not exploited by the company "Grutas da Moeda".

Regarding potentially unstable areas of the cave, the ceiling between the second and fourth galley (Presepio until Pastor) stands out due to its low thickness, as well the high development of exokarst structures (lapiés) that extend to relevant depths. The fourth gallery (Pastor) may also present higher risk due to the higher fracture intensity caused by the crossing of more than one joint set. That, together with stratification joints, could potentially give rise to blocks detachments. These in situ observations allow us to point it as the area with most subsidence risk.

Table 1 Main joint sets of the prospected area

Sets	Orientation
Set 1	N35°W
Set 2	N150°E
Set 3	N75°E
Set 4	N100°E

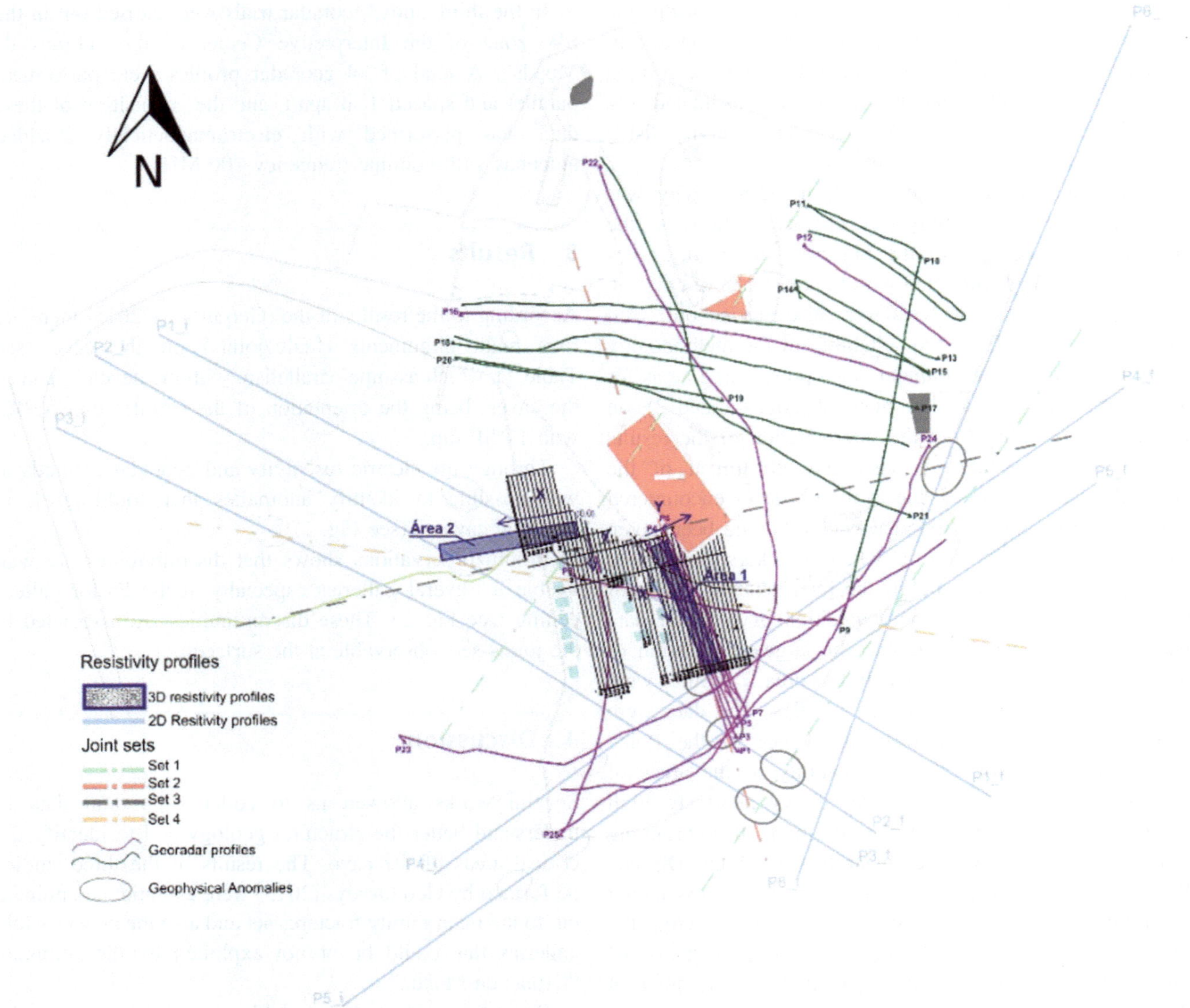

Fig. 1 Representation of the resistivity profiles and main joint sets of the prospected area. Adapted and revised from GeoSurveys (2015)

Fig. 2 Discontinuities in limestone ceiling of Pastor gallery

5 Concluding Remarks

This study gave us important data and tools to continue the project, and it is considered important to implement a geotechnical monitoring system to understand the subsidence risk associated to this area. It is proposed to use some equipment, as used by (Potherat and Duranthon 2010) and others, like vibrating wire extensometers and piezometers, in-place inclinometers, crackmeters, surface marks for surveying ground surface displacements, load cells and tiltmeters, and others that could be more convenient. Those tools are fundamental to start the monitoring phase and collect data of the rock mass behavior over time and figure it out if it is happening subsidence movements.

Knowing that there are not a lot of studies on touristic caves like this one, it is relevant to carry on with this study to guarantee the safety of thousands of people that visit it during the year as it is important too for other caves that are on the same conditions.

The continuation of these studies is considered of extreme relevance since this touristic activity involve the movement of many people inside and around the caves.

References

Azerêdo A, Wright V, Mendonça J, Cabral M, Duarte L (2015) Deciphering the history of hydrologic and climatic changes on carbonate lowstand surfaces: calcrete and organic-matter/evaporite facies association on a palimpsest Middle Jurassic landscape from Portugal. Sed Geol 323:66–91

GeoSurveys (2015) Caraterização geológica – Grutas da Moeda, São Mamede, Fátima. Prospeção Geofísica por Resistividade Elétrica e Georadar. REP42158-1; REP52175

McNeill J, Nettleton I, Webber I, Bridgeman C, Cosgrove T (2010) Living with geohazards: investigation, assessment and mitigation for a large commercial development in karstic limestone, Galway City, Ireland. 11th IAEG Cong., pp 1365–1372. Taylor and Francis Group, London

Potherat P, Duranthon JP (2010) The monitoring of unstable rock masses applied to risk management. 11th IAEG Congress, pp 1373–1380. Taylor and Francis Group, London

Geotechnical Hazards in Rocky Slopes (Northern Portugal): Focused on Methodology

José Filinto Trigo, Carlos Pacheco, João Fernandes, Pedro Ferraz, Jorge Sousa, Rui Machado, Sara Duarte, Ana Mendes, Liliana Freitas, José Teixeira, Luís Ramos, Maria José Afonso, and Helder I. Chaminé

Abstract

Interdisciplinary studies of rocky slopes stability have been performed on Northern Portugal, that allowed to establish a design methodology. This approach carries together the use of more traditional techniques, with the application of more recent technologies, such as unmanned aerial vehicles. This work presents a set of procedures that have been applied successfully in several case studies, describing the different stages, with an emphasis on the production of a ground three-dimensional model. This model is generated from photographs obtained with successive drone flights, with overlapping. It is the support for the development of subsequent stages, especially for the characterisation of the trajectories of rockfalls and modelling, the development of susceptibility mapping and of rock slope hazard. In this context, the Natural Slope Quality Index (N-SQI), an adaptation of the Slope Quality Index (SQI), orientated to natural slopes, is presented. The Block Gravity Number (BGN), created in order to hierarchise rock blocks, selecting them for modelling, is also proposed.

J. F. Trigo · C. Pacheco · J. Fernandes · P. Ferraz · J. Sousa
Division for Construction Studies (NEC), Department of Civil Engineering, School of Engineering (ISEP), P.Porto, Porto, Portugal

J. F. Trigo · R. Machado · S. Duarte · A. Mendes · L. Freitas · J. Teixeira · L. Ramos · M. J. Afonso (✉) · H. I. Chaminé
Laboratory of Cartography and Applied Geology (LABCARGA), Department of Geotechnical Engineering, School of Engineering (ISEP), P.Porto, Porto, Portugal
e-mail: mja@isep.ipp.pt

J. Teixeira
CEGOT and Department of Geography, Faculty of Arts, University of Porto, Porto, Portugal

M. J. Afonso · H. I. Chaminé
Centre GeoBioTec|UA, Aveiro, Portugal

Keywords

Rockfalls • Slope geotechnics • Natural hazards • Drones • Hydrological conditions

1 Introduction

Hazard describes any condition with the potential for causing an undesirable situation, that sometimes could incur in disasters (e.g. Wyllie and Mah 2004; Einstein and Sousa 2012). The triggering mechanism for failure involves both the presence of at least one factor (e.g. infiltration of water, earthquake, vegetation grow) and the existence or development of unfavourable conditions for stability, such as blocked drainage paths and high slope angle (Pantelidis 2009).

In Northern Portugal, rock slope instability is a common process associated with rockfalls. These mass movements occur mainly in areas where the geological, hydrological and geomorphological conditions promote the natural landscapes with the presence of high slopes and rocky boulders with different shapes and dimensions. Sometimes, these rockfalls involve significant damage and are often associated with natural phenomena, such as high-intensity and/or long-term precipitation (Wyllie and Mah 2004; Volkwein et al. 2011). Also, they may result from anthropic actions, such as the destruction by wildfires of the vegetation cover. The change in the water conditions (namely water content or pressure) is generally the major factor responsible for triggering this kind of slope instability. Water is also an important weathering factor that can play a significant role with respect to rock discontinuities and thus rock mass behaviour. The increase of precipitation (frequency and intensity) highlights new problems concerning rock mass stability (Sandersen et al. 1996).

Northern Portugal is classified, in climate terms, as Atlantic type, being temperate with a hot and dry summer. It is very influenced by the orientation of the relief, which

© Springer Nature Switzerland AG 2020

F. Fernandes et al. (eds.), *Advances in Natural Hazards and Hydrological Risks: Meeting the Challenge*, Advances in Science, Technology & Innovation,
https://doi.org/10.1007/978-3-030-34397-2_14

represents the first obstacle to the air masses from the West, after a long journey over the Atlantic Ocean. In this region, precipitation decreases from the coastal areas to inland, due to orographic barriers, and may vary by altitude control from maximum values around 3000 mm/year in the NW mountains (Gerês and Cabreira) to less than 500 mm/year in the inner Douro Valley. The NW façades have average values around 1200–1500 mm/year, and most of the precipitation is concentrated in the rain season, from October to March. This intense precipitation is the cause for the presence of water in the rocky masses and has often a double negative effect on their stability.

In fact, the percolation of water into the discontinuities induces a pressure increase on the adjacent rock surfaces and promotes the decrease of the shear strength of these surfaces. Water is also responsible for the transport of small particles that accumulate in the discontinuities, filling them with material usually poorly resistant and very deformable. Instead, in filled discontinuities, water can wash these particles.

On the isolated blocks that occur, frequently, on the slopes of Northern Portugal, water is also an important factor for the decreasing of their stability. The water that flows on the base of the blocks erodes the material, usually of residual soil origin, that constitutes its support surface, promoting a loss of support from underneath. This water flow is also responsible for the transport of residual soil material and smaller rock blocks from higher elevations, which may constitute an additional instability action on the bigger blocks. This effect of temporary water flow generated after an intense or continuous precipitation that usually flows under a torrential regime is more evident when the blocks are located close to a line of thalweg. This way, the process of blocks transportation and their fall along the slope can be triggered more easily. Finally, it is interesting to note that many rockfalls happen on summer time, corresponding to periods of time without precipitation, but with high thermal amplitudes.

2 Methodology Applied to the Study of Natural Rocky Slopes Stability in Northern Portugal

Interdisciplinary studies of rock slopes stability have been performed on Northern Portugal (e.g. S. Simão site, Amarante; S. Cristovão and Mourilhe sites, Cinfães; Santo Ovídio site, Gaia; Joane site, Famalicão) that allowed to establish a sustainable design methodology. That constraint allows always to design with nature and even design with hazards and/or rock engineering risks (e.g. Einstein and Sousa 2012; Hudson and Feng 2015). The approach requests an integrated and balanced understanding of climatology, applied geomorphology, hydrology, engineering geology, geotechnics and engineering methods. This integrative approach carries together the use of more traditional techniques, supported by a detailed mapping and fieldwork, with the application of more recent technologies, such as unmanned aerial vehicles (UAV), usually known as drones.

This methodology involved (Fig. 1), on a first stage, a topographical, hydroclimatological, morphotectonic and geostructural surveys and characterisation of the study area and, in a subsequent stage, a geotechnical, geomechanical and hydrological/hydrogeological assessment. The main geological, hydrogeological, geotechnical and geomechanical constraints were compiled and integrated along the rocky slopes. A high-precision GPS (Trimble GeoExplorer) was used for the fieldwork surveys. In addition, the use of Geographic Information Systems (GIS) on rock slope stability, using basic analysis tools for spatial management and data manipulation, has become more common.

The study of rock mass outcrops includes the description, evaluation and modelling of the mechanical properties of the intact rock and the rock anisotropies. The scanline sampling technique of discontinuities has been applied to the rocky mass faces for the rock mass description. This method includes placing a tape along the length of an outcrop and determining its orientation. All the basic geologic and geotechnical rock mass parameters are recorded and surveyed. To establish the main discontinuity sets, the structural geology data collected at the slope sites were analysed with Dips software package (Rocscience). Weathering grade (W), fracture intercept (F) and seepage conditions of rock material were used. Furthermore, uniaxial compressive strength (UCS) was estimated by Schmidt Rebound Hardness and Point Load Test (PLT). In order to classify the quality of the rocky masses, the following geomechanical classification systems and/or geotechnical indexes are the key: rock mass rating (RMR), slope mass rating (SMR) and geological strength index (GSI). SMR classification was also essential to evaluate the stability of the slopes and to offer insights into the main potential local slope failure mechanisms. In addition, other slope-based classifications systems could be applied to correct understanding of the hydrogeomechanical quality such as *Q*-slope, HP-Value and also the important inputs from Hack (2002), Pantelidis (2009), Bauer and Neumann (2011) and Pinheiro et al. (2015).

Simultaneously, a detailed visual inspection of the studied rock slopes is carried out, identifying the potential unstable blocks. This characterisation aims to establish a hierarchisation of the blocks, to be able to select the most unstable and to model their fall along the slope, using an automatic calculation program.

Therefore, a Block Gravity Number (BGN) was defined, which is a function of: (a) the weight and shape of the block, (b) the slope geometry, namely the slope angle, (c) the type

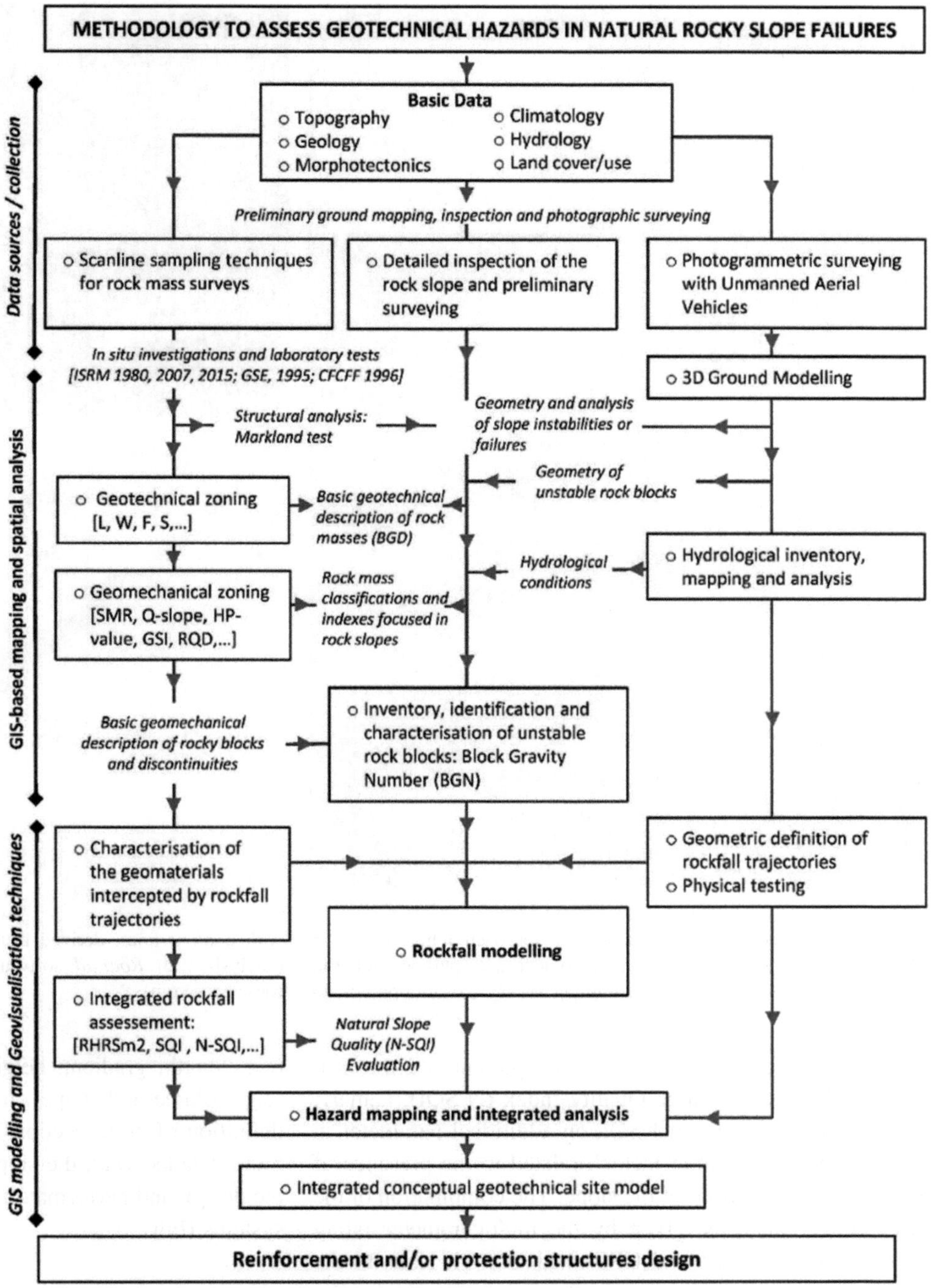

Fig. 1 General and conceptual flow chart of the on-site investigation methodology to assess geotechnical hazards in natural rock slope failures

of material that exists on its potential fall trajectory and (d) a set of factors characteristic of the block, identified as anomalies. These anomalies are related with: (a) discontinuities and their aperture, (b) the evidence of loss of support from underneath, (c) the presence of vegetation with an instability effect, (d) indicators of block displacement, (e) the existence of soil- or rock-like material acting on the surface of the block, with an instability effect and (f) the presence of water in the discontinuities or in the base of the blocks. Each of these factors has associated a score and the sum is the classification of the BGN, which permits the hierarchisation of the blocks.

To evaluate the rockfall hazard of the slopes, the Slope Quality Index (SQI) proposed by Pinheiro et al. (2015) was applied with some adaptation to rock hillslopes (e.g. S. Simão site, Amarante). This index is a practical system developed to obtain a quality index for rock slopes in highway infrastructures. The implementation of this index in

Fig. 2 Some aspects of the methodological approach: **a** high-precision GPS and UAV techniques to topographical and photogrammetric surveys; **b** geotechnical scanline sampling surveys; **c** orthophotomap with contour lines, derived from UAV survey; **d** rockfall modelling (analysis with *RocFall* software by Rocscience Inc.—https://www.rocscience.com)

natural landscape scenarios led its adaptation and the proposal of the Natural Slope Quality Index (N-SQI), namely, introducing, in the original SQI, an additional parameter in the 'environmental/traffic factor', related to the presence of buildings on the bottom of the slope. The combination of the susceptibility analysis, given by the multiparameter rating systems—RMR, SMR and GSI—with N-SQI, leads to the final step of hazard assessment, the rockfall susceptibility and hazard mapping/zoning.

The use of UAV permits a high-resolution photographic survey of the slope. Photographs are obtained with an interval of 2 s and with an overlap of more than 60%. Flights are carried out at different altitudes and, whenever possible, photographs are taken vertical to the slope. This way, it can be obtained a resolution of 1 cm/pixel. The data collected with the UAV are processed with photogrammetry OpenDroneMap software that permits to generate a high-resolution georeferenced orthophotomaps, as well as a three-dimensional model of the slope. The digital elevation model permits to calculate several parameters of the slope (e.g. length, gradient, height, etc.) and of the rock blocks (e.g. volume and shape). In addition, this model allows the definition of cross sections, which help to understand, in a more detailed way, the slope morphology. Also, it supports the design and performance evaluation of rockfall protection systems (Fig. 2).

3 Concluding Remarks

Rockfall susceptible areas can be assessed and predicted throughout integrative and multidisciplinary approaches, and thus, rockfall damage can be decreased through prevention efforts. The presented methodology was successfully applied to several case studies in Northern Portugal. This methodology puts together traditional approaches with innovative technologies, with detailed analyses both at the rocky block scale and at the slope scale. The proposal of the Block Gravity Number (BGN) is very useful on the optimisation of the rocky blocks selection to consider on the rockfall modelling. Additionally, the N-SQI index helps to the validation of the rockfall susceptibility and hazard mapping and zoning.

Acknowledgements This study was carried out under the framework of the LabCarga|ISEP re-equipment program (IPP-ISEP|PAD'2007/08). The authors would like to thank the reviewers for helpful comments and inputs.

References

Bauer M, Neumann P (2011) A guide to processing rock-fall hazard from field data. In: Vogt N, Schuppener B, Straub D, Bräu G (eds) Proceedings of the 3rd international symposium on geotechnical safety and risk, ISGSR 2011. Munich, Germany, Bundesanstalt für Wasserbau, Karlsruhe, pp 149–156

Einstein HH, Sousa LR (2012) Risk in slopes. In Sousa LR, Vargas E, Jr, Fernandes MM, Azevedo R (eds) Innovative numerical modelling in geomechanics. CRC Press, pp 201–210

Hack R (2002) An evaluation of slope stability classification. In: Dinis da Gama C, Sousa LR (eds) Eurock'2002, ISRM, Funchal, Sociedade Portuguesa de Geotecnia, pp 3–32

Hudson JA, Feng X-T (2015) Rock engineering risk. ISRM book series. CRC Press, London

Pantelidis L (2009) Rock slope stability assessment through rock mass classification systems. Int J Rock Mech Min Sci 46:315–325

Pinheiro M, Sanches S, Miranda T, Neves A, Tinoco J, Ferreira A, Correia AG (2015) A new empirical system for rock slope stability analysis in exploitation stage. Int J Rock Mech Min Sci 76:182–191

Sandersen F, Bakkehoi S, Hestnes E, Lied K (1996) The influence of meteorological factors on the initiation of debris flows, rockfalls, rockslides and rockmass stability. In: Proceedings of the 7th international symposium on landslides, vol 1, pp 97–114

Volkwein A, Schellenberg K, Labiouse V, Agliardi F, Berger F, Bourrier F, Dorren LKA, Gerber W, Jaboyedoff M (2011) Rockfall characterisation and structural protection: a review. Nat Haz Earth Syst Sci 11:2617–2651

Wyllie DC, Mah CW (2004) Rock slope engineering: civil and mining, 4th edn. Spon Press, London

The Use of Unmanned Aerial Vehicles in the Monitoring of Forest Fires

António Correia, Luis Araújo Santos, Paulo Carvalho, and José Martinho

Abstract

In this paper, a general description of the use of drones in forest management is made, addressing their various potentialities in three distinct phases: (i) prevention of forest fires; (ii) during forest fires, with particular emphasis on residential areas in rural environments and escape routes; and (iii) after the occurrence of fires. An approach will also be taken on the use of drones to combat fire and support 3D modeling of the territory. In fact, under the motto "early detection, early rescue," the versatility of this technique allows its use both for cleaning tasks on land, roads, and high and medium voltage lines, as well as collecting data on the evolution of fire or early detection of new fires, being also useful in determining the burning area.

Keywords

Wildfire • UAV • Monitoring

1 Introduction

Forest monitoring is both for ecological as well as economic and social reasons, a relevant aspect in any modern society (Ollero et al. 2006). In this field, the use of unmanned aerial vehicles, commonly called drones, has enabled the collection and processing of real-time data, essential to the efficient management of this natural resource. Particularly, after the forest fires of 2017, in Portugal, the use of this technology brings many other applications and potentialities. The increase in the use of drones is due to the high quality of the images captured by wide cameras that may have associated several functions, including thermographic (infrared) and vision (night vision). In addition, the evolution of image processing software, as well as the growth of the coverage of mobile internet networks, allows the transmission of images every second with a rapid processing. For example, the use of image processing algorithms allows the creation of 3D georeferencing models that, with the support of geographic information systems, facilitate the forest inventory. Moreover, intelligent systems are now already available which, with the use of real-time images captured by drones, accurately detect and locate the spread of fires, and enable the combat.

2 Unmanned Aerial Vehicles (UAVs)

UAVs were developed, with different sizes, different autonomies, and different equipment, in order to satisfy each of the aforementioned purposes. Figure 1a shows a Prox Dynamics Black Hornet Nano, a device of very small dimensions, weighing 16 g, equipped with three cameras.

In the category of medium-sized UAVs, Northrop Grumman MQ-8 Fire Scout is shown in Fig. 1b. With a wingspan of 4.3 m, a weight of 1430 kg, it reaches speeds rounding 130 km/h, within a radius of 190 km.

3 The Use of UAV in the Surveillance and Prevention of Forest Fires

In the activity of forest fire prevention, we can point out the forest surveillance, in periods of high fire risk, as one of the most important. This task has been carried out by security forces when patrolling and by the lookout points, which unfortunately do not have full coverage of the national

A. Correia · L. A. Santos (✉)
Department of Civil Engineering, School of Enginering (ISEC), Polytechnic of Coimbra, Coimbra, Portugal
e-mail: lmsantos@isec.pt

P. Carvalho
Poiarqui—Arquitetura e Engenharia, Lda, Vila Nova de Poiares, Portugal

J. Martinho
Faculty of Sciences and Technology, University of Coimbra, Coimbra, Portugal

© Springer Nature Switzerland AG 2020
F. Fernandes et al. (eds.), *Advances in Natural Hazards and Hydrological Risks: Meeting the Challenge*, Advances in Science, Technology & Innovation,
https://doi.org/10.1007/978-3-030-34397-2_15

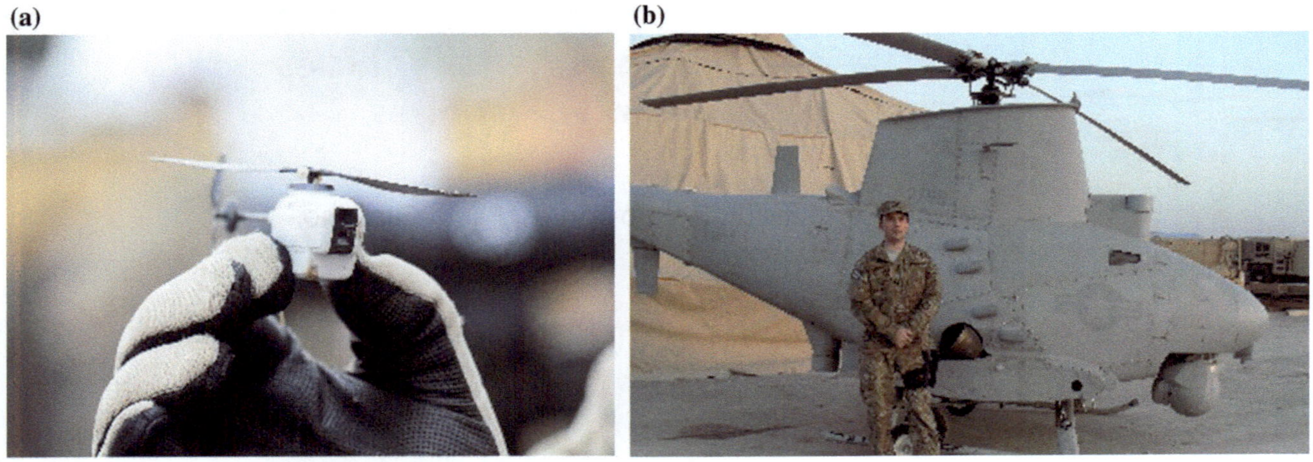

Fig. 1 Examples of UAVs: **a** Prox dynamics black hornet nano. **b** Northrop Grumman MQ-8 Fire Scout

territory. Naturally, the possibility of obtaining real-time images of territory, covering large areas of land, makes the use of UAVs, and particularly drones, of great interest for forest surveillance and forest fire prevention. In order to safeguard human lives, housing, and vehicles, it is of primary interest to have detailed knowledge of the vegetation and forest cover of the territory. As it can be seen in Fig. 2 (illustrations a and b), the use of drones allows a thorough assessment of the cleaning state of the bordering roads and all communication routes, as well as a view of the border of the villages, to analyze the conformity of fuel management bands in urban-forest interface. This issue is of extreme importance for the passive safety of populations, who, through ignorance or lack of means, invade in their urban space, which, during a fire, pose a risk to their lives and the integrity of their belongings (Correia et al. 2018).

The images depicted in Fig. 2 (illustrations a and b) intend to show that the capture of images by drones or UAVs is capable of providing valuable information on the safety or lack of safety in rural areas, concerning the invasion of the forest in rural areas. Only with "eyes in the sky," it is possible to assess large areas and to spot the spaces with correct and incorrect fuel management, and to plan future cleaning operations.

The cleaning of medium and high voltage powerlines for electric supply to populations is of extreme importance, as well as the cleaning of bands of protection of roads that serve as escape route, in alarming situations and of general panic, as it was the case of the Pedrógão Grande mega-fire, which occurred in Portugal in June 2017. As it can be seen in Fig. 3a, the ignition point of the Escalos Fundeiros fire began in contact between large vegetation and a medium tension electric powerline, which started what would become one of the most dramatic fires in the history of our country. Also, in Fig. 3b, an image of the EN 236 road can be observed, where a large number of civilians attempted to escape and lost their lives, partly due to the violence of the fire, but also because of this lack of compliance with the cleaning rules of the communication routes, which require a cleaning of 22.5 m for each side of the electric powerline

Fig. 2 Example of drone use for surveillance: **a** protection of roads; **b** protection of villages

Fig. 3 Use of drones in land recognition: **a** ignition point of Escalos Fundeiros fire (Viegas 2018); **b** road EN 236 after fire

and 10 m for each side of the axis of any road. This point is of the utmost importance, since neither the entities of authority nor the entities responsible for managing these infrastructures behave well in this process, as can easily be seen in Fig. 3.

4 The Use of UAV in Surveillance of Forest Fire Prevention

The use of UAVs in a firefighting scenario is, logically, more complex, given the aggressiveness of the environment in which planning, combat, and rescue operations are carried out, as well as safeguarding the material assets of populations. It is of utmost importance for the good conduct of all operations that the operational teams in the field have at all times a rigorous knowledge of the location, speed, and direction of advance of all the fire fronts. Obviously, in the first place, there is a danger associated with the fact that, in the airspace of the fire, there are already supposed to be another aircraft flying, such as helicopters and combat aircraft. In this sense, it should be noted that piloting drones or any UAVs in airspace of forest fires should be restricted to technically qualified and accredited personnel for this purpose (Correia et al. 2018), besides de pilot license already required in any environment.

On the other hand, it should also be noted that the existence of smoke in such airspace may make it difficult or impossible to fly these aircraft. In this sense, there must be rigid and demanding safety rules, so that the drone or UAV can perform its function, essentially monitoring, without taking risks for its integrity and continuation of its role. As shown in Fig. 4, the drone can monitor the distance of the flames from the ground without being caught by the high-level column of flames that can damage the aircraft.

Imaging by drones, in the case of large fires, where the amount of smoke is too large, can be performed by thermal imaging cameras, which easily detect flames contour through thick fumes (Fig. 5a). For the aid of communications during firefighting scenarios, drones may be equipped with radio signal repeater antennas (Fig. 5b).

Also, the appearance of secondary foci can be readily mitigated if there is a monitoring of incandescent particles projections by fire sources. These secondary outbreaks are impossible to detect by the lookout points on the ground, and when alerted by populations, it will be too late for firefighters to eliminate them in reasonable time. Figure 6a shows the appearance of secondary foci, which arise very quickly, and which can be easily detected by drones, allowing an action with great readiness (Correia et al. 2018).

In Fig. 6b, examples of an aerial image of an escape route, without safety, are presented. It should be noted that the safety or its lack on a particular escape route varies very rapidly, depending, also, on the weather conditions, which vary very rapidly during the fire. The safest place during a fire may be the area already burned, and that location may be best detected with drones.

5 The Use of UAV in the Territory Mapping and in Forest Management

This article also highlights the potential of using georeferenced images, for 3D modeling of the territory, for accurate information on urban structures, and the detection of critical points. Through the capture of images by properly georeferenced drones and their treatment, it is possible to map the terrain in question. For this, a terrain mapping software, prepared for this type of photos, is used. Among many software of this type, the one chosen for this demo is Agisoft PhotoScan (2018). Many drones, nowadays economically very accessible, have the function of storing the georeference of each photo (Fig. 7a). It consists of a *.txt file which,

Fig. 4 Use of drones for monitoring the propagations of a wildfire

Fig. 5 Drones with logistic functions: **a** drone with thermographic camera; **b** drone for communications

Fig. 6 Support of drones during fires: **a** detection of secondary foci; **b** searching for safe escape routes

in each line, contains information about the position *x* (latitude), *y* (longitude), and *z* (altitude) of the point where each photograph was taken.

Moreover, it is now possible to develop 3D models of territory using freeware software, as illustrated in Fig. 7b. DroneDeploy is a leading cloud software platform for commercial drones and provides 3D applications for territory mapping, free of charge, and for certain types of users.

Intelligent systems are currently being developed which, with the use of drones and also satellites, proceed in real time to the exact detection and location of the spread of the fire, allowing combat command to send combat or evacuation orders as well as for more effective discharges, such as the FUEGO system, developed in Berkeley, USA.

Fire Urgency Estimator in Geosynchronous Orbit (FUEGO) is a proposal for a system that, using modern

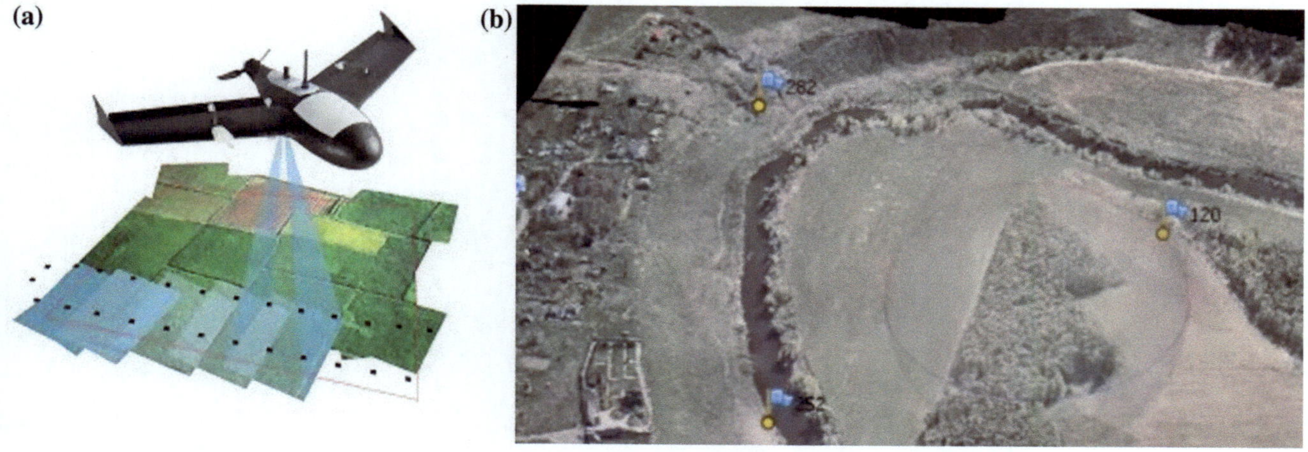

Fig. 7 Mapping of the territory with drones: **a** Laser scanning and georeferenced images; **b** 3D terrain model

digital technology (Pennypacker et al. 2013), uses platforms at various altitudes to detect forest fires, perform a rigorous quantification of the extent of fire fronts, and produce files to feed simulation software for the propagation of forest fires. This system will enable commanders of rescue operations to have the most valuable information in real time for the necessary decision-making in the theater of operations.

6 Concluding Remarks

For all the above, there seems to be no doubt that UAVs, or drones, will play, in a near future, a key role in real-time fire knowledge, planning, and monitoring of forest occupation, forest-urban interface, and also in the post-disaster period, due to all the expert support it can provide to entities involved in the forest fire management process. However, it is the responsibility of the heads of each entity to evaluate the costs, the pros and cons of this use, and to train personnel who, during the various stages of the process, can be entirely dedicated to piloting these aircraft, and collect the needed information. This paper intends to spread worldwide the potential of the use of drones and UAVs in the fight of wildfires. As for the use of satellite imagery, we will also see major developments in the process of acquiring and making data available in real time.

References

Agisoft Metashape User Manual (2018) Professional edition, version 1.5. Agisoft LLC

Correia A, Santos L, Carvalho P, Martinho J (2018) O uso de veículos aéreos não tripulados na monitorização em tempo real da floresta em situação de incêndio, 1ªas Jornadas Ibero-latino-Americanas de Proteção Civil, Coimbra, Portugal

Ollero A, Martínez-de-Dios JR, Merino L (2006) Unmanned aerial vehicles as tools for forest-fire fighting. For Ecol Manag 234(S15): S263

Pennypacker C, Jakubowski M, Kelly M, Lampton M, Schmidt C, Stephens S, Tripp R (2013) FUEGO—fire urgency estimator in geosynchronous orbit—a proposed early-warning fire detection system. Rem Sens 5(10):5173–5192

Viegas X (2018) O complexo de Incêndios de Pedrógão Grande e concelhos limítrofes, iniciado a 17 de Junho de 2017. CEIF-ADAI/LAETA, UC

Slopes Instability Phenomena and Mitigation Techniques After October 2017 Wildfires in Serra Da Estrela Region

Luis Araújo Santos, António Correia, and Paulo Coelho

Abstract

The probability of occurrence of slopes instability phenomena increases after wildfires, being usually triggered by periods of heavy precipitation or earthquakes. In order to minimize incidents or mitigate their effects, which can be delayed over time, various techniques of soil conservation and/or water infiltration may be used. This communication presents the geotechnical incidents registered in the region of Serra da Estrela (Portugal), as well as the slopes stabilization techniques employed after the October 2017 wildfires.

Keywords

Wildfire • Slope stabilization • Serra da Estrela

1 Introduction

The year 2017 will be remembered by tragic events related with Pedrogão Grande and October wildfires, in which 109 people died and more than 300 people suffered injuries. 1300 houses were destroyed, economic damage exceeded 500 million euros of losses only in Pedrogão Grande. Additionally, social costs and environmental damages must also be considered.

In this paper, only the geotechnical and geological aspects of the problems arised from the October 2017 wildfires and their consequences are analyzed. Due to the delay between wildfires and instability phenomena, geotechnical incidents are not always directly related to summer wildfires. However, the severity of October 2017 wildfires, as well as the fears of the rainy season, led to slopes stabilization works.

2 Factors Affecting the Occurrence of Instability Phenomena

For a better understanding of the various types of slope instabilities following wildfires, it is important to introduce some factors that influence, over time, the geotechnical response of burned areas. The various factors can be grouped into two sets: (i) non-fire-dependent factors and (ii) fire-dependent factors (Robichaud et al. 2010). According to these authors, the first set includes factors such as precipitation characteristics, namely their intensity, orography and anthropogenic activities. The second set of factors consists of fire severity, severity of the burned soil area, amount of soil without vegetation cover, soil repellency to water, soil erodibility and time elapsed after the fire (Robichaud et al. 2010).

2.1 Changes in Hydrological Response of Watersheds

The occurrence of instability phenomena cannot be dissociated from the conditions and hydrological response of watersheds due to changes in infiltration characteristics and erodibility of soils after wildfires deflagration (Parise and Cannon 2012). According to these authors, two processes may be the source of instability: (i) surface runoffs that are responsible for erosion, culminating in debris flows and (ii) water infiltration into the soil that triggers processes of soil mass rupture. In addition, the intensity and duration of precipitation periods are also important for natural hazards occurrence (Moody et al. 2013). It should be noted that hydrological response of watersheds depends on the removal

L. A. Santos (✉) · A. Correia
Department of Civil Engineering, School of Engineering (ISEC), Polytechnic of Coimbra, Coimbra, Portugal
e-mail: lmsantos@isec.pt

P. Coelho
Department of Civil Engineering, Faculty of Sciences and Technology, University of Coimbra, Coimbra, Portugal

© Springer Nature Switzerland AG 2020
F. Fernandes et al. (eds.), *Advances in Natural Hazards and Hydrological Risks: Meeting the Challenge*, Advances in Science, Technology & Innovation,
https://doi.org/10.1007/978-3-030-34397-2_16

of surface vegetal layer, which may alter the rate of flow, or may cause depressions that store water and give rise to new infiltration routes (Moody et al. 2013).

2.2 Modification of Soil Characteristics

The combustion of organic matter in the more superficial layer of soil can result in loss or degradation of shear strength properties. Wildfires generally generate more friable and consequently more erodible soils. This is due not only because of the disappearance of roots, but also because the absence of microbial activity which improve soil characteristics (Parise and Cannon 2012). It should also be noted that some soils physical properties, such as granulometry or plasticity, may change during wildfires. Several studies point to the reduction of clay and silt fractions in soils, giving rise to coarser and less cohesive soils that are more vulnerable to erosion (Parise and Cannon 2012). Another factor to consider is the appearance or strengthening of soil repellency to water.

2.3 Influence of Roots

The reinforcement provided by roots in soils is well known, influencing the spatial variation of soil resistance and being an important factor in the stability of steep slopes (Schwarz et al. 2015). The increase of resistance provided by the roots can be included in the Mohr–Coulomb failure criterion as an additional cohesion, as defined in Eq. (1). The resistance parameters c_r, c' and ϕ' are, respectively, the contribution of roots to soil cohesion, cohesion and angle of shear resistance (Wu 2013). Studies point to values for root contribution ranging from 1.5 kPa in deforested areas to 94.3 kPa in natural forests (Schmidt et al. 2001).

$$\tau = c_r + c' + \sigma' \tan \phi' \quad (1)$$

Wildfires can damage crowns, trunks and roots of trees and shrubs, leading to the death of this vegetation and, consequently, to the loss of resistance caused by roots (Jackson and Roering 2009) or the decrease of soil layer thickness (Benda and Dunne 1997). Root depletion can occur in two ways: by direct contact with flames, exposing them to extreme temperatures, especially in the first 5–10 cm of soil, or due to tree or shrub death caused by injuries during or after wildfires (diseases or pests) (Jackson and Roering 2009). In the first case, roots die shortly after the fire while in the second case, death can occur long after the fire. The process of reduction of root resistance reaches its minimum value eight to twelve years after the tree death, which can lead to an increase of instability phenomena (Schmidt et al. 2001; Jackson and Roering 2009).

3 Instability Phenomena Registered After October 2017 Wildfires

The post-wildfire instability phenomena are complex and depend of several factors. In this way, it becomes difficult to identify the most dominant phenomenon, being assumed that all processes of erosion or sediment transport can occur (Moody et al. 2013). In the following sections, several instability phenomena observed in Serra da Estrela region are presented.

3.1 Raindrop, Rain-Flow and Interrill Erosion

The exposition of soil without vegetation cover due to wildfires to physical action of climatic conditions, especially rainfall, is responsible for most of instability phenomena on slopes. The impact of rain drops on the soil is one of the most obvious processes of soil disintegration that, combined with surface runoff along the slope, causes the transport of sediments and/or surface landslides (Moody et al. 2013). The occurrence of these phenomena depends essentially on slope inclination as well as on precipitation duration and intensity. Figure 1a shows an example of surface erosion that can be observed on Ribeira do Gualdim (Arganil).

3.2 Rill Erosion

Rill erosion process is defined as an erosive process by which the flow accumulates in narrow channels, recurrently and, in short periods, removes the soil from this narrow area to considerable depths (Poesen et al. 2006). Figure 1b shows an occurrence of rill erosion near Penalva de Alva (Oliveira do Hospital).

3.3 Debris Flow Erosion

Debris flows are among the more destructive instability phenomena resulting from wildfires (Parise and Cannon 2012). Three initiation mechanisms may be responsible for this type of instability (Moody et al. 2013): (i) a progressive failure of eroded soil together with surface water flow and ashes; (ii) infiltration and consequent saturation of soils above the repellent soil layer, resulting in shallow debris flow; and (iii) occurrence of surface landslides induced by

Fig. 1 Instability phenomena: a surface erosion (Arganil); b rill erosion (Oliveira do Hospital); c debris flow erosion (Oliveira do Hospital)

water infiltration in soils. Figure 1c illustrates small occurrences of debris flow near Penalva de Alva (Oliveira do Hospital)

3.4 Landslides and Slope Failures

Slope stability depends on the shear resistance of soil. Wildfires destroy, in a first phase, trees and shrubs, leaving the soil uncovered, decreasing water retention at the surface. The increase of soil water contents, besides increasing the slope weight itself, reduces the effective stress in the soil and its shear resistance. Later, the deterioration and death of roots lead to an increase in probability of deep landslides (Moody et al. 2013).

There are several examples of landslide occurrences in the regions affected by October 2017 wildfires. Figure 2 shows landslides observed in different locations.

4 Slopes Stabilization Methods and Technics

Various stabilization solutions are available to promote slopes stability. A summary of the techniques adopted in Serra da Estrela region after October wildfires is presented.

4.1 Temporary Barriers for Erosion Control

Erosion control barriers, made from natural or man-made materials, have been used for decades to mitigate the effects of erosion and surface landslides (Robichaud et al. 2010). These structures are designed to: (i) promote infiltration; (ii) storing sediments and (iii) reducing sediment movements on burned slopes (Ferreira et al. 2010). When the burned areas are very wide and trunks of trees are hopelessly burned, these trunks can be used as barriers, as it was done in Penalva de Alva (Fig. 3a). In the absence of large trunks, identical solutions can be developed using smaller trunks and burned branches, as it can also be observed in Penalva de Alva (Fig. 3b).

When small amounts of sediment are to be retained or in areas with low density flows, the previous barriers can be replaced with nylon barriers with straw, such as those observed near Belmonte (Fig. 3c). These structures are permeable, retaining surface landslides and reducing flow velocity (Ferreira et al. 2010).

4.2 Mulches

Mulches (dry mulch or hydromulch) consist of material spread throughout the surface to be protected. Its application

Fig. 2 Landslides: a Penalva de Alva; b Penalva de Alva; c Oliveira do Mondego

(a) (b) (c)

Fig. 3 Erosion control barriers: **a** burned trunks (Oliveira do Hospital); **b** burned branches (Oliveira do Hospital); **c** nylon barriers (Belmonte)

(a) (b) (c)

Fig. 4 Slopes protection: **a** erosion barriers and mulch (Seia); **b** geotextile with straw (Belmonte); **c** channel barriers (Belmonte)

as an emergency treatment post-wildfire has been increasing because its stabilizing effects are immediate (Robichaud et al. 2010). This technique is often used in conjunction with seeding, enhancing soil infiltration capacity, its moisture content and the soil retention capacity of larger soil particles.

In Mata do Desterro (Seia), both stabilization with temporary barriers and mulch were carried out. As can be seen in Fig. 4a, the use of dry straw creates a vegetation cover layer and was also used in erosion control barriers together with existing burned vegetal residues. Near the locality of Santa Cruz, in Belmonte, slopes were protected by non-woven geotextile and straw (Fig. 4b).

4.3 Channels Treatment

Barriers in channels are structures made up of stones and sticks installed to reduce water flow velocity and retain sediments, especially during the first rainy season following wildfires. These barriers, should be small in size, reducing the risk of breakage and possible downstream consequences. Figure 4c illustrates one example which can be observed near Belmonte.

5 Concluding Remarks

In this communication, the main phenomena of slope instability registered in Serra da Estrela region after October wildfires and the main stabilization techniques employed are presented. It should be noted that although examples of illustrated interventions are the result of public investment, it is possible to find several stabilizations works on private land, proving that civil society is also alert to the problems associated with slope stability after wildfires. However, the lack of information is notorious, especially when compared to other countries, where several guidelines to inform population are available. For example, unlike the current practice of cutting all burned wood in order to optimize the price of this raw material, owners should be informed that, in soils susceptible to erosion, this solution is only advisable after plant regeneration to ensure the soil protection (Moreira and Vallejo 2009).

References

Benda L, Dunne T (1997) Stochastic forcing of sediment supply to channel networks from landsliding and debris flows. Water Resour Res 33(12):2849–2863

Ferreira AD, Alegre SP, Carvalho T, Silva JS, Pinheiro AQ, Coelho C (2010) Estratégias e técnicas de conservação do solo e da água após incêndios. In: Ecologia do fogo e gestão de áreas ardidas. ISA Press, pp 229–252

Jackson M, Roering J (2009) Post-fire geomorphic response in steep, forested landscapes: Oregon Coast Range, USA. Quat Sci Rev 28:1131–1146

Moody JA, Shakesby RA, Robichaud PR, Cannon SH, Martin DA (2013) Current research issues related to post-wildfire runoff and erosion processes. Earth Sci Rev 122:10–37

Moreira F, Vallejo R (2009) What to do after fire? Post-fire restoration. Living with wildfires: what science can tell us. European Forest Institute, pp 53–58

Parise M, Cannon SH (2012) Wildfire impacts in the processes that generate debris flows in burned watersheds. Nat Hazards 61:217–227

Poesen J, Vanwalleghem T, Vente J, de Knapen A, Verstraeten G, Martínez-Casasnovas JA (2006) Gully erosion in Europe. Soil erosion in Europe. Wiley, pp 515–536

Robichaud PR, Ashmun LE, Sims BD (2010) Post-fire treatment effectiveness for hillslope stabilization. General technical report RMRS-GTR-240, Forest Service, United States Department of Agriculture

Schmidt KM, Roering JJ, Stock JD, Dietrich WE, Montgomery DR, Schaub T (2001) The variability of root cohesion as an influence on shallow landslide susceptibility in the Oregon Coast Range. Can Geotech J 38:995–1024

Schwarz M, Rist A, Cohen D, Giadrossich F, Egorov P, Büttner D, Stols M, Thormann J-J (2015) Root reinforcement of soils under compression. J Geophys Res Earth Surf 120:2103–2120

Wu TH (2013) Root reinforcement of soil: review of analytical models, test results, and applications to design. Can Geotech J 50:259–274

Impact of Sea-Level Rise in the Azores Islands. Prospective Analysis Based on Current Projections

Marta Aguiar, Margarida Santos, Ana Oliveira, Luísa Magalhães, and Fernando Pereira

Abstract

Scientific projections point to an increase n the average sea-level rise from 0.25 to 2 m up to the end of this century. This increase will result in the disappearance of many beaches, low coastal areas and back retreat. Estimates indicate that more than 50 million people will be affected in the Pacific and Indian areas. The reduction of the available area and the greater pressure on natural resources, increased by the remoteness of other terrestrial areas can manifest the vulnerability in the Azores, which may affect islands or localities where the conditions of survival can become very complex. The most affected island will be São Miguel with 57.6% of the population at risk, but the analysis of the projections and impacts that may occur becomes important to define local action strategies. This analysis should be multidisciplinary, including the natural sciences and engineering, but also the social sciences, to help evaluate and dismiss incorrect perceptions of common sense.

Keywords

Climate change • Sea-level variation • Environmental impacts • Environmental education

1 Introduction

The rise of the average level of sea waters is an indicator of global warming that occurs due to natural actions (volcanic activity) and anthropogenic (industrial activity, transport, construction, agriculture, among others) and has been the object of study in recent times, being considered an imminent danger throughout the world.

Data presented in 2014 estimated an increase of the average sea level, up to 2100, from the order of 0.5–1 m, with a possibility of a significant increasing (Le Cozannet et al. 2014). Other studies seem to conclude that the average sea level is rising more rapidly due to the melting of glaciers and thermal expansion of the oceans, which can mean a rise in the order of 0.77 mm/year during the present century (Glasspool 2008).

The combination of sea-level ascent with extreme weather events can lead to more flooding and greater erosion in coastal areas, increasing risk in many regions of the planet. The Azores can be one of these regions.

2 Evidence and Projections Regarding Sea-Level Rise

The human being is a major contributor for the increase, in global temperature (0.85 °C since the end of the nineteenth century), through the burning of fossil fuels and forest cuts, leading to a greater release of gases into the atmosphere, which increases global warming, and this affects all regions, although in a differentiated way.

The temperature of sea water fluctuates over the years, however, from the 1960s, there seems to be a continuous increase that affects the heating of sea water. This increase in temperature, therefore, has effects on the stability of the polar ice masses, aggravating their melting, contributing to the rise of average sea level, but also affecting the living conditions of many species of living beings.

The combination of rising sea level and temperature will contribute to a greater number of extreme events, such as cyclones, heavy rains, marine storms and floods, and the destruction of coastal areas, affecting living conditions and causing significant economic damage. There are already

M. Aguiar · M. Santos · A. Oliveira · L. Magalhães · F. Pereira (✉)
Direção Regional das Obras Públicas e Comunicações, Secretaria Regional dos Transportes e Obras Públicas, Largo do Colégio, nº 4, 9500–054 Ponta Delgada, Portugal
e-mail: fdpereira@sapo.pt

© Springer Nature Switzerland AG 2020
F. Fernandes et al. (eds.), *Advances in Natural Hazards and Hydrological Risks: Meeting the Challenge*, Advances in Science, Technology & Innovation,
https://doi.org/10.1007/978-3-030-34397-2_17

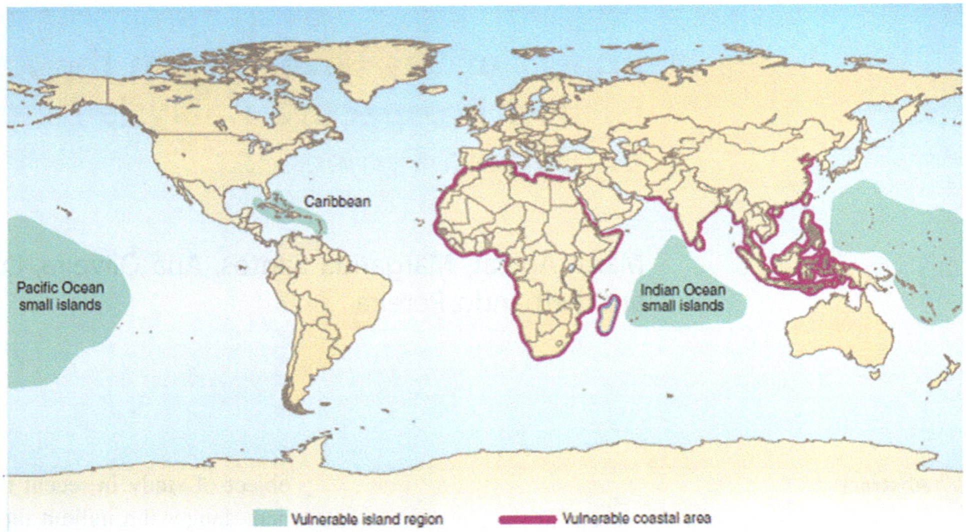

Fig. 1 Vulnerability to sea-level rise. Adapted from: Nicholls and Cazenave (2010)

regions where such phenomena are being recorded, and it looks set to worsen in the coming decades.

These changes will have very significant impacts, especially on activities linked to agricultural production, fishing and transports, whose infrastructures are increasingly exposed and vulnerable and, consequently, in living and health conditions of populations. Scientific projections point to global effects, but with distinct impacts, locating in the Indian Ocean and part of the Pacific Ocean the most affected areas (see Fig. 1), endangering many millions of people.

These projections may be much more critical if were considered estimates that by 2030 nearly half of the world's population will live in coastal areas (Bellard et al. 2013).

3 Impact in Azores Islands

3.1 Impact on Geomorphology and Human Occupation

With climate change, and the increase in sea water temperature, increases the possibility of more storms reaching the Azores islands due to its geographic location, that is, will be reflected in greater maritime agitation, greater intensity of wind and greater precipitations in more concentrated periods. Consequently, more coastal cliffs may crumble and collapse, erosion and retreat of the coastline will be more evident, with many cases already occurring as, for example, falling from a cliff in Caloura Beach (Lagoa), on Santana Beach (Ribeira Grande) and in Calhetas, which occurred in the last months on S. Miguel's Island. On the coastlines, there are mostly steep cliffs, some beaches and some platforms, known as "Fajãs," formed by lava flows or deposits resulting from mass movements of the cliffs.

The territorial dimension of each island and the resident population differ greatly from each other, as well as its orographic forms and geological age. The population agglomerates are located along the coastal line, with the exception of a few cases and due to the insular nature and orography, the main infrastructures vital to the life and economy of the islands, are also in the coastal area. This occupation leads to an overpressure in the use of territory and natural resources, contributing to an increasingly rapid environmental degradation, often due to the pressure of certain social agents in the pursuit of short-term gains, maximizing productive activities (Anderson and Kuhn 2014). This pressure alters the conditions of stability and natural evolution of the landscape and the waste that pollutes the water masses and the forests also contributes to their disappearance (Humboldt *apud* Wulf 2015).

Although, given its small size, the contribution of the economic and productive activities of the Azores to climate change is low, it is found that in almost all the islands their coasts are in danger (see Fig. 2) because the rise in the average level sea waters will interfere with human occupation in certain population clusters.

The situation is more critical on S. Miguel Island, since 57.6% of the population inhabits in vulnerable area, but it is also worrying in Graciosa Island, where about 28% of the population lives in an area of high risk of been affected by storm surges. Given the importance that the coastal area has in the archipelago (high housing occupation and economic activity), these become essential areas of action regarding the reduction of vulnerabilities and the promotion of resilience (Cavalheiro et al. 2017).

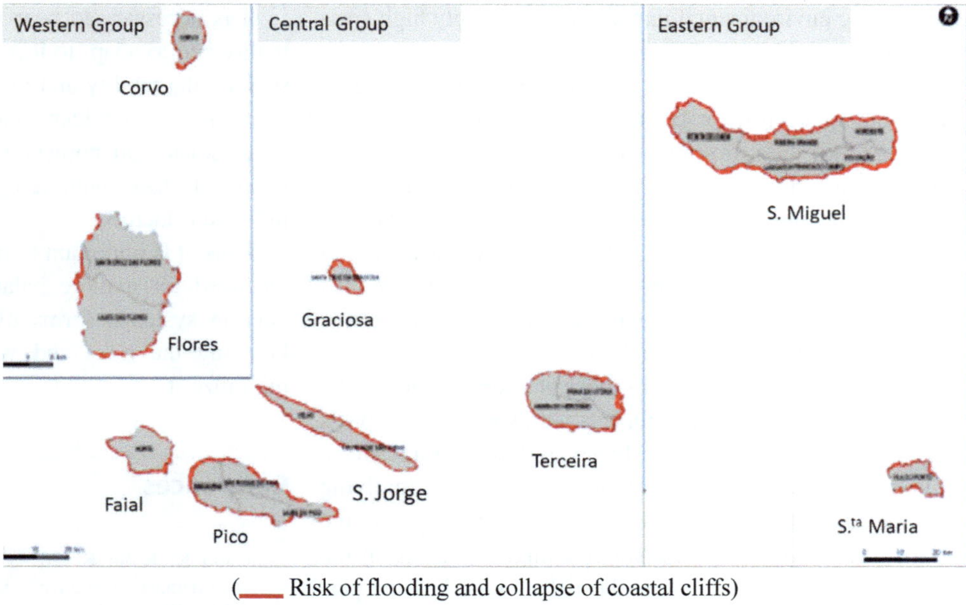

Fig. 2 Areas threatened by the sea (Cavalheiro et al. 2017)

3.2 Impacts on Economic and Social Activities

People, anchored in their social and cultural structures, have a large resilient capacity, adapting to the sudden or violent shocks. However, the changes discussed here will gradually manifest, but continuously and their impacts will be felt also in vital sectors to the survival of communities, such as water resources, biodiversity, economic activities, tourism, inhabited areas and insurance (see Table 1), aggravating situations of social stress and economic dependence.

4 Discussion

It is possible to observe that the elevation of the average level of the ocean will affect the islands and this impact will be greater than the lower island's altitude, the degree of urbanization and the amplitude of the tides.

Currently, there are interventions that aim to correct critical situations with the construction of walls and other structures of coastal protection; however, these interventions

Table 1 Threat of climate change in different sectors and examples of its impacts

Sector	Examples of impacts
Water resources	Water availability issues Saline intrusion
Biodiversity	Retreat of the cliffs Disappearance of beaches Harmful effects on flora and fauna in tidal zones
Economic activities	Fish catch reduction Infrastruture's damages Greater erosion and loss of soils Floods and flooding Production losses
Tourism	Changes in tourist options Infrastruture's damages Accidents with tourists Substantial losses and indemnities
Inhabited areas	Floods and flooding Soil instability Relocations Loss of private and social heritage
Insurance	Increase in indemnity Increase in franchises Reduction/limitation of warranties

Adapted from Tyndall Center for Climate Change Research

have strong environmental impacts and extremely high costs and possibly a limited effect in time.

Sociological studies have shown that technocracy does not contemplate respect for the environment, and populations are not safe from the ideological manipulation of economic and technological forces. Thus, the formulation and implementation of intervention policies must take into account that environmental and community sustainability will depend to a large extent on the level of risk tolerated culturally by society in general and by the social sectors directly affected (Anderson and Kuhn 2014).

Increasing coastal risks will make it necessary to implement measures to minimize impacts with significant impacts in economic and social terms. However, these measures have to be considered in the context of social realities, implying that the institutional actors have some knowledge of the public to be involved in decision-making. Knowing, in the first place, as the public faces the risks, as he has the perception of its causes, and the public awareness on this issue has to be assessed (Schmidt et al. 2012).

5 Concluding Remarks

In conclusion, the islands are vulnerable areas due to their isolation, being stricken throughout their perimeter by abrasion and marine intrusion.

The different types of orography of the islands will influence the way these will be affected, In this sense, the islands of the Azores despite having some coastal impact on the rise of the sea, this will not be so critical due to the altimetric configuration, safeguarding habitable areas.

However, the rise in sea level and changes in the precipitation regime can lead to significant reductions of resources due to the loss of soil and nutrients transported by surface runoff, less fresh water infiltration and greater saline intrusion in aquifers.

It is necessary to bear in mind that climate change is a reality and to adapt to it is necessary to reduce physical and social vulnerability and to act both locally and globally, not being able to consider raising the average level of seawater as an isolated phenomenon. This is an effect that depends on several factors, with temperature being one of the most influential factors.

Thus, it is important to inform and empower people about the need for a more balanced environment between production systems, forms of protection and quality of life, knowing the risks and ways of responding, in order to minimize it.

References

Anderson K, Kuhn K (2014) Cowspiracy: the sustainability secret. Documentary visualized in https://www.youtube.com/watch?v=3T334pdWLzc. Accessed at 19 Jan 2019

Bellard C, Leclerc C, Courchamp F (2013) Potential impact of sea level rise on French islands worldwide. Nat Conserv 5:75–86. https://doi.org/10.3897/natureconservation.@@.5533

Cavalheiro G et al (2017) Programa Regional para as Alterações Climáticas dos Açores (PRAC). Queluz, Lisboa

Glasspool AF (2008) The impact of climate change on Bermuda. Report prepared for the Bermuda National Trust, p 190

Le Cozannet G, Garcina M, Yatesa M, Idier D, Meyssignac B (2014) Approaches to evaluate the recent impacts of sea-level rise on shoreline changes. Eart Sci Rev 138:47–60

Nicholls RJ, Cazenave A (2010) Sea-level rise and its impact on coastal zones. Science 328:1517–1520. https://doi.org/10.1126/science.1185782

Schmidt L, Delicado A, Guerreiro S, Gomes C (2012) Mudanças Climáticas e económicas na costa portuguesa: perceções das comunidades, justiça social e democratização. In: VII Congresso Português de Sociologia. University of Porto, Faculdade de Letras, Faculdade de Psicologia e Ciências da Educação, Porto, pp 8–9

Wulf A (2015) A invenção da Natureza. As aventuras de Alexander Von Humboldt, o herói esquecido da ciência, Lisboa, Temas e Debates. Círculo de Leitores

The Importance of Land Cover Planning on Climatic Events: Evaluation of Peatlands' Buffer Impact on Terceira and Flores Islands (Azores, Portugal)

Dinis Pereira, Cândida Mendes, and Eduardo Dias

Abstract

The objective is to estimate the actual and potential hydrological services provided by the peatlands of Flores and Terceira islands. Peatlands were identified through the distribution of *Sphagnum,* obtained by a spectral signature from satellite image analysis. The results show an actual distribution of natural peatlands of 2766 and 2414 ha, for Terceira and Flores, respectively, which are, even so, quite lower than the potential area estimated as 8035 and 5231 ha, correspondingly. Nowadays, these peatlands have the ability to retain 72,438,317 m^3 of water. Theoretically, if all peatlands were in the natural state, this capacity would increase to 300% of the retained water. The amelioration of naturalness would increase these ecosystems' ability to act better as a buffer in extreme climate events.

Keywords

Naturalness • Spectral signature • *Sphagnum* • Water retention

1 Introduction

The most extended type of natural peatlands is forest dominated (Mendes 2010). A considerable area is still occupied by *Sphagnum*-dominated types (Mendes and Dias 2013). The main threat faced by the peatlands in the Azores is their use as pasture for livestock, leading to the degradation of the peatlands. There is a potential distribution of 35,000 ha of peatlands. Less than 30% currently persists, and of these, more than 50% are under pressure (Mendes and Dias 2017). The peatlands of Azores have numerous ecological, hydrological and biochemical functions as well as social values. These are extremely important in the regulation of the water cycle, being characterised by water retention structures that release water gradually thus acting as buffers, minimizing the effects of climate events, promoting landscape equilibrium, minimizing the impact of extreme events, such as landslides or floods, and supporting biodiversity. However, owing to peatland depletion and degradation, their natural functions have become narrowed. One of the main future challenges is the establishment of the strategies to minimize the impacts of extreme events, due to climate change, and the Azores islands are no exception. In this context, the goal of this work is to calculate hydrological services from the current distribution of peatlands in Flores and Terceira islands and compare with equal parameters estimation, considering the potential distribution of peatlands in those islands.

2 Methods

Study area: The Azores are located at about 1400 km W from Europa in the middle of the Atlantic Ocean. The archipelago has nine islands, distributed in three groups. The islands selected for this study are Flores (142 km^2, 914 m a.s.l at Morro Alto) and Terceira (402 km^2, 1023 m a.s.l at Santa Barbara mountain), as these have the largest areas of peatlands (Dias et al. 2004).

2.1 Data collection and analysis

Actual distribution of peatlands: For the definition of *Sphagnum* distribution was used: Sentinel-2 images (2018–12–07 for Terceira and 2019–01–22 for Flores) and Rapideye (2017–12–21 for Terceira and 2012–04–16 for Flores), with 2800 Terceira and 7500 Flores Ground Truths. The

D. Pereira (✉) · C. Mendes · E. Dias
FCCA, CBA Centro de Biotecnologia dos Açores, Universidade dos Açores, Rua Capitão João d'Ávila, 9700-042 Angra do Heroísmo, Portugal
e-mail: dinispereira@dinispereira.com

© Springer Nature Switzerland AG 2020
F. Fernandes et al. (eds.), *Advances in Natural Hazards and Hydrological Risks: Meeting the Challenge*, Advances in Science, Technology & Innovation,
https://doi.org/10.1007/978-3-030-34397-2_18

modulation was done in ArcGIS 10.6.1 with a supervised classification with maximum likelihood using 25 and 50% Reject Fraction (RF). For ground truth, ecological fieldwork was done. It was also integrated data from Mendes (2010), Dias et al. (2017), ATLANTIDA©GEVA database and photointerpretation of Planet Scope Images.

Potential distribution of peatlands: The potential distribution was modelled using Rasters (100 × 100 m) of Aspect E and W orientation, slope (>9°), curvature (< −0.55), TOPEX (Wilson 1984), endorheic basins, and geomorphology (as a limitation of the model in Terceira). In both situations of *Sphagnum* and forested peatland, water services values from Pereira (2015) were applied, to the extension of the peatland typology.

For actual and potential cartographic analysis purposes, two types of *Sphagnum* peatlands were assumed: (1) mixed *Sphagnum* peatlands (basin, raised and transition) and (2) hillside *Sphagnum* peatlands (hillside and blanket types) (classification of Mendes and Dias 2013) and one type of forested and shrubland peatland (in Dias 1996). Considering naturalness (based in Pereira 2015 and Mendes 2010), (1) and (2) were separated into natural (no disturbance), degraded (frequent use as pasture, *Sphagnum* present) and peat soil pasture (corresponding to wet implanted pastures, no *Sphagnum*); forested peatlands were separated in natural and peat soil forested areas (corresponding to forest production).

Hydrological services evaluation: For the actual and potential quantification of hydrological services, reference values were established. To define these reference values for water retention capacity as well as time efflux (considering peatland type and naturalness degree), we studied eight representative peatlands in Terceira island (in Pereira 2015). These were surveyed with Ground Penetrating Radar for the tri-dimensional modelling of peatlands' internal deep and layer structures. Field coring was done with peat collected by layer for bulk density, water retention capacity and water efflux determination. The obtained reference values were estimated by Pereira (2015) and applied to the different typologies defined in actual and potential peatlands, for the two islands under analysis and the services quantified.

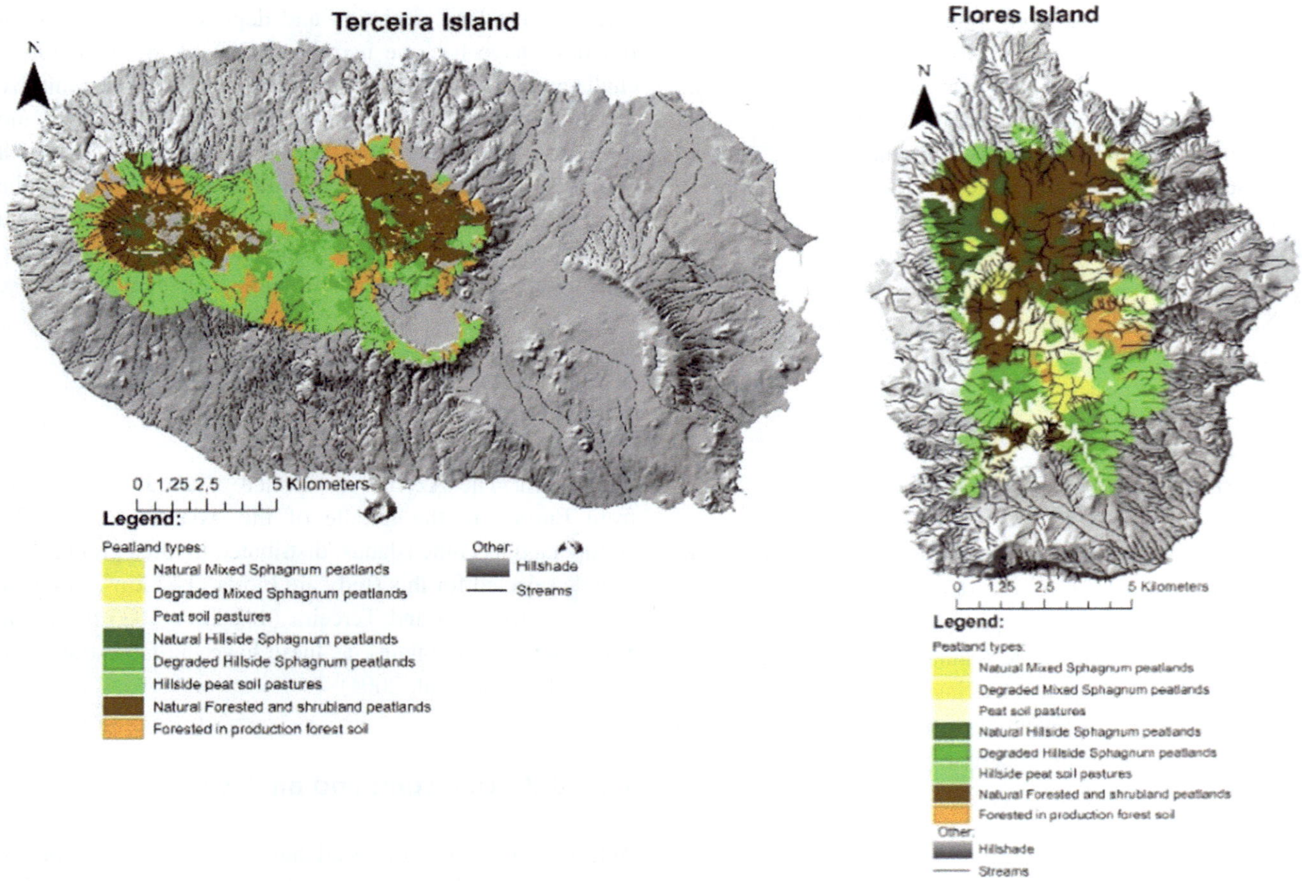

Fig. 1 Actual distribution of peatlands defined for Terceira and Flores islands. Naturalness considered in the type's definition (natural, degraded or peat soil when transformed into pasture or forest production). Modulation in ArcGIS 10.6

3 Results

3.1 Actual Distribution of Peatlands

Actual distribution of peatlands: In Terceira island, the area obtained is 909 ha (423 ha using an RF of 0.5 and 486 ha with an RF of 0.25). This distribution in Flores island is 648 ha (290 ha using an RF of 0.5 and 359 ha with an RF of 0.25). The actual distribution of peatlands is presented in Fig. 1 and Table 2. Both in Flores and in Terceira, the dominant type of peatland is *Sphagnum* dominated types (including the degraded forms). However, in potential terms, it would be the forested formations of the commonest form of Azorean peatlands.

3.2 Potential Distribution of Peatlands

The potential distribution of the peatlands (assumes all formations as natural) is expressed in Fig. 2 and Table 2. Both in Flores and Terceira, the dominant types of peatlands are forested and shrubland. *Sphagnum* types prevail in endor-

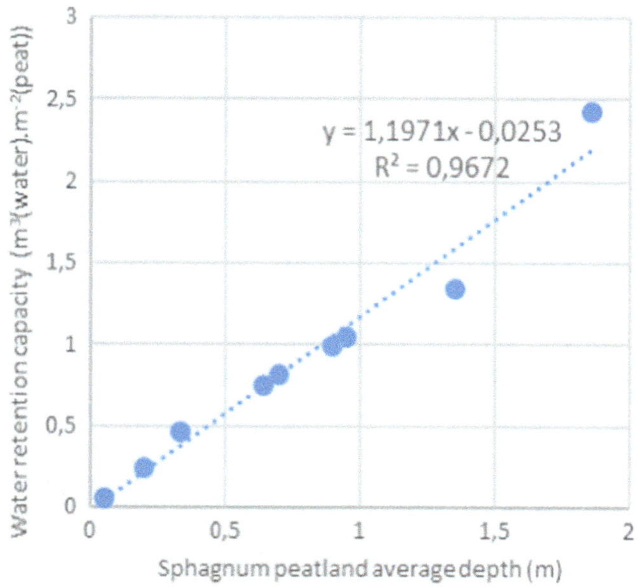

Fig. 3 Relation between *Sphagnum* peatland's depth and water retention capacity. Peat depth obtained by GPR and modulated in GIS. Water retention capacity values obtained in the lab as described by Pereira (2015)

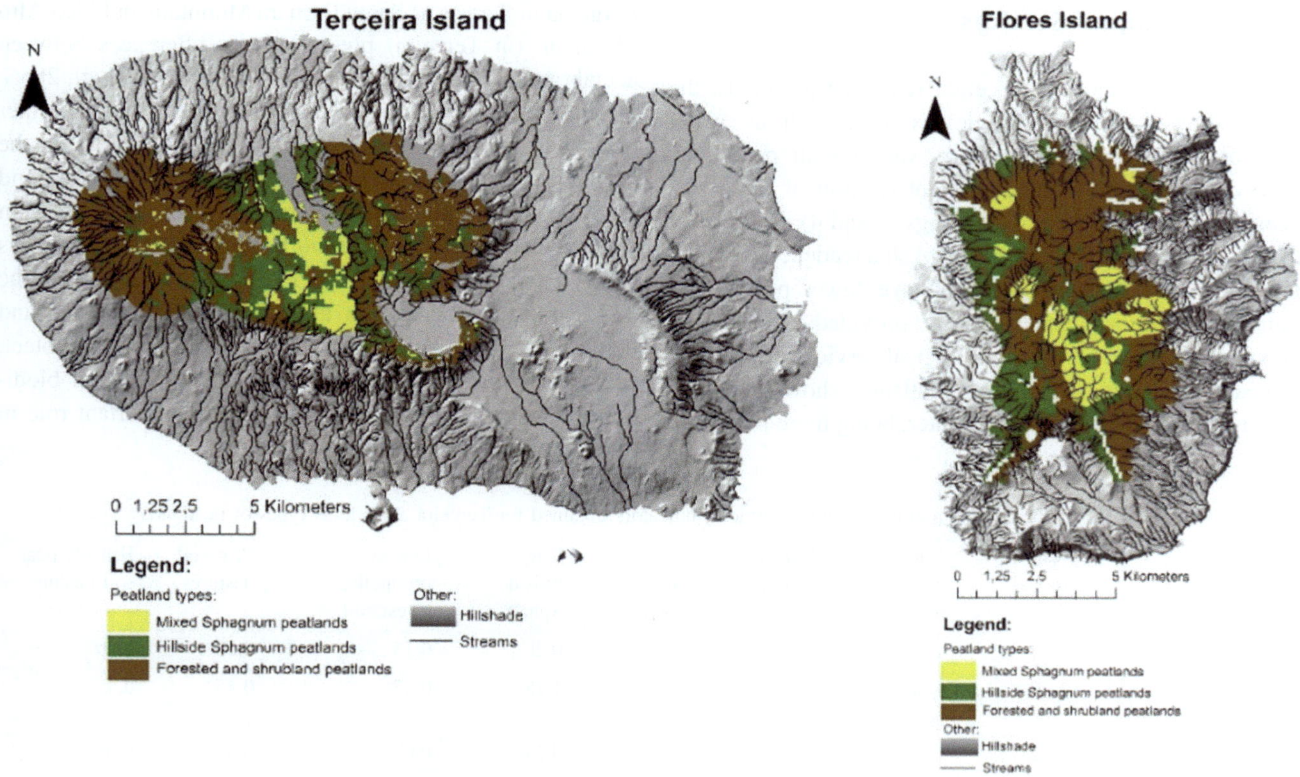

Fig. 2 Simulated potential distribution of peatlands defined for Terceira and Flores islands. Modulation in ArcGIS 10.6

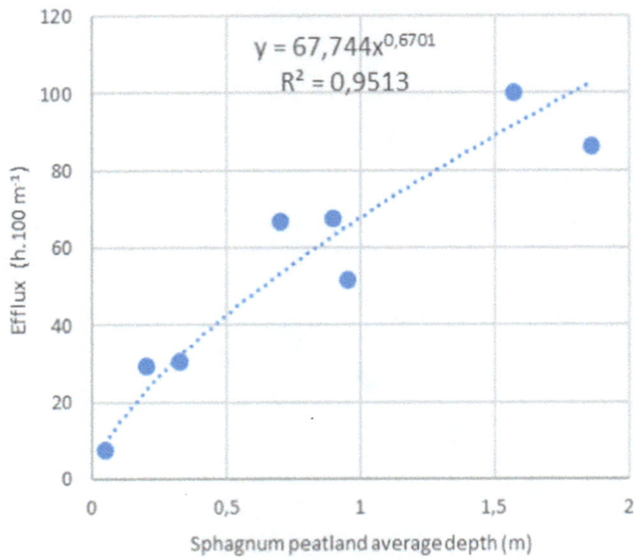

Fig. 4 Relation between *Sphagnum* peatland's depth and water efflux. Peat depth obtained by GPR and modulated in GIS. Time efflux values obtained in the lab as described by Pereira (2015)

heic valleys, and *Sphagnum* dominated hillside types monopolise extreme wind sloping areas.

3.3 Hydrological Services

Considering the study of the eight reference peatlands, the results showed that hydrologic services, such as water retention as well as water efflux, varied with peat type. However, we found a more relevant relation of these services associated with peat depth (Figs. 3 and 4). In terms of disturbances, it was possible to establish a tendency in which more degraded peatlands tend to have lower peat depth, diminishing peatland water services provided. Considering the values obtained for the hydrological services of all types of peatlands (Table 1), mixed peatlands show a greater capacity to contain gravitational water, being more important for buffering torrential regimes after large rainfall events. The same type also reveals the greatest capacity for the temporal restraint of gravitational water (efflux), working as accumulation structures.

The results show an actual distribution of natural peatlands of 2766 and 2414 ha, for Terceira and Flores, respectively, which is even so quite lower than the potential area estimated as 8079 and 5268 ha, correspondingly. These peatlands currently have the ability to retain 72,438,317 m^3 of water (Table 2). Theoretically, if all peatlands were in a natural state, this capacity would increase to 300% of the retained water.

4 Discussion

Floods accompanied by landslides are quite frequent in the Azores and could become more common, associated with climate change. As shown in this study, peatlands are extremely important landscape regulators, which are far below their potential capacity. The potential increase of an average value of 300% in water retention is relevant and can be, theoretically achieved. Additionally, we must highlight that these two islands are the ones characterized by more areas of peatlands and those in a better state of conservation. Large natural areas as Santa Barbara Mountain and Pico Alto Mountain (in Terceira) present lower differences between actual and potential hydrological services, with a 285% increase for Santa Barbara and 237% for Pico Alto. Other areas practically do not present natural peatlands, as the pasture-dominated mosaic between Serra Santa Barbara and Pico Alto have a potentially enormous increase of 1418% in water retention. The pristine north-central plateau of Flores show a potential increase of 132%, contrasting with the 432% potential increase of the southern part of the island being more disturbed. Urgent action is required to protect, sustainably manage and restore peatlands for global biodiversity protection, and it can also play an important role in reducing GHG emissions.

Table 1 Average values of peat depth and hydrological services indicators obtained for Terceira and Flores types of peatlands

	Nat. mixed Sphag	Nat. hillside Sphag	Nat. for./ shrubland	Deg. mixed Sphag	Deg. hillside Sphag	For. in production forest soil	Peat soil pastures	Hillside peat soil pastures
Peatland depth (m)	1.65	1.10	0.33	1.15	0.20	0.15	0.10	0.10
Water retention (m^3 (water) m^{-3} (peat))	1.10	1.38	0.70	1.18	1.18	0.47	0.45	0.45
Water retention (m^3 (water) m^{-2} (peat))	1.82	1.53	0.23	1.35	0.24	0.07	0.05	0.05
Efflux (h 100 m^{-1})	67.74	30.59	10.26	37.21	29.25	7.00	7.80	7.80
Perlocation velocity (m h^{-1})	0.11	0.08	0.10	0.18	0.02	0.07	0.08	0.08

Table 2 Values of actual and potential areas of peatland distribution and associated values of water retention capacity obtained for Terceira and Flores types of peatlands

	Terceira				Flores			
	Area (ha)		Water storage (m^3)		Area (ha)		Water storage (m^3)	
	Actual	Potential	Actual	Potential	Actual	Potential	Actual	Potential
Nat. mixed Sphag	71	763	1,300,615	13,841,653	123	799	2,240,017	14,505,886
Nat. hillside Sphag	197	2166	3,037,997	47,404,912	769	1327	11,875,007	29,031,787
Nat. for./shrub	2497	5150	5,775,013	11,844,863	1523	3142	3,526,759	7,226,076
Deg. mixed Sphag	379		5,137,043		148		2,002,912	
Deg. hillside Sphag	825		1,990,130		463		1,119,573	
For. in production forest soil	1864		843,477		376		170,264	
Peat soil pastures	1050		738,909		684		482,185	
Hillside peat soil pastures	1151		520,931		1146		519,303	
Total	8035	8079	19,344,115	73,091,429	5231	5268	21,936,020	50,763,749

5 Concluding Remarks

Terceira and Flores islands possess a relevant area of peatlands; however, disturbance diminishes their intervention in the hydrologic cycle control of the landscape. Restoration implementation would significantly increase the buffering capacities of peatlands in a scenario of climate change.

Acknowledgements PlanetLabs, for allowing the use of Planet and Rapideye images. Study integrated in CONNECT.GENE project (Ref. Acores-01-0145-FEDER000061) financed by FEDER and regional funds through the Azores 2020 Operational Program.

References

Dias E (1996) Vegetação Natural dos Açores. Ecologia e sintaxonomia das florestas naturais. Azores University, Department of Agriculture Sciences (Ph.D. Thesis)

Dias E, Mendes C, Melo C, Pereira D, Elias R, Elias S, Pereira F (2004) Plano de Gestão Sectoriais das áreas Terrestres da Rede Natura 2000 dos Açores. Departamento de Ciências Agrárias. Universidade dos Açores & Direcção Regional dos Serviços de Ambiente, Açores

Dias E, Pereira D, Mendes C (2017) Cartografia Ecológica para o Ordenamento e Conservação da Ramsar Central da Ilha Terceira. Geva, Angra do Heroísmo. ISBN 978-989-95707-6-4

Mendes C (2010) A Dimensão Ecológica das zonas húmidas na Gestão e Conservação dos ZEC terrestres dos Açores. Azores University, Department of Agriculture Sciences (MSc thesis)

Mendes C, Dias E (2013) Classification of Sphagnum peatlands in Azores—cases from Terceira Island. Suo Mires and Peat 64(4):147–163

Mendes C, Dias E (2017) Portugal—Açores. In: Joosten H, Tanneberger F, Moen S (eds.) Mires and peatlands of Europe: Status, distribution and conservation. Schweizerbart Science Publishers

Pereira D (2015) Avaliação do valor dos ecossistemas de turfeiras dos Açores, com recurso a modelação em Sistemas de Informação Geográfica. Azores University, Department of Agriculture Sciences (Ph.D. Thesis)

Wilson J (1984) Determining a topex score. Scottish For 384:251–256

Ponta Delgada, a Volcanic City. Main Natural Risks and Reasons that Dictate the Coexistence of Populations

Maria Anderson and Nelson Mileu

Abstract

S. Miguel is constituted by several volcanic structures, considering this island, in spite of its current quiescence, one of the most eruptive poles of the Atlantic. This is the fundamental characteristic from which other hazards, such as CO_2 emission from fumaroles, volcanic and tectonic earthquakes as well as mass movements associated with both episodes of intense precipitation and seismic activity, were developed. An exploratory study was developed to evaluate the coexistence of populations with the stated risks. Using a methodology of measurement, the application of a survey to a population sample, it is verified the low perception of the volcanic risk, both by the populations and by the institutions and authorities in the communication of the risk. However, it is noted that the population surveyed seeming to underestimate this threat, as if the memory of previous eruptions constituted only a historical heritage and, at the same time, they have a high/average perception of the other risks associated, namely seismic and hydrological risks (this includes the consequences of intense precipitation, that is, mass movements). Also, it was noted that this survey relies differently on the authorities and institutions responsible for risk communication and risk management.

Keywords

Volcanic risk • Risk perception • Coexistence of populations • Institutional confidence

M. Anderson (✉)
REDE- Associação Nacional de Voluntariado de Proteção Civil, Ponta Delgada, Portugal
e-mail: mariaanderson2@gmail.com

N. Mileu
IGOT, Universidade de Lisboa, Lisbon, Portugal

1 Introduction

The climate and the volcanic substrate that illustrate the Azores are well described in the Azorean literature, in which the geographical, meteorological and geological constraints of the inhabited space, acquire life. The harmony between man and nature is typically experienced in the islands as a precarious state that obeys a cycle of destruction—reconstruction (Pires 1979). In these communities, it is possible to observe an intrinsic coexistence with natural risks in general, presupposing an awareness, perception and acceptance of them.

There is also a recognized and deep knowledge by the scientific community present in the Azores, in particular regarding the characterization of the volcanic and seismic hazards in these islands. In the case of S. Miguel, which is the object of this exploratory study, three active central volcanoes (Sete Cidades, Fogo and Furnas) (Carmo 2013) are identified and monitored, among others. There are several studies on the characterization of volcanic and tectonic structures (França et al. 2003). The volcanic activity is therefore inherent to the island's own constitution, and historical records of several occurrences are known either on land or at sea, accompanied by seismic activity, as well as accounting for several events in which, in addition to volcanic and/or tectonic seismicity, there were mass movements (Amaral et al. 2009). More recently and already with instrumental records, it is verified in the island of S. Miguel that the main volcanic and seismic activity is associated to the volcanoes of Sete Cidades and Furnas and in the sector comprised by the volcano of Fogo-Congro allowing to affirm, based on the temporal distribution of the eruptive phenomena, for a mean period between eruptions of 21.7 years (Carmo 2013, p 47).

Regarding Civil Protection Authorities, the Regional Service for Civil Protection and Firemen of the Azores (SRPCBA), such as the University of the Azores, are institutions that play a decisive role in communication and visualization of risk and, consequently, contributes to the

© Springer Nature Switzerland AG 2020

F. Fernandes et al. (eds.), *Advances in Natural Hazards and Hydrological Risks: Meeting the Challenge*, Advances in Science, Technology & Innovation,
https://doi.org/10.1007/978-3-030-34397-2_19

perception of risks (Tavares et al. 2011). In addition, SRPCBA has been carrying out an increasing number of actions to raise public awareness of the adoption of self-protection measures.

The investigated question that is presented in this article intends to evaluate how the perception of volcanic, seismic and hydrological hazards (in this one include the consequences of intense precipitation and the mass movements) in the region denominated Fissural volcanic system dos Picos, considering the familiarity with them, and also clarifying the confidence that the institutions arouse in the sampled population.

2 Settings and Methods

The methodology used carried out a survey that allowed the interdisciplinary between the social sciences and the analysis of the main natural risks of the region. In this study, were not evaluated the social characteristics for risk construction.

The unit of sampling was the household and the unit of inquiry was the individual. The residence in the region Fissural volcanic system dos Picos (Forjaz et al. 2004) was an inclusion criterion. A sample of 17 individuals associated to "Casa do Povo da Fajã de Baixo" was residents in this region for at least 10 years.

Following the Hayes methodology (Hayes 2011), areas of potential risk were identified, using data available in diverse evaluation and characterization of S. Miguel's island as well as a survey of occurrences from regional newspapers. The geographic unit used was the volcanic region with coexistence of other risks, using the definition of the following regions: (a) Sete Cidades Volcanic System, (b) Fissural volcanic system dos Picos, (c) Fogo Volcano, (d) Furnas Volcano and (e) Povoação-Nordeste Region.

The survey questions privileged the evaluation of risk perception, referring to the history of events and, for each risk, categorized the degree of confidence in the Institutions.

The formulation of the questions encompassed risk perception according to Hayes's (Hayes 2011) method for assessing community vulnerability according to their attitudes and beliefs about risks.

The survey contains a total of 64 multiple choice and open response questions that include demographic characteristics, risk perception, knowledge of prevention measures and degree of confidence in different institutions.

Prior to the application of the survey, a session was held to raise awareness of the presence of risks and presented the main self-protection measures. Interviewers were also trained on the objectives of the questionnaire and on how they should lead the interview in a conversational manner. The interviews took place at the "Casa do Povo da Fajã de Baixo" on February 13. Each questionnaire lasted about 7 min.

To measure risk perception a five-degree scale was used for volcanic, seismic and hydrological hazards.

A scale of 4 degrees was used to measure the degree of trust in institutions, namely (a) Regional Government, (b) University, (c) media, (d) Church and (f) Family/Friends.

3 Results

Based on the sampling criteria, an elderly population was sampled. The average household size covered by the survey is 2.5 persons. Of the respondents, 82% were women, and the average age was 76 years. The sample is characterized by 64.6% of the respondents being widowers, 23.4% married and 12% being divorced. The majority, 76.5%, consider themselves healthy despite their age. Still, 52.9% live with their sons and 23.5% live with their sons and grandsons or even alone. As far as work is concerned, 88% are retired and have no connection to agriculture and 6% of the rest work in agriculture although for their own. As for forms of support for the household, 64.7% contribute with their retirement and 35.3% are dependent on their families.

Regarding the three main risks, the most frequent is the seismic (88.2%) followed by the volcanic (11.8%), not having mentioned the hydrological risk.

The volcanic risk is the least concern to them, 60.8%, because it constitutes "no risk", and 64.7% considered that "there will never be another eruption" even when faced with the fact that, at least two underwater eruptions with the emission of gases and basaltic material, in Banco do Mónaco, in July 1981 and Banco D. João de Castro, in 1997 (Madeira 2007).

Television and radio were the most important information, (58.8%) and also the family (11.8%).

As for the expected damages, the volcanic risk is the one for which major damages are expected, 47.1%, in the degree "high" and "very high" (Table 1).

The assessment of the degree of confidence of the respondents on the various Institutions was only made for hydrological and volcanic risks (Fig. 1).

4 Discussion

The sample surveyed reveals familiarity with the volcanic risk, because of the influence of television and radio (58.8%), recognizing that there are risks associated with the extrusion of lava, such as ash deposition, gasification and eventual explosions, but they are not concerned with the situation in their region. However, they admit the fact that if there were a volcanic catastrophe, the damage would be very High (17.7%) or High (29.4%). The hydrological risk is not a concern.

Table 1 Perceived damages due to risk

Risk	Very high	High (%)	Moderate (%)	Low (%)	Negligible (%)
Volcanic	17.7%	29.4	41.2	5.9	5.9
Seismic	x	23.5	58.9	5.9	5.9
Hydrological	x	29.4	29.4	5.9	23.5

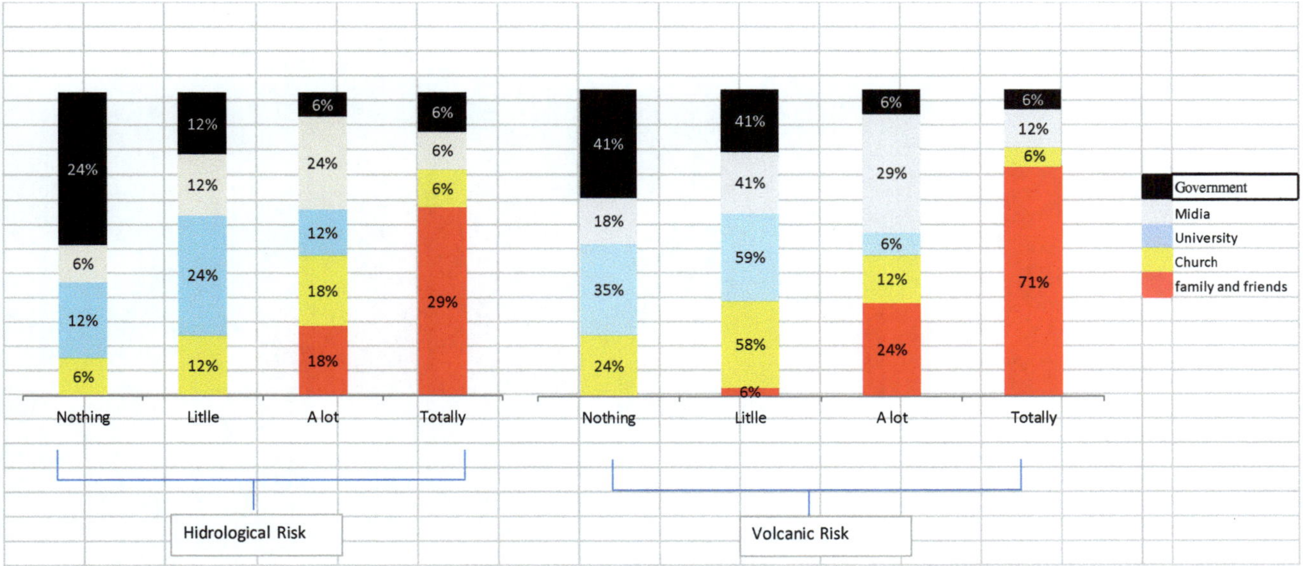

Fig. 1 A confidence shown by the respondents in two situations: affectation by hydrologic risk and by volcanic risk

Family/friends represent the highest degree of trust (much confidence or total) followed by the media, revealing their role in the information and communication of the risk. The Regional Government or the Government Institutions, was not clear, are the entities that invest more in the communication of the risk, nevertheless the results on lack of confidence.

5 Concluding Remarks

It is concluded for the elderly population that although they have knowledge about the characteristics of the volcanic risk through the media, their exposure is not perceived, inferring from this exploratory study that this fact is due to lack of confidence in the institutions/Regional Government. Another explanatory hypothesis that must be studied should evaluate the effectiveness of the degree of communication of this specific risk by governmental institutions.

It is still concluded that the sample reveals a small perception of its exposure to hydrological risk and reveals the expectation that the consequences will be insignificant or small (29.4%), with no response accounting for about 12%. The reason may be related to the fact that the annual precipitation has presented an annual deficit, although small (−9%), when compared to the normal one of reference as of 1961–1990 (Costa 2017).

Acknowledgements Thanks are due to the *Casa do Povo da Fajã de Baixo* and to the students of the University of the Azores, João Pedro Medeiros and Maria Francisco Anderson.

References

Amaral P, Marques R, Zêzere JL, Queiroz G, Goulart C (2009) Estudo comparativo de diferentes métodos probabilísticos para a avaliação da susceptibilidade à ocorrência de movimentos de vertente: um caso de estudo no vale da Ribeira Quente (S. Miguel, Açores) Volume VI, APGEOM, p 183–190

ANPC Homepage. http://planos.prociv.pt/Pages/distritos.aspx?dis=19. Last Accessed 10 Feb 2019

Carmo RL (2013) Estudos de Neotectónica na Ilha de S. Miguel, Uma contribuição para o estudo do risco sísmico no arquipélago dos Açores. University of Azores, Ponta Delgada (Ph.D. thesis)

Costa V (2017) Levantamento climatológico da precipitação em Ponta Delgada no ano de. In: Livro de Actas 2018 workshop em Ciências da Terra e do Espaço, pp 74–84, Universidade de Évora

Forjaz, Victor (Hugo) et al (2004) Atlas Básico dos Açores. 1nd edn. Observatório Vulcanológico e Geotérmico Acores, Ponta Delgada

França Z, Cruz JV, Nunes JC e Forjaz VH (2003) Geologia dos Açores: uma perspectiva actual. Açoreana 10(1):11–140

Hayes SL (2011) Volcanic risk assessment: integrating hazard ans social vulnerability analysis. University of Plymouth, Pearl (Ph.D. thesis). Accessed on http://pearl.plymouth.ac.uk

IPMA Homepage. http://www.ipma.pt Last Accessed 10 Feb 2019

Madeira J (2007) A erupção dos Capelinhos e o vulcanismo nos Açores. Boletim do Núcleo Cultural da Horta 16:29–44

Pires AM (1979) Marcas da insularidade no Mau Tempo no Canal de Vitorino Nemésio. Arquipélago, Série Ciências Humanas 1:79–90

SRPCBA Homepage. https://www.prociv.azores.gov.pt//formacao/. Last Accessed 10 Feb 2019

Tavares A, Mendes JM, Basto E (2011) Perceção dos riscos naturais e tecnológicos, confiança institucional e preparação para situações de emergência: O caso de Portugal Continental. Revista Crítica de ciências sociais 93, junho 2011:167–193

Build Back Better: Rebuild to Preserve

Fernando Pereira, Ana Oliveira, Margarida Santos, Marta Aguiar, and Luísa Magalhães

Abstract

Heritage values are socially, economically or culturally directly linked to populations and influence the static or dynamic way in which they perceive, experience and enjoy their territory. Knowledge of the risks and measures that contribute to the preservation of its livelihoods and its social and cultural environment, regardless of whether or not to change the way of organizing space and construction forms, will be important to minimize impacts and avoid social dysfunction. The Sendai Framework for Disaster Risk Reduction 2015–2030 points out for cases of damages, it will be better to rebuild, seeking to maintain communities by making use of their adaptability reinforcing their resilience. According to the Valletta Principles for the Safeguarding and Management of Historic Cities, Cities and Urban Areas (2011), also known as "The Valletta Principles", adopted by the 17th ICOMOS General Assembly on November 28, 2011 in Valletta, Malta, interventions must ensure respect for tangible and intangible heritage values, as well as the quality of life of the inhabitants. These are broad concepts, but applied to the analysis and definition of disaster risk prevention and mitigation measures, may contribute to keeping communities increasingly reflexive and resilient.

Keywords

Disaster risk reduction • The Valletta principles • Preserve social and cultural organization • Build back better • Land use re-planning

1 Introduction

The Atlantis legend, the flood in the Bible, the explosive eruption of Santorini, Pompeii's destruction by a burning cloud in 79, Lisbon in ruins due to earthquake and tsunami in 1755, San Francisco's earthquake in 1906, the town of Armero engulfed by lahars in 1985, a gas cloud released from Lake Nyos (Cameroon, Africa) by an eruption in 1986, Mount Pinatubo's eruption (Philippines) in 1991, Kobe's earthquake in 1995, and many other earthquakes in China, Iran, Chile, Italy, Greece, floods in many cities and territories outside the area of monsoon or tropical storms. And, as most remarkable in most recent times with great global and economic impact, the 2004 tsunami in Banda Aceh, with victims among tourists from all over the world, the volcanic ash clouds of Eyjafjallajokull 2010, which led to the cancellation of numerous commercial flights and closure of airports and the nuclear accident of Fukushima 2011, whose plant was hit and damaged by a tsunami. Throughout the history of humanity, there are references and descriptions of natural disasters.

Man has always been responding to disasters by rebuilding his environment, seeking to improve the safety and protection of his buildings with locally adapted solutions, even though architecture and engineering models carried from other regions.

With technological development, solutions and techniques have been globalizing, is often presented as standard, leaving populations out of the discussion and often conflicting with local geodynamic, climatic and cultural characteristics. To reduce disaster risks and improve quality of live, all around the world, the United Nations approved the

F. Pereira (✉) · A. Oliveira · M. Santos · M. Aguiar · L. Magalhães
Direção Regional das Obras Públicas e Comunicações, Secretaria Regional dos Transportes e Obras Públicas, Largo do Colégio nº 4, 9500-054 Ponta Delgada, Portugal
e-mail: fdpereira@sapo.pt

© Springer Nature Switzerland AG 2020
F. Fernandes et al. (eds.), *Advances in Natural Hazards and Hydrological Risks: Meeting the Challenge*, Advances in Science, Technology & Innovation,
https://doi.org/10.1007/978-3-030-34397-2_20

Sendai Framework for Disaster Risk Reduction 2015–2030, that aims to achieve a "*…substantial reduction of disaster risk and losses in lives, livelihoods and health and in the economic, physical, social, cultural and environmental assets of persons, businesses, communities and countries*" (UN 2015).

2 Hazards Versus Impacts

A rainy day is as beautiful as a sunny day.
Both exist; each one as he is.
Alberto Caeiro

The islands of the Azores also have in their history a series of events associated with natural disasters. It highlights here only some of the greatest impact that constitutes milestones in the social history of the Azorean population, namely, the earthquake of 1522 in Vila Franca do Campo, that was largely destroyed by the earthquake and totally buried by deposits of mass movements (Silveira 2002), the two seismic crises that became known as "Cahida da Praia" on 1614 and 1841, the earthquakes in Horta, Faial, 1926 and 1998, in Pico, 1973 and in Angra do Heroísmo, Terceira, 1980. There were also three volcanic eruptions, namely Pico Sapateiro in Ribeira Grande in 1563, a submarine, southwest of S. Miguel, which gave rise to the so-called Sabrina Island in 1811, and which survived only a few months and the most recent and impressive the Capelinhos Volcano, in Faial, in 1957–1958. Storms in Horta, Faial, 1946 and 1969, in Nortdeste/Povoação, S. Miguel 1986, on Ponta Delgada/Povoação, S. Miguel, 1996 and mass movements, on S. Jorge, 1757, when great movements of the cliffs occurred, dragging houses until the sea (Silva 2005) and Ribeira Quente, 1997.

Apart from the above-highlighted cases, there are records of natural disasters from the settlement of the islands, during fifteenth century, and covering all the islands. Although their impacts are not very significant in damages, those occurrences reveal types of hazards and vulnerable areas. It is also worth noting the various records of small tsunami, including that of 1755, associated with the Lisbon earthquake.

The Global Assessment Report on Disaster Risk Reduction 2013 uses definitions as Natural or Physical Hazard for hazardous phenomena such as floods, storms, droughts, and earthquakes, in general, but considering the scale defines Major Hazard for global or regionally important hazards such as earthquakes, tsunamis, flooding in large river basins and tropical cyclones and Local Hazard for small scale hazards such as flash or surface water flooding, fires, storms and landslides, which affect particular localities. Considering the social aspects, Exposure refers to the location and number of people and assets in risk-prone areas, Vulnerability to the degree of susceptibility of these assets to suffer damages and losses and Resilience is about systems ability to absorb or buffer losses, and recover. It also defines the Emerging Risk to describe the risk of extremely low probability disasters associated with new patterns of risk and vulnerability, such as increasing dependence on modern societies in vulnerable energy and telecommunication networks.

The Azores Archipelago consists of nine small-inhabited islands, so the natural hazards are essentially of local scale, except for some earthquakes or storms that may affect several islands having a major scale. The islands essentially result from submarine and subaerial volcanic activity, with central volcanoes being associated with sub-Plinian and hydromagmatic explosive eruptions, but there are also fissural-type volcanic systems with effusive or moderately explosive eruptions of Hawaiian and Strombolian type (Pacheco et al. 2013). There are also many scoria cones and surface materials are essentially pyroclastic deposits and some basaltic lava flows.

Considering that the vast majority of population settlements are located at the foot of the slopes of volcanoes, near the coastline, with only a few settlements located in the interior of the islands, two of them in the bottom of volcanic caldera, the exposure and vulnerability are significantly high. Therefore, the most considerable natural hazards in Azores Islands are earthquakes, volcanic eruptions, landslides, cyclones, sea storms, floods, and flashfloods (see Table 1) and the increasing dependence on electricity and communications vulnerable networks.

3 Approach to the Valletta Principles and Sendai Framework for Disaster Risk Reduction

There is a date on the balcony… which reminds us
of the time when the house was rebuilt.
Agustina Bessa-Luís

Physical forms of landscape and manmade constructions are tangible elements, whereas cultural practices and references, symbols, traditions, and activities developed are intangible elements (see Table 2). All together, they constitute the values of a community.

The Valletta principles establish that "*Historic towns and urban areas are made up of tangible and intangible elements.*" and defines the spirit of place as "*… the physical and the spiritual elements that give the area its specific identity, meaning, emotion and mystery. The spirit creates the space and at the same time the space constructs and structures this spirit.*"

Is the "*spirit of place*" that unites people living locally and those who in any way feel that there are their roots or good moments of life. The loss of these places, even a little

Table 1 Natural hazards types in the Azores Islands

Hazard type	Examples
Geological	Earthquake, volcanic eruption, landslides, subsidence
Meteorological	Cyclones, extratropical storms, drought, hail storms, lightning
Oceanographic	Tsunami, sea storm, coastal and submarine landslides
Hydrological	Flood, flashflood, landslides

Adapted from Schneiderbauer and Ehrlich (2004)

Table 2 Tangible and intangible elements highlighted in The Valletta Principles. Adapted from UN (2011)

Tangible elements	Intangible elements
Urban structure	Activities
Architectural elements	Symbolic and historic functions
Landscapes within and around the town	Cultural practices
Archaeological remains	Cultural references
Panoramas	Traditions
Skylines	Memories
Viewlines	
Landmark sites	

idealized may represent the disorganization of a community and a sense of orphanhood for those not living there but feel it as belonging to them.

The Sendai Framework for Disaster Risk Reduction 2015–2030, has as expected outcome and goal, "*The substantial reduction of disaster risk and losses in lives, livelihoods and health and in the economic, physical, social, cultural and environmental assets of persons, businesses, communities and countries.*". Action priorities focus on understanding disaster risk, strengthening risk governance, investing in population resilience, improving preparedness for an effective response, and "Build Back Better" in recovery, rehabilitation, and reconstruction.

4 Discussion

«Build back better» is a concept that aims to sensibilize social agents to the necessary, in case of disaster, to rebuild by reducing existing risks, preventing the creation of new risks and ensuring the continuity of population communities.

Many geographical and geological processes, including coastal erosion in many areas may be seen as only a natural hazard, however, in areas with anthropogenic pressure, it is a risk (Borges et al. 2009).

Normally, the risk is quantified by the number of people affected, but there are physical elements as built-up areas, transport networks or types of infrastructure that, are important to the support and social and emotional recovery of these populations. Seismically safe buildings withstand only a certain standard and even an earthquake-resistant building can be vulnerable to flooding and mass movements, so vulnerability depends on severity and type of risk. In this way, risk assessment involves knowing the hazards, the elements at risk and their spatial distribution (Schneiderbauer and Ehrlich 2004).

A single process can be triggered by more than one mechanism and there are no objective criteria to distinguish the triggering mechanism (Shanmugam 2016), and also different mechanisms can trigger the same type of process, making the assessment of risk and of vulnerability more complex.

"*Attitudes, behaviors and actions of persons are habitually influenced by religious beliefs and social and cultural milieu.*" (Fabiansson 2013), aspects that are often forgotten by scientific experts and governments when promoting guidelines for safety and wellbeing.

Human behavior, sense of security and risk assessment are complex aspects whose assessment requires a multidisciplinary knowledge. Developing knowledge and best practices on disaster risk reduction, including education and training, requires the involvement of social actors, including government officials at all levels, civil society, community groups, as well as the private sector, with a view to the common interest.

To use climate data to make planning implies considering the climate in a dynamic perspective, introducing an interpretative factor based on the probability of occurrence of certain values taking into account the historical evolution of the trend. (Azevedo and Gonçalves 1993).

5 Concluding Remarks

On ne savait pas! (We did not know!)
A sentence heard at the end of the Second World War. Today, we know.
Prefiguration Association of the Patrick Chauvel Foundation

Disaster is a critical opportunity to "Build Back Better" by integrating solutions to reduce existing disaster risk and preventive measures to avoid creating new types of risk, safeguarding the cultural and social identity of communities, enhancing their resilience.

Recovery, rehabilitation, and reconstruction must be prepared before an event occurs. Empowering the maximum number of people, regardless of gender and degree of disability, to publicly lead and promote approaches to response, recovery, rehabilitation, and reconstruction will be crucial.

In order to "we did not know" no longer be an excuse, scholars, and technicians involved in scientific investigation of natural risks disasters and all those who somehow work with this data, have a duty to clearly and comprehensibly disclose the information and projections with which they work so that the general population can adapt to changes.

In addition, the decision-makers and disseminators of information should assume the responsibility to ensure that information flows appropriately between them and for people.

Acknowledgements We would like to thank Ana Cunha, Regional Secretary for Transport and Public Works of the Government of the Azores, Frederico Sousa, Regional Director for Public Works and Communications, for the conditions provided to develop this work and all the co-workers who supported us and, indirectly, contributed.

References

Azevedo E, Gonçalves D (1993) Alguns aspectos da evolução da temperatura do ar e da precipitação na Ilha Terceira Açores desde 1874, presented at the Seminar "Water Resources and Environment" promoted by the Municipality of Angra do Heroísmo, Angra do Heroísmo

Borges P, Lameira G, Calado H (2009) A erosão costeira como factor condicionante da sustentabilidade, 1° Congresso de Desenvolvimento Regional de Cabo Verde, Cabo Verde

Fabiansson C (2013) Risk perception—the visible and invisible risks—the risk discourse from a multidisciplinary perspective, College of Arts, Victoria University, Melbourne Australia. http://www.riskanduncertainty.net/thought-pieces/risk-perception-the-visible-and-invisible-risks-the-risk-discourse-from-a-multidisciplinary-perspective. Consulted at 7 Feb 2019

Pacheco JM et al (2013) Notas sobre a geologia do arquipélago dos Açores, In: Dias R, Araújo A, Terrinha P, Kullberg JC (eds) (2013) Geologia de Portugal, vol 2, Escolar Editora, 595–690

Schneiderbauer S, Ehrlich D (2004) Risk, hazard and people's vulnerability to natural hazards. A review of definitions concepts and data. European Commission. Directorate general. Joint research centre 40p

Shanmugam G (2016) Slides, slumps, debris flow, turbidity currents, and bottom currents. University of Texas at Arlington, Arlington, TX, USA

Silva M (2005) Caracterização da sismicidade histórica dos Açores com base na reinterpretação de dados de macrossísmica: Contribuição para a avaliação do risco sísmico nas Ilhas do Grupo Central, University of Azores, Ponta Delgada (MSc thesis)

Silveira D (2002) Caracterização da sismicidade histórica da Ilha de S. Miguel com base na reinterpretação de dados de macrssísmica: Contribuição para a avaliação do risco sísmico, University of Azores, Ponta Delgada (MSc thesis)

UN (2011) The Valletta principles for the safeguarding and management of Historic Cities, Towns and Urban Areas. Adopted by the 17th ICOMOS General Assembly on 28 November 2011

UN (2013) The global assessment report on disaster risk reduction 2013–2015. www.preventionweb.net/english/

UN (2015) The Sendai framework for disaster risk reduction 2015–2030. www.unisdr.org

Vulnerability to Mass Movements' Hazards. Contribution of Sociology to Increasing Community Resilience

Margarida Santos, Marta Aguiar, Ana Oliveira, Luísa Magalhães, and Fernando Pereira

Abstract

There is yet an incorrect perception of common sense about environmental issues, mainly about the increase and rapid environmental degradation due to the misuse of natural resources. The pressure on the territory, changing the form by the intensification of the human occupation and the overexploitation of the resources can generate local phenomena that, combined with events of local or regional scale, can cause significant losses in the communities and their livelihoods. Mass movements are short return periods hazards, highly localized, mainly weather-related, but also triggered by seismic or volcanic activity or even technological accidents. The generalization or axiomatization of the idea that innovation is synonymous with greater efficiency, greater security, and greater sustainability, can lead to the technocracy of the solutions, devaluing and fragmenting the cultural and social organization of communities. Knowledge of human capacity for adaptation and analysis of the impacts of climate change forecasts on the social structure of communities, especially those that will first experience the effects, is important for the adoption of measures that contribute to its preservation. Sociology, in one of its aspects, seeks to recover and reveal the materiality of the structure and social life of communities.

Keywords

Mass movements • Natural hazards • Climate change • Environmental degradation • Environmental sociology • Human resilience

1 Introduction

In the dawn of October 22, 1522, an amazing earthquake was felt throughout the Island of S. Miguel and Vila Franca do Campo, the main population center of the time, was largely destroyed by the earthquake and totally buried by deposits of mass movements (Silveira 2002).

On July 9, 1757, around 11 o'clock on the night, on S. Jorge Island occurred an earthquake of catastrophic characteristics, which to date was the one has caused a greater number of direct deaths throughout the archipelago, causing also great mass movements with greater relevance to the north coast of the island, where several islets were formed that later disappeared. These movements dragged houses up to the sea ripped apart by the foundations and the Fajãs de S. João and the Vimes were submerged under a deposit of rough rock, with loss of all that was in them (Silva 2005).

An earthquake occurred on June 15, 1841 at 3:30 in the morning, and completely destroyed the Vila da Praia da Vitória and Fontinhas Village, causing severe damage in other locations on Terceira Island. It also caused a superficial rupture, originating a crack that comes from the sea, crosses the entire beach and extends for not less than a league (Silva 2005).

On December 18, 1987, a large landslide destroyed, in Ponta da Fajã Grande, Flores Island, several houses, forcing the rehousing of eight families and buried the Chapel of Our Lady of Fatima. Because of it the site was considered of "high geological risk", and the settlement was evacuated and later it was also subject to legal restrictions that expressly forbid to build any type of construction in that area and to inhabit existing properties.

At dawn on October 31, 1997, in Ribeira Quente, S. Miguel Island, much-localized landslides brought death to 29 people, wounding and dislodging many more. The tragedy was due to natural events such as the saturation of the

M. Santos · M. Aguiar · A. Oliveira · L. Magalhães · F. Pereira (✉)
Direção Regional das Obras Públicas e Comunicações, Secretaria Regional dos Transportes e Obras Públicas, Largo Do Colégio, nº 4, 9500-054 Ponta Delgada, Portugal
e-mail: fdpereira@sapo.pt

© Springer Nature Switzerland AG 2020
F. Fernandes et al. (eds.), *Advances in Natural Hazards and Hydrological Risks: Meeting the Challenge*, Advances in Science, Technology & Innovation,
https://doi.org/10.1007/978-3-030-34397-2_21

soil and the high vulnerability represented here by the presence of inhabited houses on the slope's base in an area where there was no memory of similar events (Raposo 1998).

After the end of World War II, a process of economic development and world population growth began, which also led to an increase in mobility, both in tourism and in business, resulting in the expansion of urban areas, needing structured territorial management (Marques 2013). Environmental sociology seeks to incorporate variables, perspectives, and scientific paradigms to produce knowledge that contributes to solving the problem of this pressure on the territories and communities in order to recover and value the structure and materiality of social life.

Since the nineteenth century, it has been observed that the environment undergoes changes and that man has an active role in that. It was Humbolt who warned against this reality and found that the natural world is related to the "political and moral history of humanity" (Wulf 2015). This idea of natural oneness also represents the descent of man from the pedestal where he considered himself above all things. (Le Buffonn *apud* Wulf 2015). From here, the human being becomes an apprentice to understand that the dominion over nature is not sustainable. It is the sociologist's job to construct a line of thought capable of providing a comprehensive perspective on man's connection to nature. (Catton; Dunlap, 1978; Redclift; Woodgate *apud* McReynolds 1999).

2 Methodology

Mass movements commonly are classified base on the mechanism of movement and speed or considering the morphology of the phenomenon or its lithological setting (Glade and Alexander 2013). However, in another perspective, "*… slopes movements corresponds to a movement of a rock mass and/or unstable soil, whose center of gravity of the affected material progresses downstream and to the outside*." (Marques 2013), an explanation maybe more understandable for non-experts.

The data of the Geological Hazard Table in the documents submitted to the public discussion of the Regional Program for Climate Change of the Azores (PRAC) were the basis used to analyze the potential risk areas of mass movements in each island of the Azores Archipelago and understanding how social sciences can be a partner in the solutions.

3 Social Background

3.1 World Events

Mass movements are the greatest natural hazards in subaerial environments; affecting thousands of people around the world (see Table 1). The main force that contributes to the movement is gravity, but characteristics as the angle of inclination, the type of material, the water content and the anthropic activities can contribute to the effect of gravity, regardless of the type of process that triggers the movement (see Table 1).

Considering the number of events and the number of people affected in Europe, the number of deaths and, above all, the value of damages is highly disproportionate. This could be interpreted by the fact that despite the level of development and quality of life, there are many vulnerabilities associated with high population density, industrialization, and economic activities. The Asian continent is the most devastated over time, the most important being India and China, countries with large territorial areas, many of them densely populated, and strong pressure on vulnerable zones.

The case of Europe can be paradigmatic, in fact, it is one of the continents with one of the best living standards, many years of spatial planning and a high level of knowledge of the population, yet vulnerability and damage are significant.

3.2 Situation on Azores

All the islands in general, have extensive areas highly vulnerable to the occurrence of landslides (Cavalheiro et al. 2017). Coastal areas are the most threatened, and are doubly threatened, on the one hand, because they are located on the

Table 1 Hydrological landslides and dry mass movements in different continents between 1900 and 2019

Continent	Events	Death toll	People affected	Damages ('000 US $)
Africa	50	2939	80,752	58,036
Americas	197	23,374	5,583,104	4,043,727
Asia	423	26,537	8,570,281	3,622,216
Europe	82	16,939	49,006	3,111,489
Oceania	20	632	21,315	2466

Source EM-DAT: The Emergency Events Database—Université catholique de Louvain (UCL)—CRED, D. Guha-Sapir. www.emdat.be, Brussels, Belgium

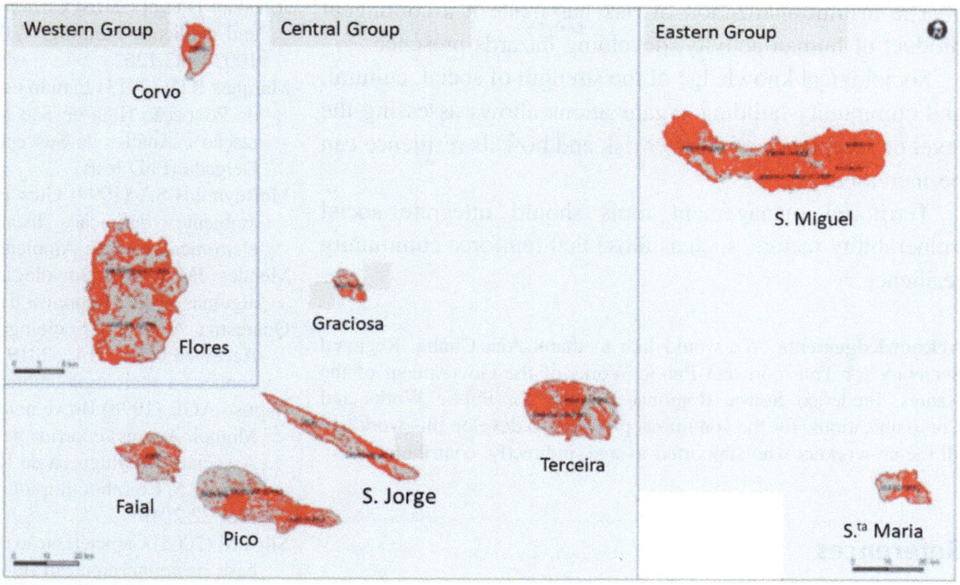

Fig. 1 Areas with high susceptibility to mass movements (Cavalheiro et al. 2017) (■— High susceptibility areas)

base of volcanic structures usually with sharp slopes and on the other, because they are located on cliffs under the influence of the sea and marine storms (see Fig. 1). Most of the population settlements and major infrastructure and economic activities are located in coastal areas.

The island of São Miguel, home to more than 50% of the population of the archipelago, is highly vulnerable throughout most of its territory. Given the destructive power of slope movements, this risk is a major concern for the safety of goods and persons (Cavalheiro et al. 2017).

4 Discussion

While human activity was much less than the self-regenerative capacity of living systems, ecological concerns did not have much space in ecological thinking (Quaresma 2017). However, the modernization that began with industrialization in the nineteenth-century institutionalized risk and gained unpredictable contours, making it difficult for sociologists to develop a systematic assessment of human ecological issues (Giddens 1991).

Associated with modernity comes globalization, and there is still some undervaluation of environmental concerns and reduced awareness of the limitations of technical and scientific knowledge and forces that evolve beyond human rules.

Societies only evolve if becoming reflexive (Beck *apud* Mendes 2016) and sociology places the risk in social contexts, attending to the activities of individuals, social groups and communities (Lidskog, Sundqvist *apud* Mendes 2016: 19), which requires a reflection on the structural factors of social vulnerability and should lead to its explicit integration in territorial planning.

The development of the internet and mobile communications, as well as various technological applications such as sensors and applications for mobile phones, have enabled many citizens to develop their own observations and analyzes of natural phenomena (Jonathan et al. 2018). The proliferation of mobile phones in all inhabited areas, even geographically remote areas, and across all segments of society, allows new ways of collecting data and transmitting environmentally relevant information. Smartphones can be equipped with sensors or simply be used to transmit physical observations and user measurements, but also, to receive information about natural phenomena and measures to implement or react to risk situations.

Despite all the history of catastrophic events, people resist, rebuilding their habitats and ways of life, seeking to preserve their cultural, and social heritage. The social sciences, among them environmental sociology, can contribute to that by the integration and valorization the citizen science, empowering these amateur scientists as social and cultural actors who contribute with their knowledge to strengthen social cohesion and increase community resilience.

5 Concluding Remarks

Risk assessment and acceptance depend on the social and cultural context of communities.

Scientific and industrial development results in a set of risks that are not confining, neither in space nor in time.

The institutionalization of risk has made it a contingent product of human activity, devaluing hazards presence.

Sociological knowledge of the strength of social, cultural, and community-building organizations allows assessing the level of acceptance of disaster risk and how its resilience can be increased.

Territorial management tools should integrate social vulnerability factors, such as those that reinforce community resilience.

Acknowledgements We would like to thank Ana Cunha, Regional Secretary for Transport and Public Works of the Government of the Azores, Frederico Sousa, Regional Director for Public Works and Communications, for the conditions provided to develop this work and all the co-workers who supported us and, indirectly, contributed.

References

Cavalheiro G et al (2017) Programa Regional para as Alterações Climáticas dos Açores (PRAC). Queluz, Lisboa

EM-DAT: The Emergency Events Database—Universite catholique de Louvain (UCL)—CRED, D. Guha-Sapir. www.emdat.be. Brussels, Belgium. Accessed 4 Apr 2019

Giddens A (1991) As consequências da modernidade. Celta Editora, Oeiras

Glade T, Alexander DE (2013) Classification of Natural Disasters. In: Bobrowsky PT (eds) Encyclopedia of Natural Hazards. Encyclopedia of Earth Sciences Series. Springer, Dordrecht

Jonathan D et al (2018) Citizen science for hydrological risk reduction and resilience building. WIREs Water 5:e1262. https://doi.org/10.1002/wat2.1262

Marques RTF (2013) Estudo de Movimentos de Vertente no Concelho da Povoação (Ilha de São Miguel, Açores: Inventariação, Caraterização e Análise da Susceptibilidade. University of Azores, Ponta Delgada (PhD tesis)

McReynolds SA (1999) Guia para o iniciante em sociologia do meio ambiente: definição, lista de jornais e bibliografia [versão eletrónica], revista: Ambiente & Sociedade, São Paulo

Mendes JM (2016) Sociologia do Risco: uma breve introdução e algumas lições. Imprensa da Universidade de Coimbra, Coimbra

Quaresma A (2017) Sociologia – Conheça a sociologia ambiental. Accessed Jan 14 2019. http://sociologiacienciaevida.com.br/conheca-a-sociologia-ambiental/

Raposo AGB (1998) Breve nota sobre a tragédia da Ribeira Quente (S. Miguel, Açores) ocorrida na madrugada de 31 de Outubro de 1997. Associação Portuguesa de Riscos, Prevenção e Segurança, Territorium nº 5, Coimbra. http://hdl.handle.net/10316.2/40136. Accessed: Mar 22 2019

Silva M (2005) Caracterização da sismicidade histórica dos Açores com base na reinterpretação de dados de macrossísmica: Contribuição para a avaliação do risco sísmico nas Ilhas do Grupo Central, University of Azores, Ponta Delgada (MSc thesis)

Silveira D (2002) Caracterização da sismicidade histórica da Ilha de S. Miguel com base na reinterpretação de dados de macrossísmica: Contribuição para a avaliação do risco sísmico, University of Azores, Ponta Delgada (MSc thesis)

Wulf A (2015) A invenção da Natureza. As aventuras de Alexander Von Humboldt, o herói esquecido da ciência, Lisboa, Temas e Debates – Círculo de Leitores

Natural Hazards, Hydrodynamics and Engineering Design

Flooding Risk Assessment in the Middle Tiber River Valley with Remediation Proposal

Giuseppe Sappa and Flavia Ferranti

Abstract

Flooding is a natural process that can have significant impacts on people, economy and environment, in the area it occurs. It is influenced by climate changes, river dynamics, urban drainage, and land use modifications. Climatic variability and climate change are shifting rainfall patterns, making heavy rain less frequent worldwide, but very often more intense. The aim of this work is to contribute to flood risk management in the Middle Tiber River Valley (MTV), in Latium Region (Italy), presenting a GIS-supported methodology to identify risk-control measures to prevent, defense, mitigate, and recover from river flooding. This study deals with the results of the hydraulic modeling of the Tiber River section, from Orte and Castel Giubileo, for flooding risk estimation and with the aim of design a flooding risk reduction, using HEC-GeoRAS and HEC-RAS hydraulic modeling tools, supported by a Geographic Information System (ArcGIS). With the aim to encourage the use of mitigation measures, in this paper, a series of structural defense strategies have been identified as reducing flooding risk to residents and businesses in the study area.

Keywords

Flooding risk management • Middle Tiber River Valley • GIS • HEC-RAS

1 Introduction

Climate change and global warming are placing additional pressure on flooding risk, increasing the frequency of high-intensity rainfall and consequently the likelihood of river and surface water flooding (Piniewski et al. 2017). River floods are generally generated by rainfall that ranges from short-duration and high-intensity to long-duration and moderate-intensity. Most floods in the Middle Tiber River Valley (MTV) are caused by long-lasting rainfall of moderate intensity. Even if along the Tiber River there were frequent floods in the first decade of the millennium, none of them was a major event and losses were economic in nature (Calenda et al. 2005). Existing drainage infrastructure and public sewers are not designed to cope with such volumes of rainwater. In fact, even if no single flood event can be directly attributed to global warming, the recurrence of devastating 100-year and 200-year floods throughout in recent years may foreshadow a future where repeated disasters of this magnitude are much more common (Sun et al. 2018; Akter et al. 2018; Cloete et al. 2018). Moreover, the human alteration of the land use, like the rivers engineering, the destruction of natural protective systems, and the expanding construction on floodplains, may be related to the increased flooding risk worldwide (Akter et al. 2018). By managing land and river systems, the chance of flooding from rivers can be significantly reduced.

Flood management is ensured through national directives and there are regional actions to implement the necessary measures. Legislative Decree no. 49/2010 transposed the EU Floods Directive (2007/60) into national law on assessment and management of flood risk. In this framework, the Tiber River Basin Authority, according to Italian Law 183/89 on soil protection, had anticipated the EU directive contents developing a planning tool, the District River Basin Management Plan, for the identification of flood-prone areas and the level of risk exposure. This work presents the results of the updating of the Tiber River hydraulic model, developed for the river stretch between Orte and Castel Giubileo, and the description of the methodologies applied for the identification of the flooding areas, as reported in the Tiber River Basin Management Plan (P1) in 2009.

G. Sappa (✉) · F. Ferranti
Department of Civil, Building and Environmental Engineering, Sapienza University of Rome, Rome, Italy
e-mail: giuseppe.sappa@uniroma1.it

© Springer Nature Switzerland AG 2020

F. Fernandes et al. (eds.), *Advances in Natural Hazards and Hydrological Risks: Meeting the Challenge*, Advances in Science, Technology & Innovation,
https://doi.org/10.1007/978-3-030-34397-2_22

2 Study Area

The Tiber River Basin, the largest river basin in Central Italy (the third largest river one in Italy), lies in the Central Apennines District and drains towards the Mediterranean Sea covering land area of 17,500 km^2, crossing six administrative regions. The study area is the Tiber River section from Orte to Castel Giubileo, in the Middle Valley of the Tiber River (MTV), in the central part of Latium Region, few kilometers in the North of Rome (Fig. 1).

The MTV structurally corresponds to the "Paglia-Tevere Graben" (Funiciello and Parotto 1978), an NNW-SSE trending extensional basin developed since the late Early Pliocene (Barberi et al. 1994). This basin is mostly filled by Plio-Pleistocene marine terrigenous deposits. However, coeval non-marine and transitional terrigenous and carbonatic deposits extensively outcrop along the western margin of the Mt. Peglia-Lucretili Mts. ridge (Fiseha et al. 2013; Mancini et al. 2004), while Pleistocene volcano-sedimentary successions are well-exposed west of the Tiber River (Mazza and La Vigna 2011).

3 Materials and Methods

The aim of this paper is to present the results of the application of a methodology for flooding risk management by the proposal of flood mitigation and protection in the Middle Tiber River Valley. For this reason, HEC-RAS (version 3.1.3) (U.S. Army Corps of Engineers 2005) and ArcGIS (ESRI, ArcGIS Desktop 2007) software packages were combined to identify the potential flooding areas in a Tiber River stretch, between Orte and Castel Giubileo. HEC-RAS software has been used to identify, for a stream discharge with a 200-year return period, the flow exceeding the riverbank levels. The 200-year return stream discharge comes from the elaboration of 1000 years of historical precipitations series, generated by specific hydrological software, as precipitations are the main parameter, from which is possible to distinguish run-off, evapotranspiration and groundwater recharge in a basin (Sappa et al. 2015). In particular, HEC-RAS software has allowed the calculation of water levels, depths and flow velocities for the different flow configurations and different cross-sections in the study area. The Tiber River hydraulic

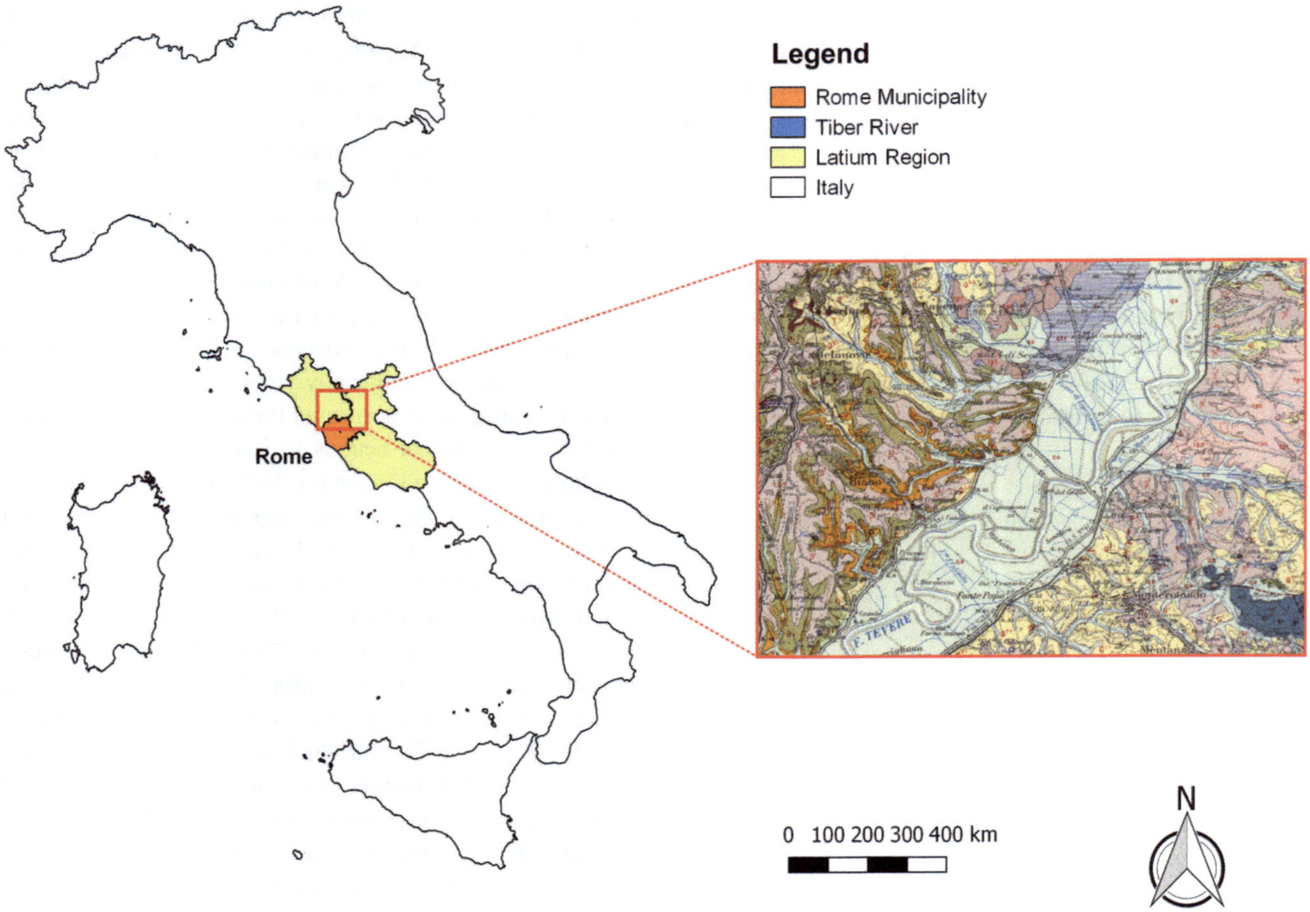

Fig. 1 Location and geology setting of the study area

model has been set up by HEC-RAS software, based on topographical maps, orthophotographic images and digital elevation models (DEM), which, adequately integrated, where the input for ArcGIS software application. The Tiber River cross-section profiles were densified to meet specific needs for spatial accuracy. The areas have been identified and countered in ArcGIS, and used for the design of the water reservoirs according to a three-step methodology:

Step 1: Insertion of new 67 cross-sections obtained from a 2008 year Multibeam survey, integrated along the floodplains with points extracted by DEM Aquater. Fourteen of these cross-sections are so long to cover the entire over-flowed track, while all the rest are only the riverbed. The locations of these cross-sections have been chosen according to the necessity of obtaining a more detailed local mapping of the fluvial conformation in the areas of greatest interest (embankments, expansion tanks, riverbed variations) and to individuate the presence of phenomena of deepening and engraving of the riverbed.

Step 2: Replacement of the cross-sections, obtained from the previous survey, made in 1997–1998, by means of the linear interpolation function available on HEC-RAS software. The results of these interpolations have been used to detect erosion phenomena along the Tiber River and consequently to update the existing topographic survey. With this procedure, nineteen Tiber River cross-sections have been updated in the area of Campo Grande, in the municipality of Fiano Romano, and in the Bacche area, in the municipality of Orte.

Step 3: Replacement of the riverbed obtained from the previous survey by extrapolation from the Multibeam survey, available only for the stretch between Ponte del Grillo and Castel Giubileo. The digital elevation model (DEM), obtained from the Multibeam survey, has been previously treated, carrying out a mosaic of the 18-grid elements making the LEG II (Ponte del Grillo—Castel Giubileo section), and afterwards carrying out the georeferences from the UTM WGS84 reference system to the UTM ED50 system, fused 33. Subsequently, along tracks specially arranged along the shaft, the dimensions were extracted according to a 10 m pitch in the floodplain area and 0.5 m in the riverbed by means of the tool "Data Management Tool" by ArcGIS Map.

4 Results and Discussion

The identification of the flooding areas has been performed by intercepting the water surface generated by the hydraulic model with the surface of the ground according to two phases: (i) an automatic procedure using HEC-GeoRAS and (ii) the result's validation according to criteria of reliability and safety.

HEC-GeoRAS, using ArcGIS Desktop, has been applied to post-process the results of the hydraulic model for flood damage computations and overflooding mapping. Using HEC-GeoRAS, the water surface (TIN surface) has been reconstructed by the envelope of the maximum water levels in the various sections. The water surface has been superimposed with the ground surface (DEM Aquater), identifying the intersecting line representative of the wet contour. The overlapping of the two digital models has provided (i) the flooding areas contours, obtained as the intersection of the digital elevation model and the water surface, and (ii) the GRID with the data of the difference between the DEMs.

The overlap of the water surface with the digital elevation model has been followed by a refinement phase aimed to reduce the errors related to this automatic procedure and to the representative limits of the various supports. In summary, the following procedures have been applied:

(1) evaluation and correction of altitude errors using topographical maps;
(2) evaluation, through sensitivity verification, of "uncertain" situations in the presence of anthropization.

The areas, where the floodwater reservoirs are placed are generally flat areas, adjacent to the river, without significant anthropic settlements. The sizing and numbering of these areas are related to the needed water volume storage and it has been determined following this iterative procedure:

(1) calculation of non-overflowing volumes due to the defense measures for the protection of the inhabited settlements (V_1);
(2) identification of flat and non-urbanized upstream areas;
(3) calculation of the overflow curve and available capacity in relation to the water table;
(4) calculation of non-overflowing volumes naturally due to the flood water reservoir (V_2);
(5) comparison between the flood water reservoir capacity and the sum of non-overflowing volumes (V_3);
(6) if capacity is sufficient, the procedure is completed, otherwise, we return to point 2 identifying a more suitable area.

The non-overflowing volumes have been calculated from the results of the simulation carried out with Hec-Ras and HecGeo-Ras, comparing the configuration without defense works with that relating to the addition of embankments. The overflow curve has been obtained with the use of GIS software able to calculate the volume between a plane at a given altitude and the underlying digital terrain model. The results of the application of this methodology are reported in Table 1.

Table 1 Summary of the volumes involved in the Middle Tiber River Valley flooding risk management

Area	V_1 (10^3 mc)	Flood water reservoir	V_2 (10^3 mc)	V_3 (10^3 mc)
Monterotondo	1893.82	N°1	2033.95	3699.82
Castelnuovo di Porto	1589.71	N°2	2865.98	4702.23
Procoio	601.04	N°3	279.39	2382.59
Marcigliana	115.64	N°4	8086.17	9016.19
Settebagni	2527.62	N°5	2209.41	2493.23
Total	6727.83	–	15,474.9	22,294.06

Fig. 2 Updating map showing the flood-prone areas in the study area

5 Concluding Remarks

This paper proposes a GIS-supported procedure for river flood risk reduction in the Middle Tiber River Valley. Hydraulic modeling and flood inundation mapping have been performed to provide important information from a flood event including the level of inundation and water surface elevations within the study area (Fig. 2).

References

Akter T, Quevauviller P, Eisenreich SJ, Vaes G (2018) Impacts of climate and land use changes on flood risk management for the Schijn River, Belgium. Environ Sci Policy 89:163–175

Barberi F, Buonasorte G, Cioni R, Fiordelisi A, Foresi L, Iaccarino S, Laurenzi MA, Sbrana A, Vernia L, Villa IM (1994) Plio-Pleistocene geological evolution of the geo thermal area of Tuscany and Latium. Mem Descr Carta Geol d'Italia 49:77–134

Calenda G, Mancini CP, Volpi E (2005) Distribution of the extreme peak floods of the Tiber River from the XV century. Adv Water Resour 28(6):615–625

Cloete G, Benito G, Grodek T, Porat N, Enzel Y (2018) Analyses of the magnitude and frequency of a 400-year flood record in the Fish River Basin, Namibia. Geomorphology 320:1–17

Fiseha BM, Setegn SG, Melesse AM, Volpi E, Fiori A (2013) Hydrological analysis of the Upper Tiber River Basin, Central Italy: a watershed modelling approach. Hydrol Process 27(16):2339–2351

Funiciello R, Parotto M (1978) Il substrato sedimentario nell'area dei Colli Albani: considerazioni geodinamiche e paleogeografiche sul margine tirrenico dell'Appennino centrale. Geol Romana 17:233–287

Mancini M, Girotti O, Cavinato GP (2004) Il Pliocene e il Quaternario della Media Valle del Tevere (Appennino Centrale). Geol Romana 37:175–236

Mazza R, La Vigna F (2011) Hydrogeology of the southern Middle Tiber Valley (Central Italy). AQUA Mundi 2:93–102
Piniewski M, Mezghani A, Szcześniak M, Kundzewicz ZW (2017) Regional projections of temperature and precipitation changes: robustness and uncertainty aspects. Meteorol Z 26(2):1–12
Sappa G, Trotta A, Vitale S (2015) Climate change impacts on groundwater active recharge in coastal plain of Dar es Salaam (Tanzania). Engineering geology for society and territory, vol1: Climate change and engineering geology. pp 177–180
Sun P, Wen Q, Zhang Q, Singh VP, Sun Y, Li J (2018) Nonstationarity-based evaluation of flood frequency and flood risk in the Huai River basin, China. J Hydrol 567:393–404

The 2017 Flash Flood of Livorno (Italy): Lessons Learnt from an Exceptional Hydrologic Event

Chiara Arrighi and Fabio Castelli

Abstract

Between 9th and 10th September 2017 a catastrophic, very intense rainfall event struck the city of Livorno in the Tyrrenian coast of central Italy causing eight fatalities. The event was characterized by very high rainfall intensity (>100 mm h^{-1}) that persisted on a small portion of territory causing huge rainfall accumulations (>260 mm). The aim of the work is to reconstruct the flash flood to understand how much the event was exceptional under the hydrologic and impact aspects. The rainfall event and the inundation have been reconstructed by analyzing the rain gauge data of the regional hydrologic service. The discharges have been simulated using MOBIDIC, a continuous distributed hydrologic model already operational in the region for hydrologic predictions. The exposure and damages to buildings have been analyzed and compared with the official claims. The results show that the flash flood was exceptional since most of the river reaches in the area have overcome of about 50% the 200 years return period discharge, used to design the flood defenses according to the national regulations.

Keywords

Flood risk • Flood vulnerability • Flood damages • Risk management • Pedestrians

1 Introduction

Rainfall with high intensity persisting on small catchments is increasing in Italian coastal areas (Silvestro et al. 2016). These localized events are difficult to predict and might have a high destructive potential including fatalities, especially when the inundation propagates fast, with high flow velocities capable of sweeping away vehicles (Fig. 1) and pedestrians (Arrighi et al. 2017, 2019; Salvati et al. 2018). The 2017 flood that struck the city of Livorno caused 8 fatalities and highlighted the following weaknesses of the local flood risk management strategy: (i) the ineffectiveness of the mitigation system; (ii) the scarce risk awareness of citizens; (iii) the poor management of residual risk.

This work aims at reconstructing the event from the precipitation up to the impacts to better understand what went wrong and how exceptional was the event with respect to the recorded time series. Section 2 describes the methodology, Sect. 3 presents the results and Sect. 4 discusses the main findings of the work here limited to the most significant hydrologic and impact aspects.

C. Arrighi (✉) · F. Castelli
Department Civil and Environmental Engineering, University of Florence, Florence, Italy
e-mail: chiara.arrighi@dicea.unifi.it

2 Materials and Methods

2.1 Hydrologic Model

The precipitation data recorded during the flood event by the regional hydrologic service have been used as input of the hydrologic model. Table 1 shows the three rain gauges insisting in the catchments of the Livorno area.

The rainfall-runoff simulation has been carried out using MOBIDIC (MOdello di Bilancio Idrologico DIstribuito e Continuo) a fully distributed, raster-based hydrological balance model (Yang et al. 2014; Castelli et al. 2009; Campo et al. 2006).

MOBIDIC simulates the water balance in a series of reservoirs and fluxes between them (Fig. 2). Once the surface runoff (Q_h and R_d) and baseflow are calculated, three different methods can be used for river routing, i.e. the lag method, the linear reservoir method, and the Muskingum–Cunge method. For details on the equations and assumptions consult the references (Yang et al. 2014; Castelli et al. 2009).

© Springer Nature Switzerland AG 2020
F. Fernandes et al. (eds.), *Advances in Natural Hazards and Hydrological Risks: Meeting the Challenge*, Advances in Science, Technology & Innovation,
https://doi.org/10.1007/978-3-030-34397-2_23

Fig. 1 The aftermath of Livorno flood with the evidence of significant vehicle entrainment. *Credits* Repubblica.it and LaPresse

Table 1 Cumulate rainfall recorded in the tree rain gauges insisting on the catchments of Livorno area on the 10th September 2017

Station	Rainfall 15 min (mm)	Rainfall 60 min (mm)	Rainfall 120 min (mm)	Rainfall 180 min (mm)
Valle Benedetta	38.4	120.8	210.2	235
Quercianella	42.4	121.8	188.2	206.2
Livorno	27.8	63.4	69.1	79.2

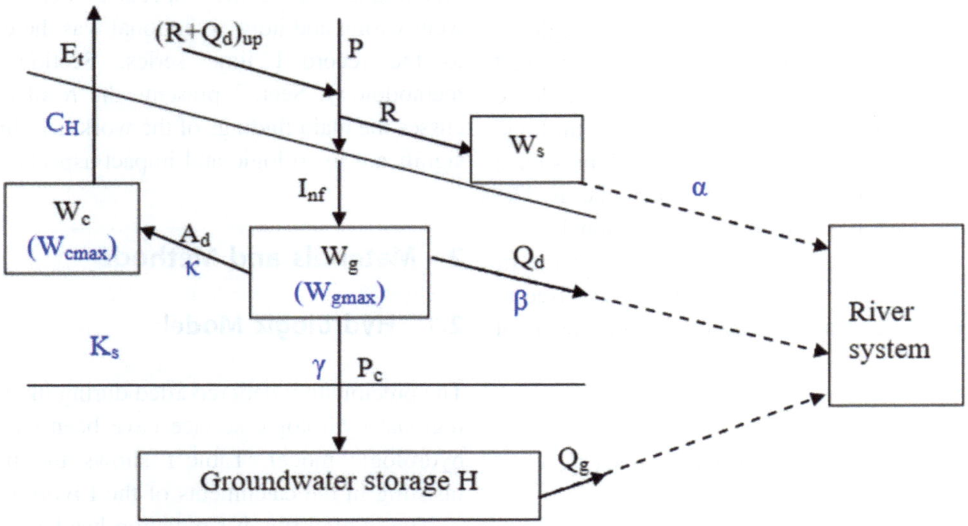

Fig. 2 A schematic representation of MOBIDIC by Yang et al. (2014). Boxes denote different water storages (gravitational storage W_g, capillary storage W_c, groundwater storage H, surface storage W_s, and the river system), solid arrows fluxes (evaporation E_t, precipitation P, infiltration I_{nf}, adsorption A_d, percolation P_c, surface runoff R, interflow Q_d, groundwater discharge Q_g, and surface runoff and interflow from upper cells $(R + Q_d)_{up}$), dashed arrows different routings, and blue characters major model parameters

Besides rainfall data, the most relevant input data to the model are the river network, the digital terrain model (10 m resolution) and the regional soil map (scale 1:250,000).

The initial soil moisture state has been assumed equal to 80% of saturation to be compared with previous analyses.

2.2 Exposure and Vulnerability

Exposure analysis considers buildings (source: regional cartography 1:10,000) and population (source: 2011 census data) located in the perimeter of the area inundated by Rio

Maggiore. The monetary value of the exposed building is calculated based on surface area and official unitary market values.

Vulnerability is evaluated based on piece-wise linear stage-damage functions which assign a percent physical loss to the flood depth in the building (Arrighi et al. 2018).

3 Results

3.1 Hydrology

Rio Maggiore and Ardenza, i.e. the two main streams responsible for the inundation can be identified in panel b of Fig. 3 (Rio Maggiore is the northernmost one). The maximum discharges evaluated for Rio Maggiore and Ardenza are 141 and 373 m^3/s, respectively. These discharges are quite significant for the two streams which have a catchment extension of 8 and 21 km^2, respectively. The resulting areal contribution to the flood peak is about 17.6 $m^3/s\ km^2$ for both streams.

The peak discharges of the 2017 flood event have been compared to the design discharge which, in Italy, coincides with the 200 years recurrence interval. The results of this comparison are depicted in Fig. 3 (panel a) as the ratio between the 2017 peak Q and the 200 years return period Q_{200}. The map shows that in most of the network the Q_{200} has been exceeded by 30–50% with local discharges 2–3 times the design discharge.

This is also confirmed by the return period evaluated for the cumulative precipitation at the three rain gauges of Table 1 which resulted higher than 500 years.

The 2017 rainfall event has been thus used to reassess the intensity–duration–frequency curves. If the event is accounted for the new design peak discharges are 101 and 282 m^3/s for Rio Maggiore and Ardenza streams, respectively.

3.2 Exposure and Vulnerability

As a preliminary hydraulic evaluation, the perimeter and water depths of the inundated area have been reconstructed based on a survey carried out immediately after the event and a LiDAR-derived digital terrain model (1 m resolution). Although the flooded areas were not densely populated, the inundation of the Rio Maggiore (areal extent 1.15 km^2) affected 364 buildings and the inundation of Ardenza (areal extent 1.8 km^2) affected 386 buildings. In terms of exposed built area, Rio Maggiore affected around 83,140 m^2 of structures and Ardenza affected around 46,500 m^2 of structures. In monetary terms, the value of exposed buildings, based on official market values is around 324 Mio €.

The population affected is estimated around 6000 people based on 2011 census.

Assuming as recovery cost a fraction of the market value (20%) and using stage damage functions for buildings without cellars (Arrighi et al. 2018) the monetary losses to building inundated by the Rio Maggiore sum up to about 6 Mio €, which is 2.9% of the exposed asset in the area. Figure 4 shows a detail of the monetary losses for the downstream area inundated by Rio Maggiore.

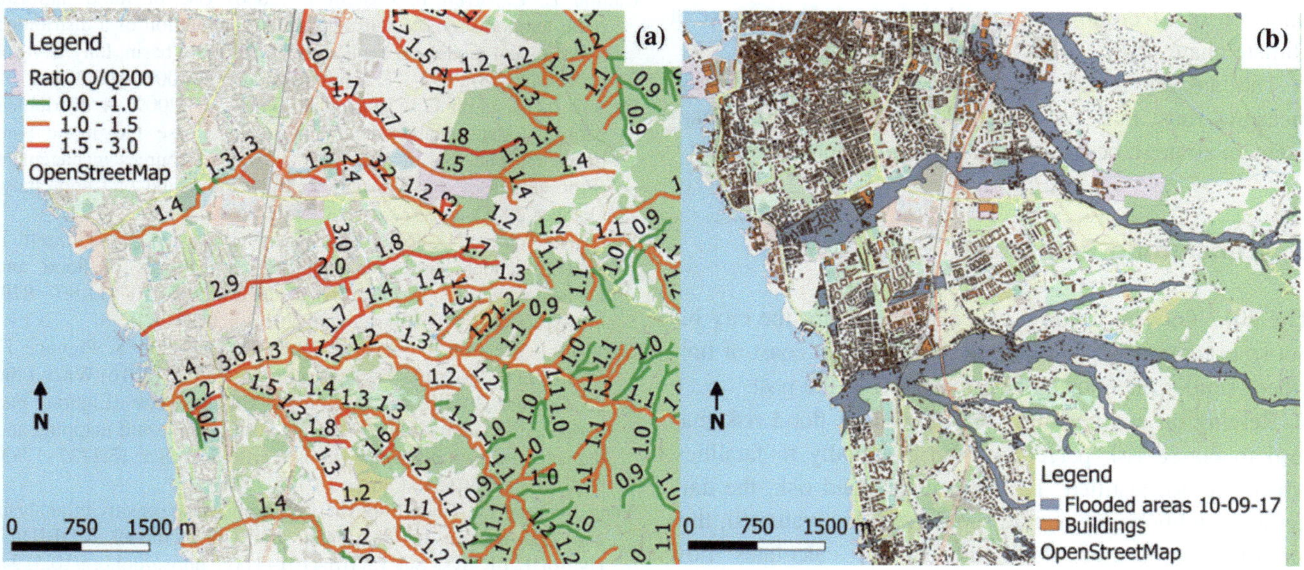

Fig. 3 Ratio between the simulated maximum discharge of the 2017 event Q and the 200 years return period discharge Q_{200} (panel a); inundated areas of the 2017 event and position of buildings

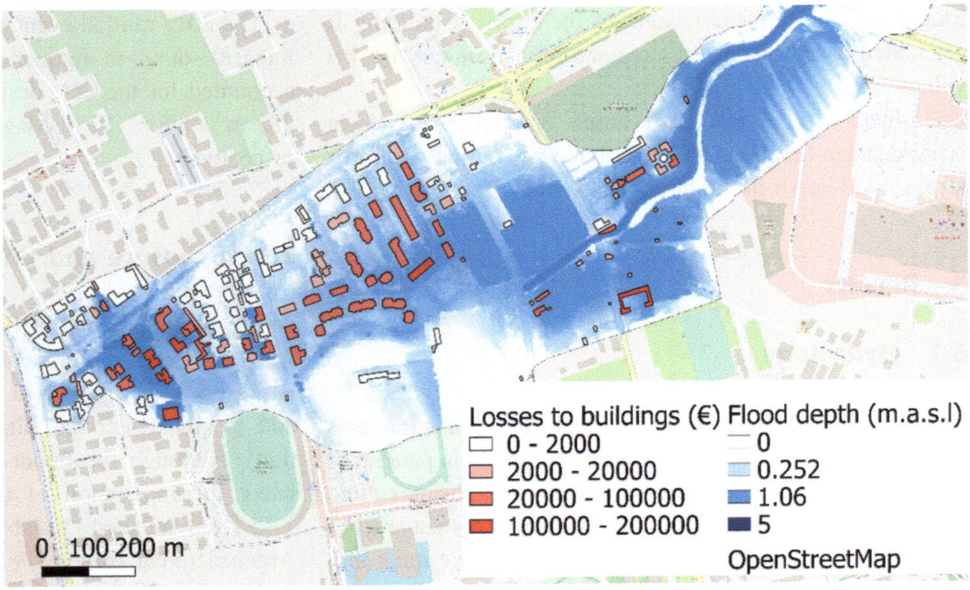

Fig. 4 Flood depths and monetary losses to buildings for a detail of the Rio Maggiore stream, close to the mouth

4 Discussion

Based on the results of the hydrologic model and on the analysis of the rainfall, the flood event of September 2017 in Livorno was exceptional. Events of similar intensity were never recorded before according to the pluviometry data. River discharges largely exceeded the design discharge of the mitigation measures on the Rio Maggiore, which were ineffective during the event. About 6 Mio € of monetary losses were estimated only for Rio Maggiore inundation, almost 3% of the total exposure.

This analysis raises two main points of reflection: (1) How reliable is the determination of design hydrograph based on 'short' time series? (2) How meaningful is the selection of a certain design hydrograph in the context of climate change?

The answer is not trivial and obviously, prevention measures must only be a portion of the overall risk management strategy besides mitigation and preparedness.

5 Concluding Remarks

The 2017 flood in Livorno was exceptional for the city but not so exceptional if considering the Tyrrenian coast of Italy where similar phenomena have occurred in the past.

Relying on engineering works to reduce flood risk may lead to biased risk perception and eventually to fatalities, which are not justified. To better understand risk, the dangerous conditions for people should also be evaluated, thus requiring a detailed flood model that calculates flow velocities.

Moreover, the design procedure should be reframed in view of changing climate projections and residual risk better managed to avoid human losses.

References

Arrighi C, Oumeraci H, Castelli F (2017) Hydrodynamics of pedestrians' instability in floodwaters. Hydrol Earth Syst Sci 21:515–531. https://doi.org/10.5194/hess-21-515-2017

Arrighi C, Brugioni M, Castelli F, Franceschini S, Mazzanti B (2018) Flood risk assessment in art cities: the exemplary case of Florence (Italy). J Flood Risk Manag. https://doi.org/10.1111/jfr3.12226

Arrighi C, Pregnolato M, Dawson RJ, Castelli F (2019) Preparedness against mobility disruption by floods. Sci Total Environ 654 (2019):1010–1022. https://doi.org/10.1016/j.scitotenv.2018.11.191

Campo L, Caparrini F, Castelli F (2006) Use of multi-platform, multi-temporal remote sensing data for calibration of a distributed hydrological model: an application in the Arno basin, Italy. Hydrol Process 20(13):2693–2712. https://doi.org/10.1002/hyp.6061

Castelli F, Menduni G, Caparrini F, Campo L (2009) A distributed package for sustainable water management: a case study in the Arno basin. In: The role of hydrology in water resources management (Proceedings of a symposium held on the island of Capri, Italy, October 2008), vol 327. IAHS Publ., pp 52–61

Salvati P, Petrucci O, Rossi M, Bianchi C, Pasqua AA, Guzzetti F (2018) Gender, age and circumstances analysis of flood and landslide fatalities in Italy. Sci Total Environ 610–611:867–879. https://doi.org/10.1016/j.scitotenv.2017.08.064

Silvestro F, Rebora N, Rossi L, Dolia D, Gabellani S, Pignone F, Trasforini E, Rudari R, De Angeli S, Masciulli C (2016) What if the 25 October 2011 event that struck Cinque Terre (Liguria) had happened in Genoa, Italy? Flooding scenarios, hazard mapping and damage estimation. Nat Hazards Earth Syst Sci 16:1737–1753. https://doi.org/10.5194/nhess-16-1737-2016

Yang J, Castelli F, Chen Y (2014) Multi-objective sensitivity analysis and optimization of distributed hydrologic model MOBIDIC. Hydrol Earth Syst Sci 18(10):4101–4112. https://doi.org/10.5194/hess-18-4101-2014

Evaluating and Mapping the Hazard and Risk of Vehicle Instability Within a Flood Prone Area

Ricardo A. Bocanegra and Félix Francés

Abstract

River overflows can significantly affect vehicles that are circulating or parked in flood-prone areas. Vehicle dragging has not only an economic impact but also the potential loss of human lives. Due to this, proper flood management requires the identification of risk and safe areas for vehicles during these types of events. Loss of stability of a vehicle can be generated by the hydrodynamic mechanisms of floating, sliding and/or toppling. The case study was Rambla del Poyo, which can flood parts of the metropolitan area of Valencia (Spain). To establish the hazard, we used maximum velocity and depth inundation maps for the return periods of 50, 100 and 500 years and different types of vehicles. For the estimation of the final risk, it was needed the car density in the area and the histogram of the types of vehicles. Using the selected stability model, it was possible to map the hazard for each type of vehicle individually and the risk for all vehicles that circulate or are parked in the case study. Especially interesting for evacuation plans was the mapping of the safe areas. Also it was found that vehicles would lose their stability mostly by flotation.

Keywords

Vehicle stability • Hazard and risk mapping of vehicle instability • Vehicle safe areas • Floods

1 Introduction

Floods negatively affect the territory generating social and economic losses. Among the elements that can be affected by floods are vehicles, which in turn can intensify the effects of floods when they are washed away by the flow (Teo et al. 2012). In cities, the highest number of deaths during floods occurs inside vehicles (Fitzgerald et al. 2010). Additionally, the rescue of people from vehicles in flooded areas can be quite expensive (Smith et al. 2017).

Despite the serious impacts that the drag of vehicle can have during floods, very few studies have been done to date to establish the hazard and risk to which vehicles are subject during these events. This research identified the risk of vehicle instability due to overflowing of the Rambla del Poyo, which is located on the East coast of Spain.

2 Materials and Methods

2.1 Vehicle Flood Hazard

Loss of stability of vehicles during floods is caused by the phenomena of flotation, sliding and toppling (Shand et al. 2011). In this study, the stability of the vehicles was determined using the model proposed by Arrighi et al. (2016). In this method the stability of a car is established by means of a hazard index, which is calculated as the ratio between a critical mobility parameter $\theta_{V_{cr}}$ and a mobility parameter θ_V for a predefined reference vehicle, according to the following expression:

$$\frac{\theta_{V_{cr}}}{\theta_V}\begin{cases} > 1 & \text{Vehicle in motion by sliding} \\ = 1 & \text{Start of sliding of the vehicle} \\ < 1\ y \geq 0 & \text{Vehicle at rest} \\ < 0 & \text{Vehicle in motion by flotation} \end{cases} \quad (1)$$

The mobility θ_V and critical mobility $\theta_{V_{cr}}$ parameters are calculated using the following equations, which are defined based on Fig. 1:

R. A. Bocanegra · F. Francés (✉)
Research Institute of Water and Environmental Engineering (IIAMA), Universitat Politècnica de València, Valencia, Spain
e-mail: ffrances@upv.es

© Springer Nature Switzerland AG 2020
F. Fernandes et al. (eds.), *Advances in Natural Hazards and Hydrological Risks: Meeting the Challenge*, Advances in Science, Technology & Innovation,
https://doi.org/10.1007/978-3-030-34397-2_24

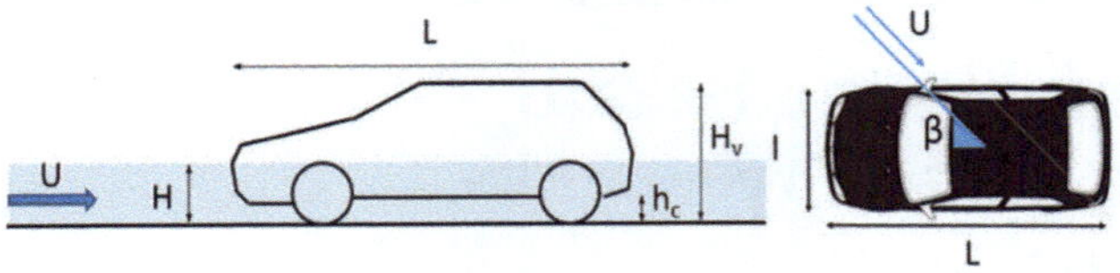

Fig. 1 Geometry of a flooded car. *Source* Arrighi et al. (2016)

$$\theta_V = \frac{2L}{(H_v - h_c)} * \frac{l}{l * \cos\beta + L * \sin\beta} * \left(\frac{\rho_c * (H_v - h_c)}{\rho * (H - h_c)} - 1\right) \quad (2)$$

$$\theta_{V_{cr}} = 8.2 * \mathrm{Fr}^2 - 14.1 * \mathrm{Fr} + 5.4 \quad (3)$$

where ρ_c = car density, ρ = water density and Fr = Froude number.

2.2 Vehicle Flood Risk

Flood risk of vehicles is defined as the combination of the probability of a vehicle being swept away or floated (hazard) and the density of traffic or vehicles parked in the flooded area (vulnerability). Solving the general risk equation in a discrete form, the risk of flooding for vehicles is calculated as follows for one pixel in mean number of vehicles washed away per year per unit area:

$$\text{Risk} = \sum_{i}^{K} \sum_{j}^{N} d * f[i] * D[j] * \Delta P_j \quad (4)$$

where K = number of types of vehicles through which the vehicle fleet is represented; f [i] = proportion of type car i; d = density of vehicles in the pixel; N = number of flood hazard maps, each of them for a given return period T_j; D [j] = stability of the vehicle given by Eq. (1) for event j, it is equal to 0 (stable) when $0 \leq \theta_{Vcr}/\theta_V < 1$ and equal to 1 otherwise; ΔP_j = occurrence probability between two considered events j and $j - 1$.

2.3 Characteristics of the Study Area

The case study in this paper is the estimation of the hazard and risk of vehicles due to overflows of the Rambla del Poyo in the urban centers of the municipalities of Massanassa and Alfafar, which are located in the south of the metropolitan area of Valencia, in Spain (Fig. 2). Rambla del Poyo is an intermittent stream that begins at an approximate elevation of 1023 masl, flows into L'Albufera lake, has a catchment of 430 km^2, a mean annual precipitation less than 500 mm and a slope that fluctuate between values higher than 16% in the upper part and lower than 2% in the lower part, which corresponds to the flood-prone area. It is strongly anthropized and is characterized by presenting severe flash floods time to time during Autumn.

The distribution of cars in the study area was determined through data published by ANFAC (the Spanish Association of Truck and Automobile Manufacturers) in 2018 (ANFAC 2018). The vehicle fleet was classified into four types of vehicles: (i) utility and smaller cars; (ii) compact; (iii) small SUVs; and (iv) medium-sized SUVs and larger cars.

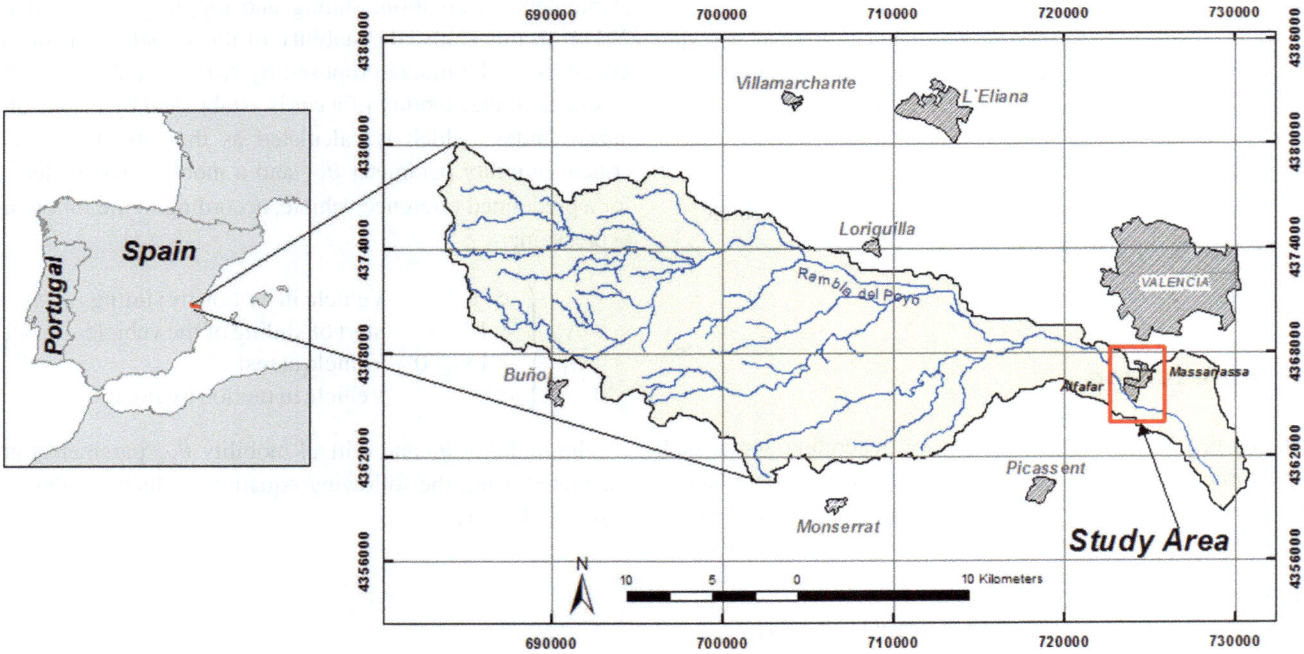

Fig. 2 Location of Rambla del Poyo catchment and the analyzed study area

Table 1 Characteristics of studied vehicles

	Utility Seat Ibiza	Compact Seat León	Small SUV Peugeot 2008	Medium SUV Volkswagen Tiguan
Length (m)	3.68	4.18	4.16	4.43
Width (m)	1.61	1.74	1.74	1.81
Height (m)	1.42	1.44	1.56	1.67
Ground clearance (m)	0.12	0.12	0.17	0.18
Density (kg/m^3)	108.00	125.86	104.41	115.26
Proportion	0.26	0.32	0.15	0.27

According to sales in 2018, it was considered that the representative cars of these types of vehicles were the Seat Ibiza, the Seat Leon, the Peugeot 2008 and the Volkswagen Tiguan, respectively. Table 1 shows the proportion of each of these types of cars in the vehicle fleet and the main dimensions of the analyzed cars.

3 Results

The hazard and risk of vehicle instability in the case study were established considering the flood hazard maps with return periods of 50, 100 and 500 years, provided by the Jucar River Water Authority (CHJ 2011). Velocity and water depth in the inundation area were obtained using the hydraulic model Infoworks 2D.

Calculations of hazard and risk of vehicle instability were done assuming that cars were completely water-tight and oriented in the normal direction to the flow. Figure 3 shows the hazard indexes obtained for the car Volkswagen Tiguan. These indexes were obtained using Eq. (1) for the three considered floods. Figure 4 shows the total risk map, which corresponds to the sum of the risk maps obtained for each type of car using Eq. (4). These maps were obtained based on the maps of the hazard indexes, the proportion of each type of car and the density of cars in the flood-prone area of

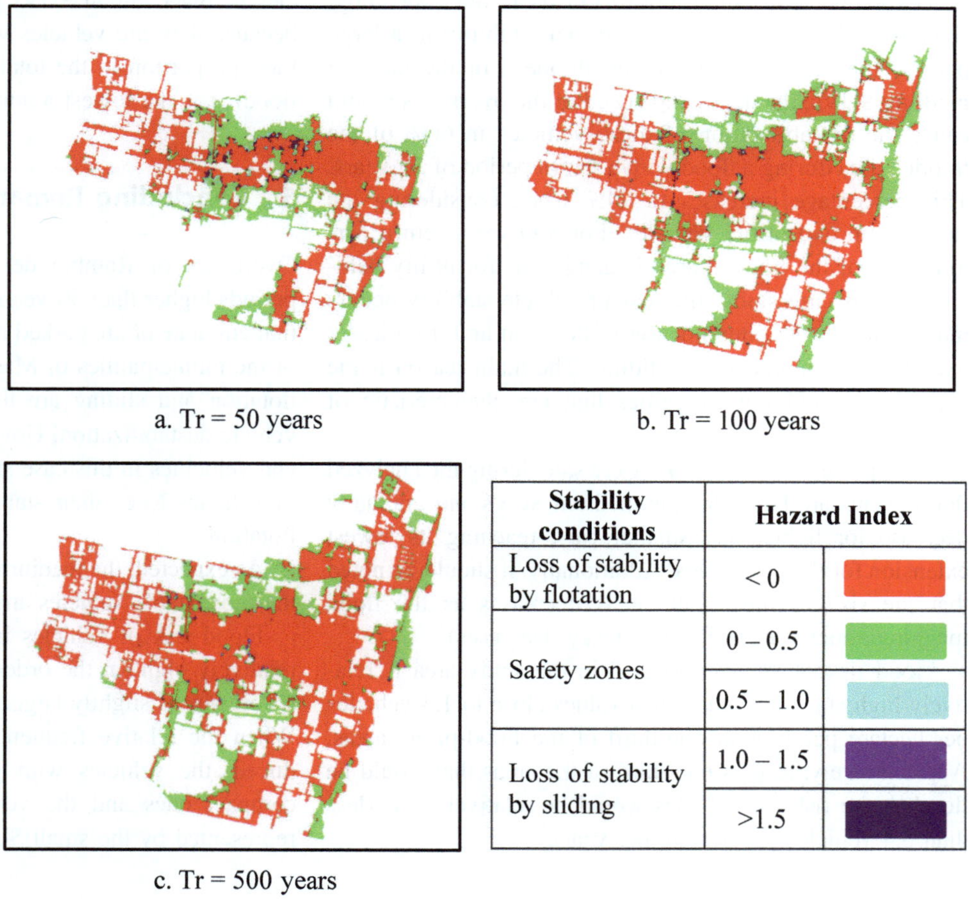

Fig. 3 Hazard index for the Volkswagen Tiguan in the case study with return periods of 50, 100 and 500 years

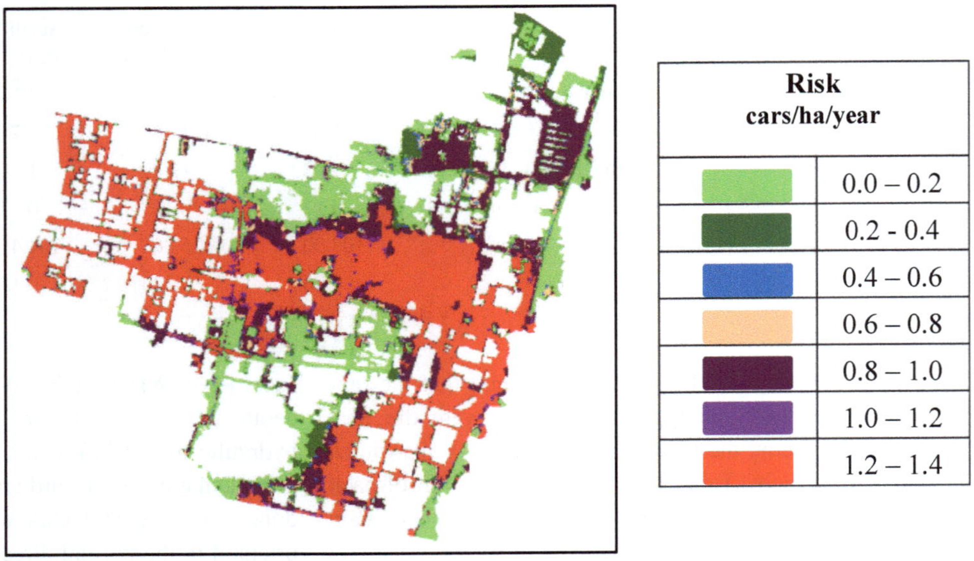

Fig. 4 Map of flood risk of vehicles that are circulating or parked in the case study

Rambla del Poyo, which, according to Francés et al. (2008), is 0.0054 cars/m^2.

4 Discussion

Floods of Rambla del Poyo with a return period of over 50 years represent a high hazard to cars in the case study, since even larger cars would lose their stability in a large part of the areas that would be flooded. In the case of medium SUVs, the depth and speed of the overflows would cause the destabilization of these vehicles in 66% of the flooded area during a flood with a return period of 50 years. This percentage increases to 74% when considering the same event and utility vehicles. For 500 years return period, 81% of the flooded area is dangerous for utility vehicles. In this case study, the loss of vehicle stability occurs mainly due to the phenomenon of flotation and, to a lesser extent, the phenomenon of sliding. The main reason is the high depths and low velocities that are characteristic of floods in the area.

As expected, the zones that were safe during the analyzed flood events are larger for the medium SUVs and gradually decrease for lighter and smaller cars, reaching its lowest extension for the utility cars. Additionally, it should be noted that the size of these safe areas decreases as the flood magnitude increases (flood frequency decreases).

Flood risk of vehicles located in the study area is relatively high since it would reach values close to 1.4 vehicles per hectare per year in one-third of the flood-prone areas. Approximately, also in one-third of the areas that would be flooded, the risk for vehicles would be relatively low (less than 0.4 vehicles per hectare per year).

The analysis of each type of vehicle indicates that the compact cars present the greatest risk to floods (in other words, potential damages), since, despite not being the most unstable, they represent the highest percentage among the vehicles located in the studied area. Utility vehicles, which are the most unstable, present a risk slightly lower than compact vehicles because their proportion among all vehicles is lower. Small SUVs present the lowest risk to floods because they are vehicles with relatively high stability and their proportion in the total number of vehicles exposed to flooding is the lowest among the types of cars analyzed.

5 Concluding Remarks

Overflows of Rambla del Poyo corresponding to return periods higher than 50 years constitute a high hazard to cars that circulate or are parked in large areas of the urban centers of the municipalities of Massanassa and Alfafar. In general, flotation and sliding are the two possible mechanisms of vehicle destabilization. However, the combination of depths and velocities in this case study results in a high percentage of vehicles loses their stability due to the phenomenon of flotation.

As expected, the medium SUV vehicles were shown to be the most stable vehicles among the types of analyzed cars.

Flood risk of vehicles in the analyzed municipalities is relatively high (in the order of 1.4 cars/ha/year) in a sector whose size is slightly larger than a third of the flooded areas. When the relative frequency of each type of car is introduced, the vehicles with greatest risk correspond to the compact ones and the vehicles with the lowest risk are represented by the small SUVs.

Finally, we would like to underline that determining hazard and risk levels of vehicle stability during floods would help the implementation of actions in urban planning and civil protection that can decrease the negative effects of this type of events.

References

ANFAC, Spanish Association of Truck and Automobile Manufacturers (2018) http://www.anfac.com/estadisticas.action. Accessed 22 Dec 2018

Arrighi C, Huybrechts N, Ouahsine A, Chaseé P, Oumeraci H, Castelli F (2016) Vehicles instability criteria for flood risk assessment of a street network. Proc IAHS 373:143–146

CHJ (Confederación Hidrográfica del Júcar) (2011) Asistencia técnica para el desarrollo del sistema nacional de cartografía de zonas inundables en la Demarcación Hidrográfica del Júcar

Fitzgerald G, Du W, Jamal A, Clark M, Hou X (2010) Flood fatalities in contemporary Australia (1997–2008). Emerg Med Australas 22:180–186

Francés F et al (2008) Efficiency of non-structural flood mitigation measures: "room for the river" and "retaining water in the landscape". ERA-NET CRUE Research Report No I-6. General Directorate for Research in the European Commission

Shand T, Cox R, Blacka M, Smith G (2011) Appropriate safety criteria for vehicles. Australian rainfall and runoff, revision project 10: report number: P10/S2/020. Australian Rainfall and Runoff (AR&R), Sydney, Australia

Smith GP, Modra BD, Tucker TA, Cox RJ (2017) Vehicle stability testing for flood flows. Technical report. University of New South Wales, Sydney, Australia

Teo Y, Xia J, Falconer R, Lin B (2012) Experimental studies on the interaction between vehicles and floodplain flows. Int J River Basin Manag 10(2):149–160

Definition of Flood-Prone Areas: A Comparison between HEC-RAS and Iber Software Results

Márcia Martins, Pedro Gonçalves, Alberto Gomes, and José Teixeira

Abstract

The availability of hydraulic modelling software allowing dynamic analysis of hydraulic variables behaviour, such as the water depth and flow velocity, is important for the assessment and management of flood-prone areas and, consequently, to reduce the associated risks. This study presents a comparison between the modelling results of a 1D (one-dimensional) and 2D (two-dimensional) hydraulic models—HEC-RAS and Iber software—in the definition of flood-prone areas for a sector of the Leça River (Northern Portugal). The main comparison refers to the maximum extent of the flood and the water depth, obtained from the mathematical simulation with these two free softwares, for return periods of 10, 50 and 100-years. Despite the two softwares consider different hydraulic approaches, it was recorded small differences in the extent of the flooded area and in the maximum water column height.

Keywords

Hydraulic modelling • Flood-prone areas • HEC-RAS • Iber • Leça River

1 Introduction

The increase use of hydraulic modelling software's is related with the fact that many of them are free, like the software's used in this work, but also, due to an higher demand in a global context of climate change where it is expected, for certain areas, an intensification of the frequency and impacts of extreme hydrologic events (EEA 2017). Thus, tools are needed that provide reliable and accurate results to support the scientific analysis and management of flood-prone areas in order to mitigate and reduce associated risks (Martins et al. 2019; Santos 2009).

In this study, the main goal was to define and compare flood-prone areas using an experimental sector of the Leça River (Northern Portugal), named the Santa Cruz do Bispo, with the application of different hydraulic modelling free software. The study area is part of the Leça River watershed, comprising a segment of ~4.4 km with altitudes ranging from 1 to 77 m. Some sectors of the river bank are urbanized areas, which can be affected during a flood event, but there are also agricultural areas that can be flooded causing significant economic losses by the crops destruction (Gonçalves et al. 2015).

The key interest of our research is to evaluate and compare the simulation results for a 1D and 2D hydraulic models, using for that the HEC-RAS and Iber software's, in order to obtain the maximum extent of the flood and the water depth, for return periods of 10, 50 and 100-years. These return periods were chosen to provide an analysis covering a frequent flood (10 years), a flood whereupon the flooded area boundaries are still in the memory of local residents (50 years) and a centenary flood (100 years), as is defined by the European legislation (European Union 2007).

2 Materials and Methods

The methodological process adopted is divided into three main steps: (i) base data acquisition and preprocessing, (ii) simulation by the hydraulic modelling software, and (iii) flood-prone area and water depth definition, validation and comparison of results.

In the first step, it was necessary to collect and preprocess the input data, i.e. topographical, meteorological, hydrological data and land use information, which supported the hydraulic simulation (Fig. 1). For the construction of the digital surface model (DSM), it was processed all the

M. Martins (✉) · A. Gomes · J. Teixeira
CEGOT, Faculty of Arts, University of Porto, Porto, Portugal
e-mail: marciam_castro@hotmail.com

P. Gonçalves
Laboratório Nacional de Energia e Geologia, I.P. Unidade de Recursos Minerais e Geofisica, Aljustrel, Beja, Portugal

© Springer Nature Switzerland AG 2020
F. Fernandes et al. (eds.), *Advances in Natural Hazards and Hydrological Risks: Meeting the Challenge*, Advances in Science, Technology & Innovation,
https://doi.org/10.1007/978-3-030-34397-2_25

topographical data, namely, the contour lines (1 m contour interval) and the elevation points (mean density of 1.5 points/m^2) as well as other vector input data, such as buildings, roads, walls and fluvial network (scale 1: 1000). This step is extremely important since floods are strongly conditioned by the terrain features, thus, to obtain better results it is necessary to have a detailed DSM (1 m pixel size grid, approx.). The values of peak flood discharge were calculated by applying the cinematic formula of *Giandotti* for the return periods of 10, 50 and 100 years. Then, the roughness values were assigned, using the Manning n coefficient for each type of land use, based on the official land use map (COS2007) (Geoportal of DGT 2018).

After these initial data preparation, with HEC-RAS 1D model, the process started by the creation of a geometric file corresponding to (Fig. 1): geometric center of the channel; river bank and water flow direction; cross-sections (260 cross-sections with an average distance between them of 16 m and an average length of 496 m); obstructions to water flow such as walls, weirs and buildings, location of the bridges, non-contributory areas and Manning n values for each type of land use. In addition to the geometric information, HEC-RAS also requires the definition of boundary conditions, the values of the flood peak discharges and the shape and model of the bridges.

In turn, the simulation with the Iber 2D model (Fig. 1) is structured in three main stages: pre-process, process and post-process. In the first stage, defined/imported the geometry, determined the conditions of the problem, assigned the roughness values, created the calculation mesh, etc. are done The second stage refers to the calculation process and, in the third stage, the results are displayed and analyzed. For this, it is necessary to use the initial input data, such as a detailed DSM, the values of the peak flood discharge and the

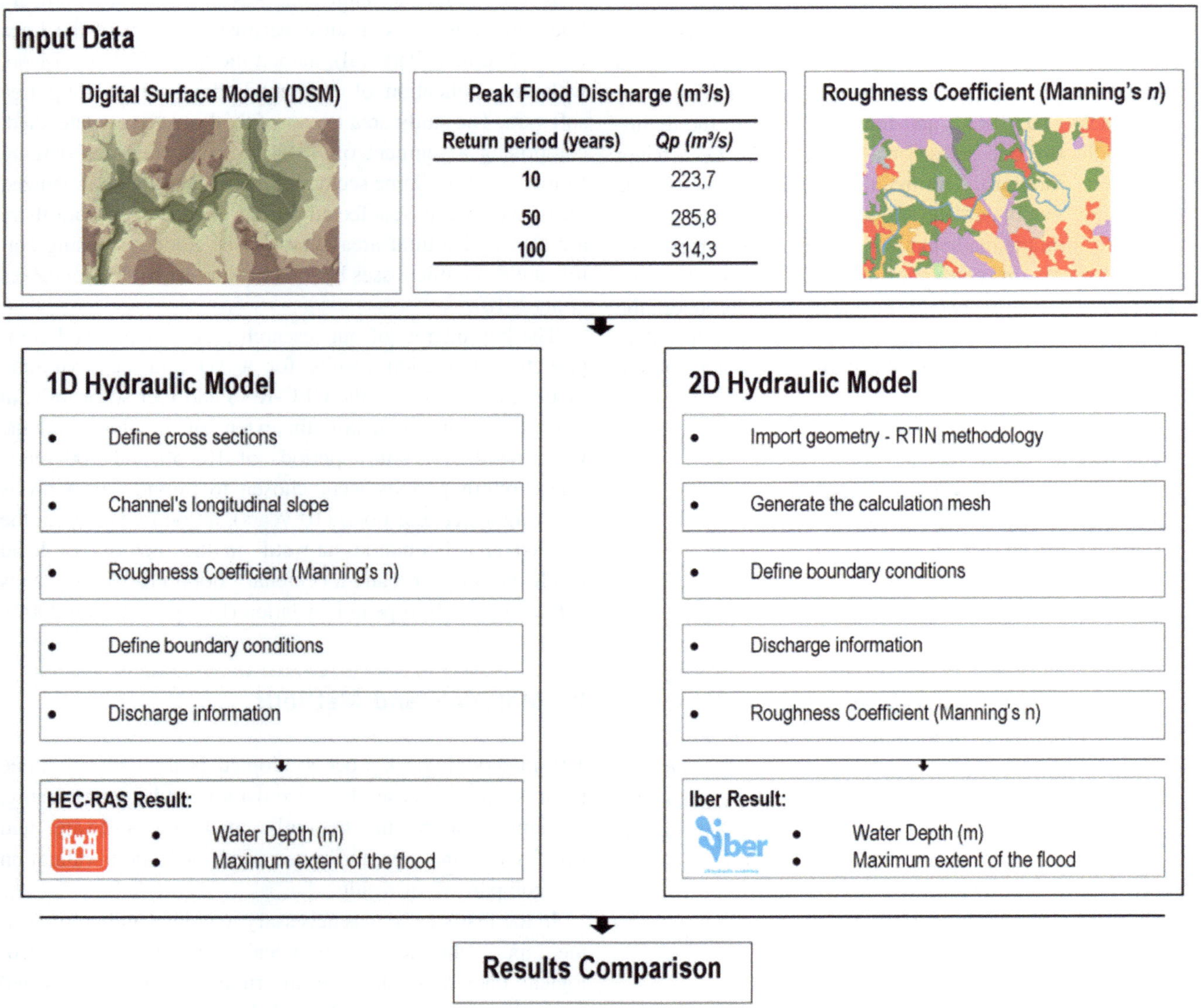

Fig. 1 Conceptual framework and main steps followed in this work

assignment of the Manning n coefficient values for each type of land use. Since the Iber is a 2D model it allows importing geometries automatically from the DSM. Although there is a great diversity of formats allowed to import geometries, we choose the RTIN methodology (Right-Triangulated Irregular Network) that divides the terrain surface into triangles rectangles of various sizes. Then, the calculation mesh was created, presenting the triple of the elements of the initial geometry to properly represent the terrain topography used for the flood simulation. The boundary conditions were defined, assigning the values of the peak flood discharge for different return periods and defining for the whole model the initial condition of the water depth. The roughness of the surface was assigned automatically through the Manning roughness coefficient.

Despite both hydraulic models provide other results, e.g. flow velocity, in this study we chose to compare only the maximum extent of the flood and the water depth.

3 Results

Both software's allows the calculation of distinct hydraulic parameters, resulting in different maps showing the flood characteristics Moel et al. (2009) argument that maps representing the maximum extent of a flood for a specific event are the most common flood hazard maps, describing historical episodes or a particular/extraordinary events.

Analyzing the simulation results (Fig. 2 and Table 1) there are differences between 1D and 2D hydraulic models in terms of flooded area, maximum water column height and number of affected buildings. Differences related to the extent of the flooded areas are around 21.8% for a 10-years return period, 20.5% for a 50-years return period and 19.3% for a flood event with a 100-years return period (Table 1).

The analysis of the results shows differences in the maximum water column height and, generally, the Iber hydraulic simulation reaches higher values. Presenting a larger flooded area as well as higher water column height values, the Iber simulation results also affect more buildings compared to HEC-RAS simulation scenario (almost the double for each considered return period). Concerning the computational time spent by each software, the HEC-RAS took $\sim$10 min and the Iber spent $\sim$22 min to perform the simulation.

Figure 3 shows two transversal profiles, displaying the water column height in two distinct sectors, *Rua das Carvalhas* and *Ponte do Carro*, allowing a more accurate comparison of the performance of hydraulic models. Profile 1 considers a wide and flat valley and for a flood event with a 100-years return period, HEC-RAS simulation reaches a maximum height (water column) of $\sim$5.8 m, while the Iber reaches $\sim$6.6 m, almost to 1 m of difference, which it's substantial when we are dealing with valley flat sectors. Profile 2 shows an entrenched and narrow valley representing one of the most critical sectors to the occurrence of floods, where it is observed that both hydraulic simulations present the same water column height, corresponding to $\sim$5.4 m.

Based on the information and marks provided by the local inhabitants regarding the significant flood event of 2001, the presumable water column height in the *Ponte do Carro* sector was estimated, reaching approximately 4.5 m above the river thalweg.

Hydraulic simulations also reveal that the banks of the Leça valley more affected by floods are mainly occupied by agricultural lands, involving complete/partial crop destruction.

The observed differences between the simulations can be related with two main reasons: the fact that each software is based on distinct equation, as well as the topography of the river channel/valley is modelling by different ways. The Iber topographic mesh that covers all the topography and surface of the valley instead of the cross-sections of HEC-RAS (simplifying the topographic model) can explain the higher values of water column and the extent of the flooded area from Iber simulations.

4 Discussion and Concluding Remarks

The comparison of the hydraulic simulations achieved with the two hydraulic models (HEC-RAS and Iber software's), gave some important differences, especially in wide and flat valley sectors. HEC-RAS software allows one-dimensional hydraulics calculations for a steady or unsteady flow river, while Iber software solves the 2D shallow water equations for the determination of hydraulic parameters. The one-dimensional models consider the river as a line associated with a sequence of cross-sections that allows the definition of the channel/valley geometry, while the two-dimensional models are based on a calculated mesh composed by polygonal elements (Guaya Caraguay and Montalván Alcívar 2018). Two-dimensional modelling requires longer computation time and entails a greater computational capacity (Guaya Caraguay and Montalván Alcívar 2018). Besides, a more thorough comparison and validated by good and enough field data would be needed to determine whether two-dimensional software which involves more complexity, is actually more accurate. The results of both models are very dependent on the quality and accuracy of the input data introduced in the software,

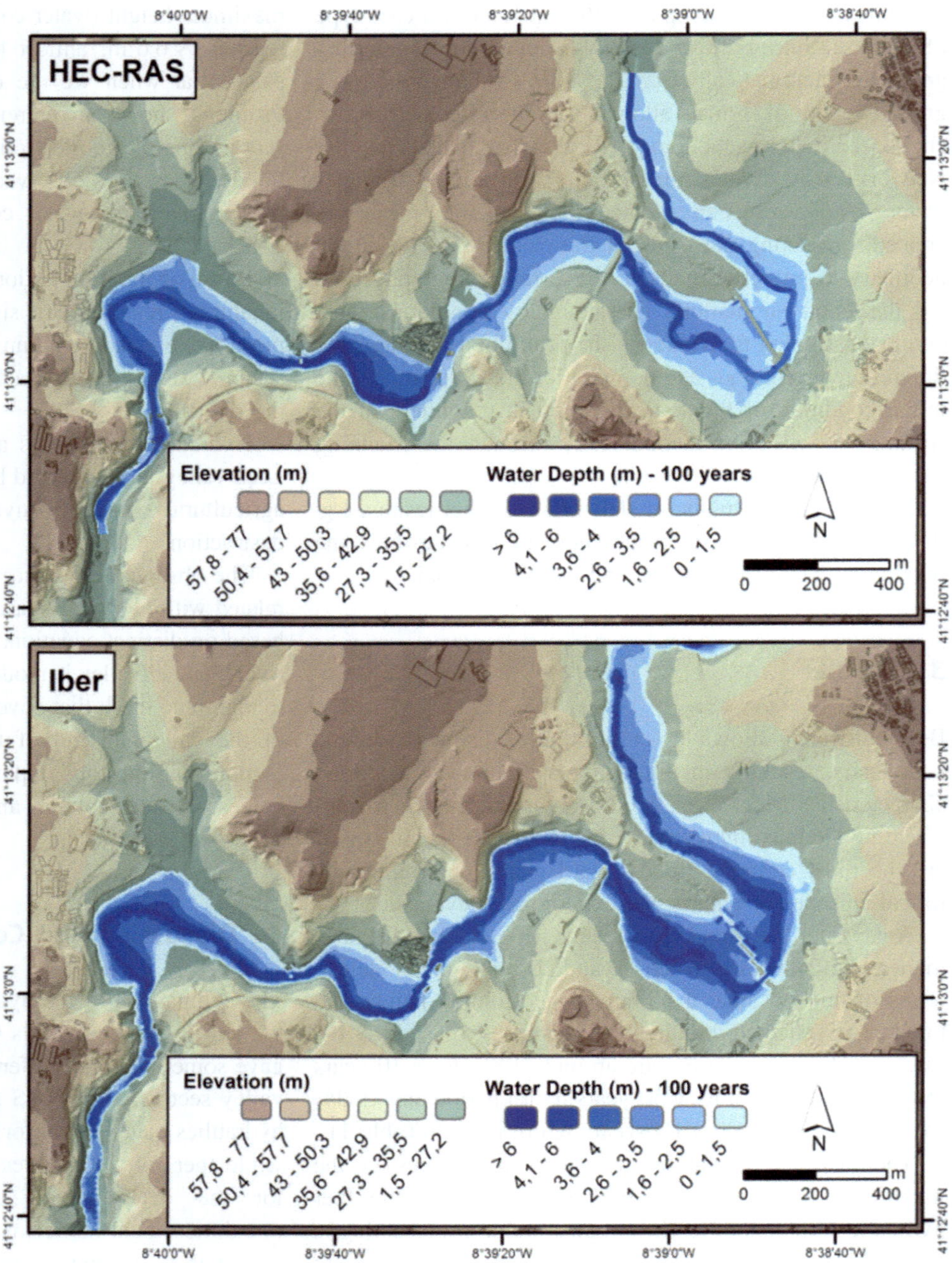

Fig. 2 Water depth obtained by hydraulic modelling performed with HEC-RAS and Iber software for a return period of 100 years

Table 1 Comparison between the results obtained (flooded area, max. water column height and number of affected buildings) with hydraulic modelling in HEC-RAS and Iber software

Return period	HEC-RAS			Iber		
	Flooded area (m^2)	Maximum water column height (m)	Affected buildings	Flooded area (m^2)	Maximum water column height (m)	Affected buildings
10 years	417,141.8	6.8	25	508,135.4	7.6	41
50 years	444,255.6	7.4	28	535,450.3	8	49
100 years	458,379.9	7.7	29	546,925	8.2	49

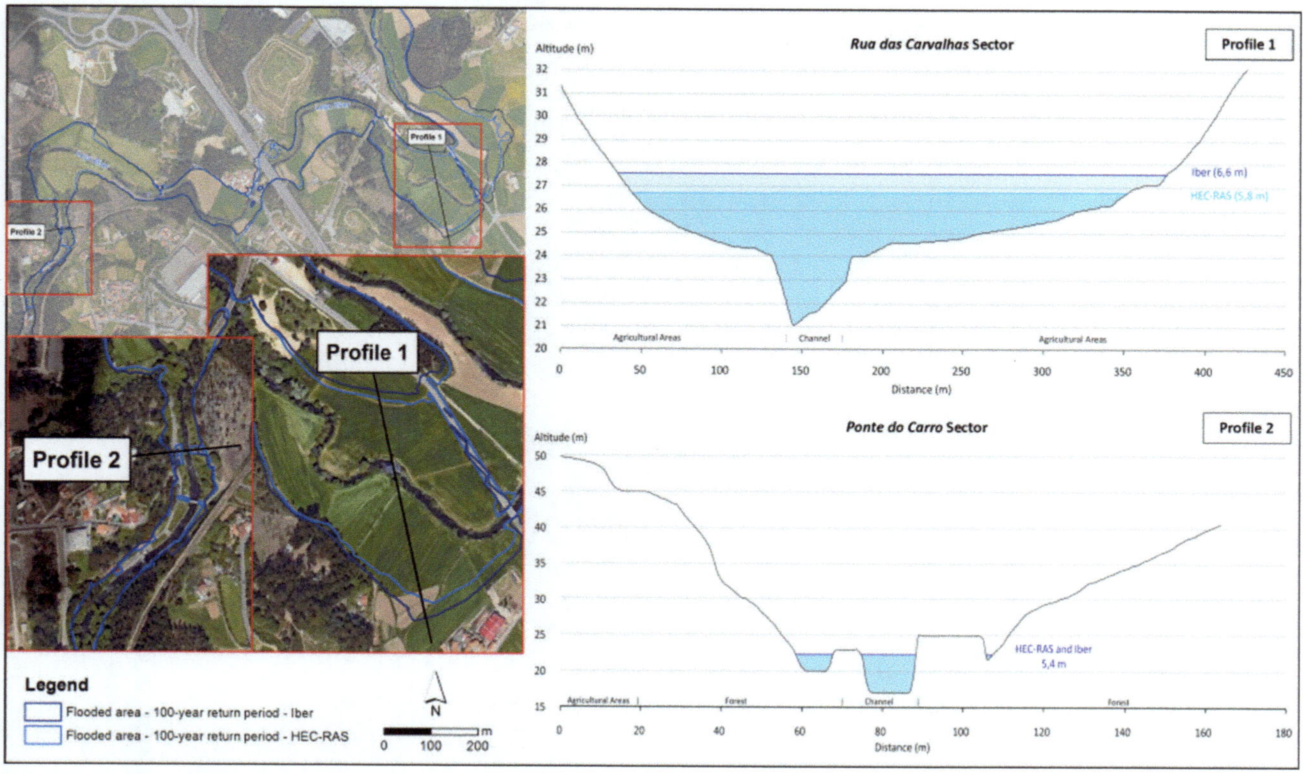

Fig. 3 Transversal profiles comparing the results of the hydraulic modelling made by HEC-RAS and Iber software's for a return period of 100 years

particularly the digital surface model (DSM) that needs detailed topographical data to accurately represent the characteristics of the terrain.

Despite some differences between the results of 1D and 2D hydraulic simulations, both are powerful tools that allow a good comprehension and reliable results for the simulation of flood events, helping the assessment and management of the flood risk.

References

EEA (2017) Climate change, impacts and vulnerability in Europe 2016. EEA Report No 1/2017. European Environment Agency, 419 pp

European Union (2007) Directive 2007/60/EC of the European Parliament and of the Council of 23 October 2007 on the assessment and management of flood risks. Off J Eur Union 27–34

Geoportal of DGT (2018) http://mapas.dgterritorio.pt/geoportal/catalogo.html. Accessed 08 Oct 2018

Gonçalves P, Marafuz I, Gomes A (2015) Flood hazard, Santa Cruz do Bispo Sector, Leça River, Portugal: a methodological contribution to improve land use planning. J Maps 11(5):760–771

Guaya Caraguay CI, Montalván Alcívar EA (2018) Modelación numérica bidimensional del río Garrapata en la provincia de Manabí aplicando el programa Iber. EPN, Quito

Martins M, Gomes A, Santos P (2019) Delimitation of flood-prone areas in the urban center of Amarante, North of Portugal, using the Iber software. Rev Bras Geomorfol 20(1):185–200

Moel H, Van Alphen J, Aerts JCJH (2009) Flood maps in Europe—methods, availability and use. Nat Hazards Earth Syst Sci 9(2):289–301

Santos P (2009) Cartografia de áreas inundáveis a partir do método de reconstituição hidrogeomorfológica e do método hidrológico-hidráulico: estudo comparativo na bacia hidrográfica do rio Arunca. Dissertação de Mestrado, Universidade de Coimbra, Coimbra

Azores Assessment and Management of Flood Risks

Dina Pacheco, Sandra Mendes, and Raquel Cymbron

Abstract

The European Directive on the assessment and management of flood risks, which entered into force in November 2007, establishes the regulatory framework for the assessment and management of flood risks within the European Union (EU), in order to reduce the negative consequences of flood for human life and health, the environment, cultural heritage and economic activity, as well as to take adequate measures to identify flood hazard and flood risk areas. Among other assets, this Directive requires EU Member States to first carry out a preliminary assessment based on the following instruments: flood hazard assessment maps; risk maps and flood risk management plans (FRMP). Portugal transposed this directive in 2010, marking a new attitude towards flood risk management by the integration of figure flood risk management plans (FRMP), in its water management and planning policies, structured in 6-year cycles. The first flood risk management plan for the Azores RBMP was concluded at the end of 2015, and further approved as a policy instrument by Regional Legislative Decree no. 20/2016/A of October 10. The plan aims an integrated river basin risk management in order to reduce the consequences associated with floods in sensitive areas covered by Territorial Management Instruments and to decrease susceptibility risks associated with local and regional particularities. The first cycle resulted in the selection of five critical zones characterized by event reoccurrence, loss of human and number of people affected, on the islands of São Miguel, Terceira, and Flores. The next planning cycle (2022–2027) will reevaluate the current critical high-risk flood areas, and identify new critical areas for both river basins and coastal areas at risk of flooding.

Keywords

Floods • Risk • River basins • Azores

1 Introduction

Directive 2007/60/EC (Flood Directive), of the Parliament and the Council, of October 23, transposed into national law, through Decree-Law no. 115/2010, dated October 22, established the regulatory framework for the assessment and management of flood risks in the European Union (EU), in order to reduce the consequences associated with the occurrence of these phenomena for human health, the environment, cultural heritage, and economic activities. Among other provisions, this Directive requires EU Member States to undertake a preliminary assessment of flood risks and to prepare flood hazard maps, flood risk maps and flood risk management plans for areas with significant flood risk.

The Flood Risk Management Plan (PGRIA) currently in force refers uniquely to fluvial flooding. In the regional context, this specificity is related to the torrential nature of most watercourses, to its morphology, topographical patterns, and steep basin slopes. The high drainage density in the Region, and the small size of the river basins, characterized by a reduced concentration-time, entails a generalized risk of sudden fluvial floods, with difficult predictability of its location and magnitude.

The intrinsic physical characteristics of the river basins contribute to magnify the adverse consequences of floods. The examples of these situations are common throughout the Region, gaining significant importance due to the socio-economic impacts caused by recent events in the islands of São Miguel, Terceira, and Flores.

D. Pacheco (✉) · S. Mendes · R. Cymbron
Direção de Serviços de Recursos Hídricos e Ordenamento do Território, Direção Regional do Ambiente, Secretaria Regional da Energia, Ambiente e Turismo, Ponta Delgada, Portugal
e-mail: Dina.MD.Pacheco@azores.gov.pt

© Springer Nature Switzerland AG 2020

F. Fernandes et al. (eds.), *Advances in Natural Hazards and Hydrological Risks: Meeting the Challenge*, Advances in Science, Technology & Innovation,
https://doi.org/10.1007/978-3-030-34397-2_26

2 Methods

A total of 727 river basins from the Azores Archipelago (DROTRH, 2001 and DRA, 2012), contemplating the total extension of the regional watercourses (hydrographic network with 7000 km) were analyzed for the critical zone selection.

For the classification of flood risk, all river basins were classified using ESRI technology and GIS handling information technology, based on the combination of three criteria:

1. Historical floods information (Marques, 2013);
2. Watercourses referred in the Municipal Emergency Plans as potentially population hazard;
3. Watercourses that intersect urban areas demarcated in the Municipal Plans.

Based on the cumulative combination of the above three criteria, a subsequent three-level flood risk hierarchy was developed: Low, Moderate and High. Furthermore, the river basins with high risk of flood were further scrutinized for three more criteria to determine population hazard, namely:

- Reoccurrence;
- Loss of human lives;
- Number of people affected.

3 Results

High-risk river basins are present in five islands: Santa Maria, São Miguel, Terceira, São Jorge and Flores (Fig. 1).

Crossing the criteria that determined risk and vulnerability for populations resulted in the selection of five critical zones. Table 1 lists all the systemized information, and their location in Fig. 2.

Flood hazard and flood risk maps were delimited for the five critical zones assessed that correspond to geographic areas that can be flooded in case of high, medium and low probability. Furthermore, the flood risk maps reflect the potential flood impact, with an indication of both population and the economic activity types affected, as exemplified in Fig. 3.

In the second cycle of the Flood Directive implementation (2022–2027), the PGRIA will focus on the reassessment of the flood risk areas identified in the previous cycle, as well

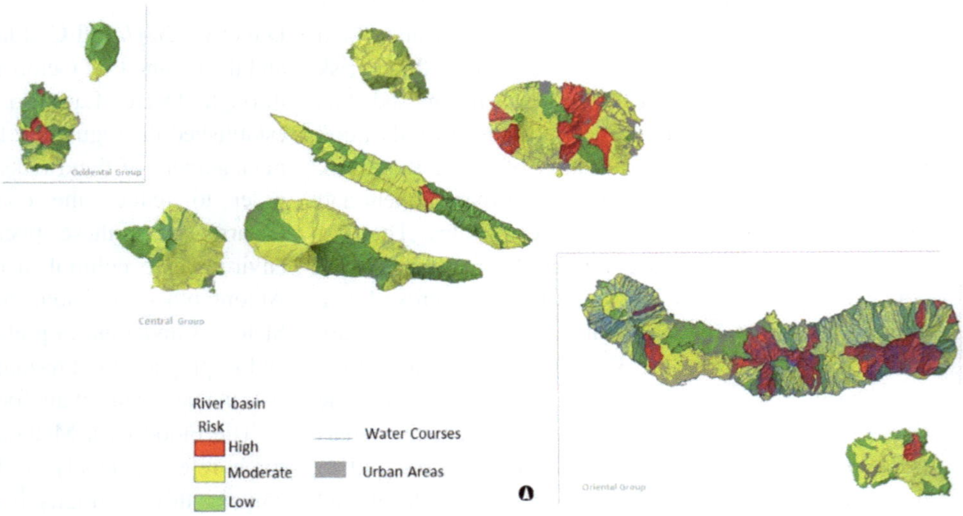

Fig. 1 Example of flood risk river basin maps. Adapted from PGRIA (2015)

Table 1 Criteria for critical areas selection

Island	River basin	N	Loss human	People affected	Origin	Factor
São Miguel	Ribeira Grande	3	1	–	River	Precipitation
	Ribeira da Povoação	10	79	–	River	Precipitation
Terceira	Ribeira da Agualva	4	9	100	River	Precipitation
	Ribeiras de Porto Judeu (Ribeiras do Testo e Grota do Tapete	2	–	30	River	Precipitation
Flores	Ribeira Grande	3	–	202	River	Precipitation

N—Events with adverse impact or damage that occurred between 1588 and 2013. Loss human–human lives, dead and/or missing. People affected–people evacuated and/or displaced

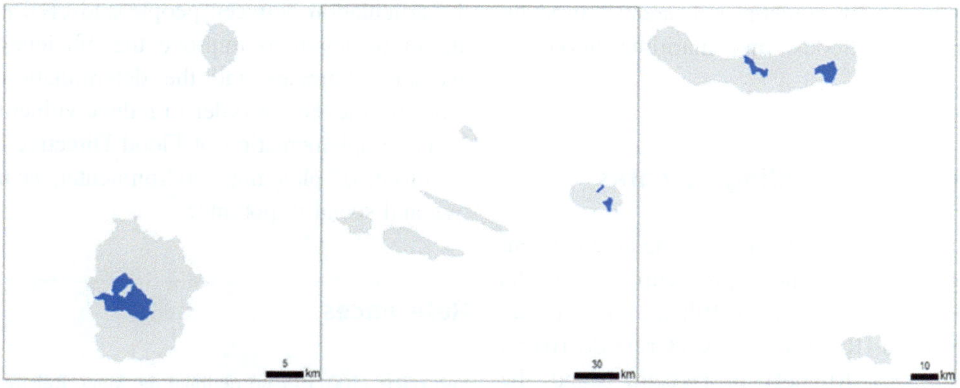

Fig. 2 Location of river basins selected as critical areas in Autonomous Region of Azores

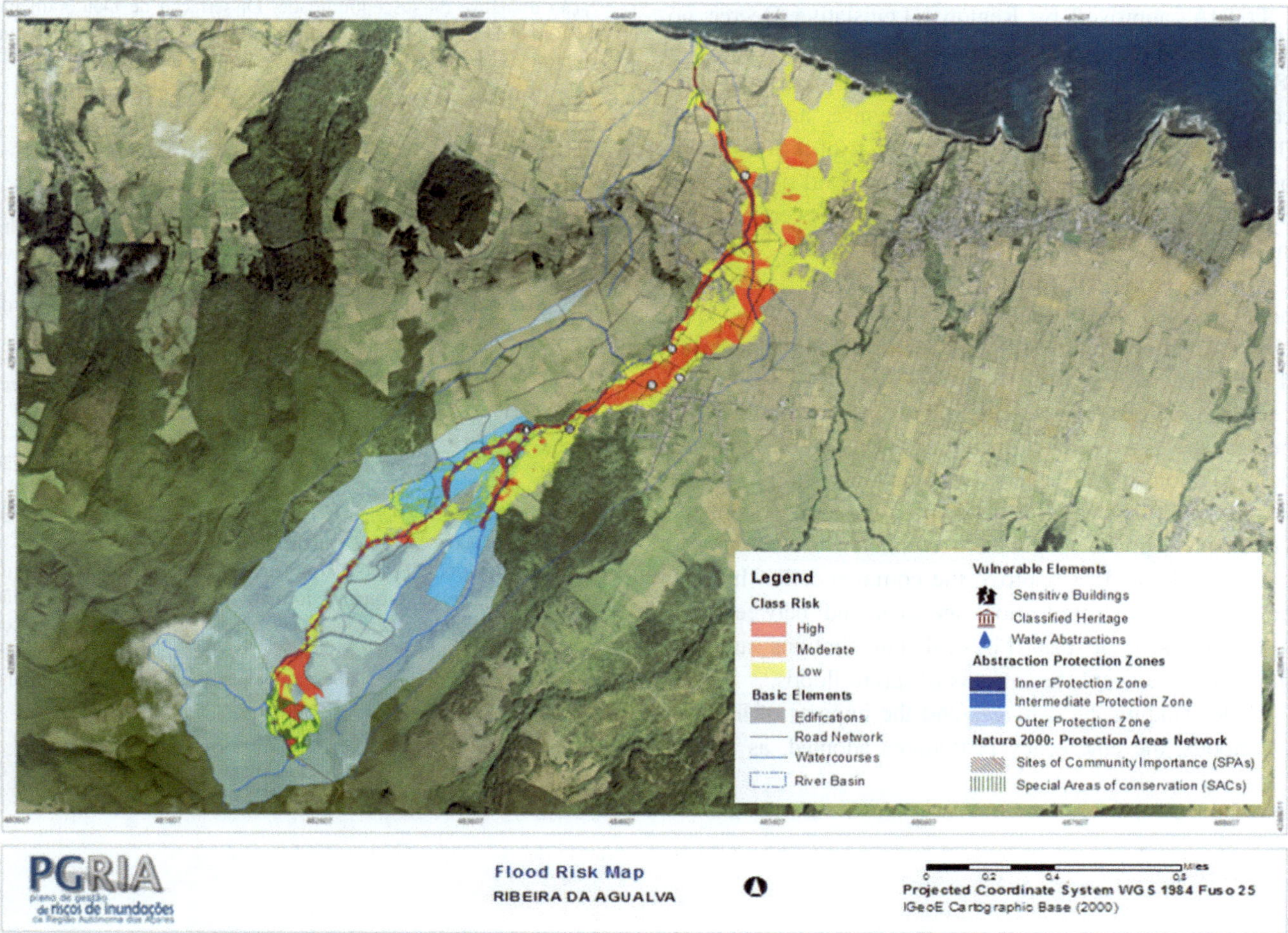

Fig. 3 Example of flood risk maps

as in the evaluation of new potential risk areas caused by fluvial floods, torrential floods, and maritime floods in coastal areas.

4 Discussion and Concluding Remarks

In the Azores the population is the most vulnerable element to flood risk, whereas the most preoccupancy results occurred in the Ribeira da Povoação, Ribeira da Agualva, Ribeiras do Porto Judeu (Ribeira do Testo/Grota do Barril) and Ribeira Grande (São Miguel). In Terceira Island, the road network was identified as the second most vulnerable element to flood risk thus, presenting the higher percentages, the water springs is another affected element to high flood risk exposure. In Povoação sensitive buildings, also assume great vulnerability often due to exposure and susceptibility.

PGRIA, approved by Regional Legislative Decree no. 20/2016/A of 10 October, comprises a program composed of 28 measures grouped into five objectives: prevention, protection, preparedness, emergency response, and recovery. The measures include the implementation of a flood monitoring and warning system, awareness of flood risks and the implementation of intervention projects in several streams and infrastructures, among others.

The measures were prioritized as critical: very high; high; moderate, and low. Regional authorities and the local/municipal authorities are responsible for the implementation of the Plan measures.

Due to regional characteristics, prevention will always be the best strategy to prevent new risks. This implies not only the regulation of land use and occupation in flooding areas but also an efficient flood disaster information system that will enable a greater population to understand the exposure to flood risks and to improve the communication between the different civil protection agencies and between civil protection and citizens. Physical interventions are also important to prevent the impacts of future floods.

Understanding the damages and the impacts of flood on livelihoods, the emergency procedures adopted, as well as the articulation between people and civil protection agents are fundamental to improve the efficiency of flood management strategies and the determination of appropriate protection levels, in order to reduce vulnerability and costs.

The implementation of Flood Directive in the Azores is, therefore, of planning, environmental, economic, informative and social importance.

References

DROTRH (2001) Plano regional da Água. Relatório Técnico, Versão para consulta pública, Ponta Delgada, p 414

DRA (2012) Plano de Gestão de Região Hidrográfica dos Açores. Relatório Técnico da Versão para consulta pública, Secretaria Regional do Ambiente e do Mar

Marques R (2013) Estudo de movimentos de vertente no concelho da Povoação (ilha de São Miguel, Açores): Inventariação, caracterização e análise da susceptibilidade. Dissertação de Doutoramento no Ramo de Geologia, especialidade Riscos Geológicos. Departamento de Geociências, Universidade dos Açores, pp 456

Law

Decreto-Lei no. 115/2010, de 22 de outubro—Estabelece um quadro para a avaliação e gestão dos riscos de inundações, com o objetivo de reduzir as suas consequências prejudiciais, e transpõe a Diretiva no. 2007/60/CE, do Parlamento Europeu e do Conselho, de 23 de outubro

Decreto Legislativo Regional no. 1-A/2017/A, de 6 de fevereiro—Aprova o Plano de Gestão da Região Hidrográfica dos Açores 2016-2021

Decreto Legislativo Regional no. 20/2016/A, de 10 de outubro—Aprova o Plano de Gestão de Riscos de Inundações da Região Autónoma dos Açores

Diretiva 2000/60/CE do Parlamento Europeu e do Conselho, de 23 de outubro de 2000—Diretiva Quadro da Água

Diretiva no. 2007/60/CE, do Parlamento e do Conselho, de 23 de outubro - Estabelece o quadro normativo para a avaliação e gestão dos riscos de inundações no espaço da União Europeia, a fim de reduzir as consequências associadas à ocorrência destes fenómenos aos níveis da saúde humana, do ambiente, do património cultural e das atividades económicas

Incorporating Apparent Shear Stress in the Roughness to Improve the Discharge Prediction in Overbank Flows

João Nuno Fernandes

Abstract

During flood events, the overbank flow is rather common in alluvial valleys. The so-called compound channel flow is characterized by the interaction between the deeper and faster main channel flow and the shallower and slower floodplain flow. The flow structure becomes much more complex than in a single channel. In the shear layer formed in a region near the interface, a lateral momentum exchange between the main channel and the floodplains and the secondary currents due to the non-isotropic turbulence lead to a 3D flow structure. Despite the availability of 2D and 3D flow models that may solve this question, 1D methods are often preferred due to the less amount of parameters required and to the shorter processing time. In the present paper, a new method to compute flow discharge in compound channels is proposed namely by incorporating the apparent shear stress in the Manning roughness. The method is calibrated and validated taking into account stage–discharge relationships in several compound channel facilities.

Keywords

Floods • Compound channel • Floodplain

1 Introduction

During flood events, natural rivers in alluvial valleys comprise one main channel and one or two lateral floodplains (i.e. compound channels). The latter are normally prone to inundation, causing serious disasters with considerable environmental, economic or human losses. The current study is related to the prediction of the flow discharge in these channels, which is considered of paramount importance for flood management and protection.

When the flow occurs simultaneously in the main channel and in the floodplains, a much more complex flow structure is observed. Momentum transfer between the flows in each subsection is observed (e.g. Sellin 1964). The difference between the flow velocities and water depths in these subsections leads to the generation of a mixing layer with interaction of vortices with vertical and horizontal axes, affecting the channel conveyance or transport capacity (Nezu and Nakayama 1997).

The main flow mechanisms are identified in Fig. 1. Due to this complexity, the flow discharge in a compound channel at a given flow depth is not as easily predicted as for single channels (Prinos and Townsend 1984).

The Divided Channel Method (DCM) is the traditional procedure to predict the flow discharge of a straight compound channel. It proposes the conceptual division of the compound channel into the main channel and lateral floodplains. The total discharge, Q, is calculated by the sum of the discharges of the subsections:

$$Q = \sum_i Q_i = \sum_i \frac{A_i R_i^{2/3}}{n_i} s_0^{1/2} \quad (1)$$

where A, R and n are the area, the hydraulic radius and the manning roughness for each subsection, i (i.e. main channel, *mc*, and floodplain, *fp*); s_0 is the streamwise slope.

In order to account for the head loss generated by the interactions between the flows in the main channel and in the floodplains, the Apparent Shear Stress Method, ASSM (Moreta and Martin-Vide 2010) proposes to include additional shear stress, τ_a, acting on the vertical interface between the main channel and the floodplains. The flow discharge may be calculated by the sum of the subsections discharges assuming the additional shear stress (N_f and h stand for the number of floodplains, 1 or 2 and for the water depth):

J. N. Fernandes (✉)
National Laboratory for Civil Engineering, Lisbon, Portugal
e-mail: jnfernandes@lnec.pt

© Springer Nature Switzerland AG 2020
F. Fernandes et al. (eds.), *Advances in Natural Hazards and Hydrological Risks: Meeting the Challenge*, Advances in Science, Technology & Innovation,
https://doi.org/10.1007/978-3-030-34397-2_27

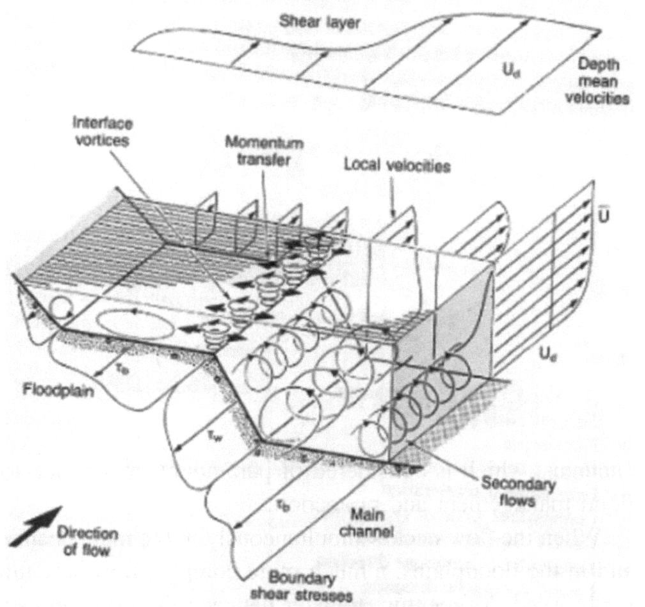

Fig. 1 Flow mechanisms in compound channel (adapted from Shiono and Knight 1991)

$$Q = Q_{mc} + Q_{fp} = A_{mc}\frac{R_{mc}^{2/3}}{n_{mc}}S_0^{1/2}\sqrt{1-\left(\frac{N_f\tau_a h_{fp}}{\rho g A_{mc} S_0}\right)} + A_{fp}\frac{R_{fp}^{2/3}}{n_{fp}}S_0^{1/2}\sqrt{1+\left(\frac{\tau_a h_{fp}}{\rho g A_{fp} S_0}\right)} \quad (2)$$

The inclusion of the apparent shear stress leads to an increase of the flow discharge in the floodplain and a decrease of the flow discharge in the main channel and in the whole cross-section as compared with the results obtained through the DCM.

In the present paper, a method to incorporate this apparent shear stress in the roughness Manning coefficient is developed in order to improve the discharge prediction in overbank flows. The calibration and validation of the proposed method are performed taking advantage of a large dataset of experimental data on compound channel flows collected in Fernandes et al. (2015).

2 Experimental Data

The dataset was gathered in Fernandes et al. (2015). All experiments were conducted under uniform flow. Figure 2 shows schematic representations of compound channels and the most relevant cross-section geometrical parameters.

The experimental data are presented in Table 1. N stands for the number of experiments.

3 Methodology

Taking into account that not whole data have information of the subsection flow discharge it was decided to divide it in: ***Dataset 1*** (for calibration)—data with experimental measurement of the flow discharge in the main channel and floodplain and ***Dataset 2*** (for validation)—Experimental data with measurement of only total discharge.

The proposed method was calibrated and validated according to the following steps:

(i) Calculation of the discharge flow in the main channel and floodplain using Divided Channel Method for the experiments of Dataset 1 using Eq. (1);

(ii) Using the following equations:

$$Q_{mc} = A_{mc}\frac{R_{mc}^{2/3}}{\phi_{mc} n_{mc}}S_0^{1/2} \quad (4)$$

$$Q_{fp} = A_{fp}\frac{R_{fp}^{2/3}}{\phi_{fp} n_{fp}}S_0^{1/2} \quad (5)$$

and the experimentally measured flow discharges in the main channel and in the floodplains, Q_{mc} and Q_{fp}, calculate the correction coefficients for the *mc* and *fp* roughness coefficients, ϕ_{mc} and ϕ_{fp} for each experiment of dataset 1.

(iii) Assuming the average of the correction coefficients, calculate the total flow discharge by the sum of the subsection discharges for experiments in dataset 2.

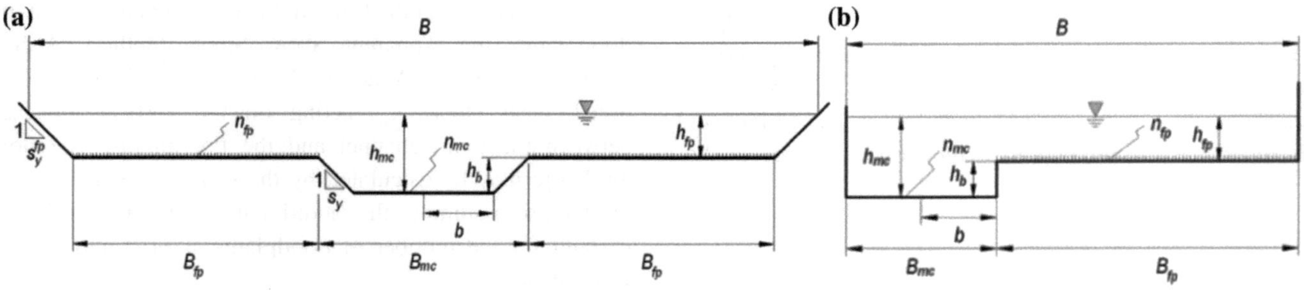

Fig. 2 Schematic cross-section of a **a** symmetric and **b** symmetric compound channel

Table 1 Summary of the experimental data

Reference	B_{fp} (m)	B_{mc} (m)	h_b (m)	s_y (−)	s_y^{fp} (−)	s_O (× 10^{-3})	h_r (−)	N
Symmetric geometry and smooth floodplains (284 experiments)								
Knight and Demetriou (1983)	0.076–0.229	0.152	0.076	0	0	0.96	0.108–0.503	18
Myers (1984)	0.18–0.3	0.16	0.08–0.12	0	0	0.93	0.067–0.535	33
James and Brown (1977)	0.572	0.279	0.051	1	1	1–3	0.002–0.311	50
James and Brown (1977)	0.191	0.279	0.051	1	1	1–3	0.008–0.383	42
James and Brown (1977)	0.191	0.279	0.051	1	1	1–3	0.020–0.423	38
James and Brown (1977)	0.502	0.381	0.069	1	1	1–3	0.011–0.315	19
Noutsopou and Hadjipanos (1983)	0.225–0.425	0.150	0.075	0	0	1.5	0.187–0.479	16
Prinos and Townsend (1984)	0.381	0.244–0.367	0.102	0.5	0	0.3	0.089–0.329	10
Wormleaton and Merret (1990)	2.25	1.80	0.15	1	0	1.03	0.041–0.500	23
Flow database FCF s8	2.25	1.80	0.15	0	0	1.03	0.050–0.500	8
Flow database FCF s10	2.25	2.10	0.15	2	0	1.03	0.051–0.464	8
Atabay (2001)	0.407	0.398	0.05	0	0	2.02	0.071–0.490	13
Fernandes (2013)	0.70	0.60	0.10	1	0	1.1	0.100–0.380	6
Symmetric geometry and rough floodplains (181 experiments)								
James and Brown (1977) T13	0.502	0.381	0.069	1	1	1–3	0.044–0.385	18
Wormleaton et al. (1982)	0.46	0.288	0.12	0	0	0.4–1.8	0.111–0.429	40
Knight and Hamed (1984)	0.076–0.229	0.152	0.076	0	0	0.97	0.104–0.518	48
Prinos and Townsend (1984)	0.381	0.244–0.367	0.102	0.5	0	0.3	0.089–0.329	30
Wormleaton and Merret (1990)	2.25	1.80	0.15	1	0	1.03	0.038–0.505	8
Hu et al. (2010)	0.35	0.30	0.06	0	0	1.0	0.341–0.528	5
Fernandes (2013)	0.70	0.60	0.10	1	0	1.1	0.150–0.300	3
Tang (1999)	0.41	0.398	0.05	0	0	2.0–2.1	0.172–0.603	29
Asymmetric geometry and smooth floodplains (33 experiments)								
Myers (1978)	0.356	0.254	0.102	0	0	0.26455	0.086–0.394	10
Flow database FCF s6	2.25	1.8	0.15	1	0	1.03	0.052–0.503	8
Atabay (2001)	0.407	0.398	0.05	0	0	2.04	0.165–0.499	8
Bousmar (2002)	0.40	0.40	0.05	0	0	0.9	0.081–0.366	4
Proust (2005)	0.80	0.40	0.05	0	0	1.8	0.219–0.412	3
Asymmetric geometry and rough floodplains (108 experiments)								
James and Brown (1977)	0.1905	0.279	0.051	1	1	1–3	0.002–0.444	36
James and Brown (1977)	0.368	0.279	0.051	1	1	1–3	0.048–0.413	43
James and Brown (1977)	0.502	0.279	0.051	1	1	1–3	0.008–0.378	29

4 Results

The Dataset 1 was used to evaluate the correction coefficients for the floodplain and main channel Manning roughness. A total of 184 experiments collected from the literature were included in this dataset. The results for the correction coefficients compared to the relative depth of each experiment are presented in Fig. 3.

It was not observed a strong correlation between the relative depth and the correction coefficients. The same type of correlation was tried for the relationship between the correction coefficients with other characteristics of the compound channel like the geometrical dimensions or the aspect ratio or cross-section shape. As the results did not reveal strong correlations with any compound channel characteristic, it was decided to assume constant correction coefficients. Their values, obtained by averaging all values, were $\Phi mc = 1.15$ and $\Phi fp = 0.91$.

The validation dataset was then used to check the advantage of considering these coefficients in comparison to the results of DCM. The results are presented in Fig. 4.

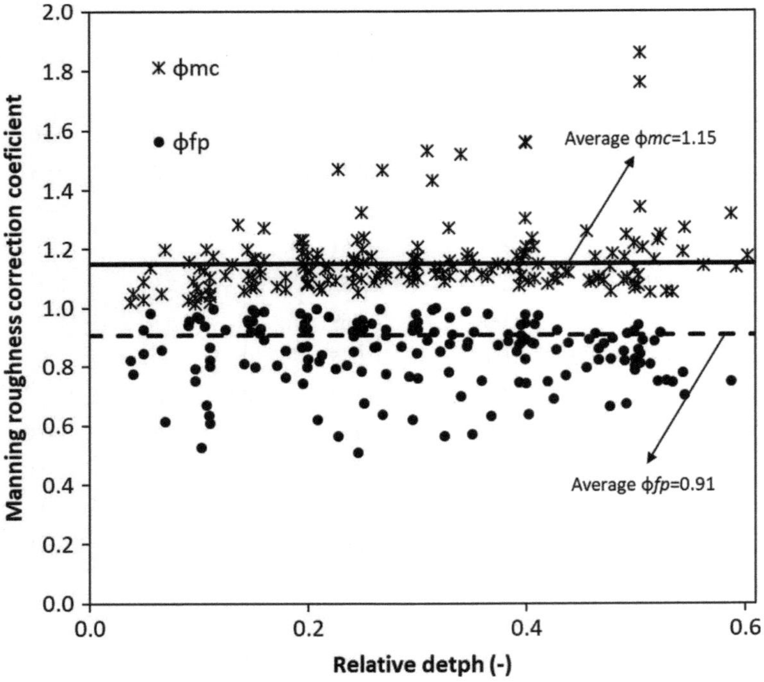

Fig. 3 Correction coefficients for main channel and floodplain manning roughness

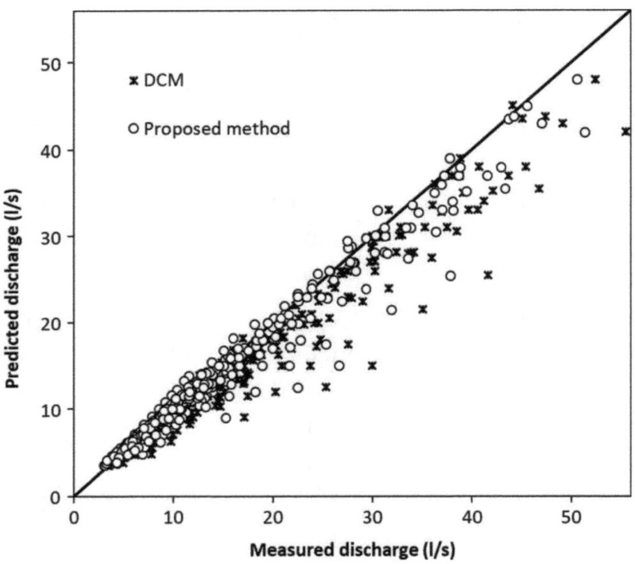

Fig. 4 Results of DCM and proposed method for dataset 2

The performance of this method in comparison to DCM is also evaluated taking into account the deviation, Δ, between each pair of calculated and measured discharge:

$$\Delta\,(\%\,) = 100 \times (Q_c - Q_m)/Q_m \tag{6}$$

where Q_m and Q_c are, respectively, the measured and the calculated flow discharge.

The result for DCM and for the proposed method were, 11.3 and 1.6%, respectively.

5 Concluding Remarks

Overbank flows are rather common in alluvial valleys during flood events. It is well known that under these conditions the flow discharge is not as easily predicted as for single channels. A new and simple method was proposed to evaluate this flow discharge in compound channels by incorporating the apparent shear stress as a correction of the Manning coefficients.

The method allowed a considerable improvement in the traditional method. Despite the promising results, further research will be conducted to understand the influence of the characteristics of the compound channel in the correction coefficients.

Acknowledgements The author acknowledges the support to the grant with reference SFRH/BPD/114768/2016 and to the Project *MixFluv—Mixing Layers in fluvial systems* (reference Lisboa-01-0145-Feder-031771) by FEDER and Science and Technology Foundation.

References

Atabay S (2001) Stage-discharge, resistance and sediment transport relationships for flow in straight compound channels. Ph.D. thesis, University of Birmingham, U.K.

Bousmar D (2002) Flow modelling in compound channels, momentum transfer between main channel and prismatic or non-prismatic floodplains. Ph.D. Thesis, Université catholique de Louvain, Louvain-la-Neuve, Belgium

Fernandes J (2013) Compound channel uniform and non-uniform flows with and without vegetation in the floodplain. Ph.D. Thesis, Universidade de Lisboa, Portugal

Fernandes J, Leal J, Cardoso A (2015) Assessment of stage-discharge predictors for compound open-channels. Flow Meas Instrum 45:62–67

Hu C, Ji Z, Guo Q (2010) Flow movement and sediment transport in compound channels. J Hydraul Res 48(1):23–32

James M, Brown B (1977) Geometric parameters that influence floodplain flow. Research report H-77-I, U.S. Army Engineer Waterways Experiment Station, Hydraulic Laboratory, Vicksburg, Missouri, U.S.A., p 141

Knight D, Demetriou J (1983) Flood plain and main channel flow interaction. J Hydraul Eng 109(8):1073–1092

Knight D, Hamed M (1984) Boundary shear in symmetrical compound channels. J Hydraul Eng 110(10):1412–1430

Moreta P, Martin-Vide J (2010) Apparent friction coefficient in straight compound channels. J Hydraul Res 48(2):169–177

Myers W (1978) Momentum transfer in a compound channel. J Hydraul Res 16(2):139–150

Myers W (1984) Frictional resistance in channels with floodplains. In: Proceedings 1st international conference channels and channel control structures, Computational Mechanics Centre, Germany

Nezu I, Nakayama T (1997) Space–time correlation structures of horizontal coherent vortices in compound channel flows by using particle-tracking velocimetry. J Hydraul Res 35(2):191–208

Prinos P, Townsend R (1984) Comparison of methods for predicting discharge in compound open channels. Adv Water Resour 7 (12):180–187

Proust S (2005) Ecoulements non-uniformes en lits composés: effets de variations de largeur du lit majeur. Ph.D. Thesis, INSA de Lyon, Lyon, France (in French)

Sellin R (1964) A laboratory investigation into the interaction between the flow in the channel of a river and that over its floodplain. La Houille Blanche 7:793–802

Shiono K, Knight DW, (1991) Turbulent open-channel flows with variable depth across the channel. J Fluid Mech 222(1):617

Tang X (1999) Derivation of the wave speed-discharge relationship from cross section survey for use in approximate flood routing methods. Ph.D. Thesis, University of Birmingham, U.K.

Wormleaton P, Merrett D (1990) An improved method of calculation for steady uniform flow in prismatic main channel/floodplain sections. J Hydraul Res 28(2):157–174

Wormleaton P, Allen J, Hadjipanos P (1982) Discharge assessment in compound channel flow. J Hydraul Div 108(9):975–994

Dispersive Effects During Long Wave Run-up on a Plane Beach

Ahmed Abdalazeez, Ira Didenkulova, and Denys Dutykh

Abstract

Dispersive effects during long wave run-up on a plane beach are studied. We take advantage of experimental data collection of different wave types (single pulses, sinusoidal waves, bi-harmonic waves, and frequency modulated wave trains) and simulate their run-up using two models: (i) nondispersive nonlinear shallow water theory and (ii) dispersive Boussinesq-type model based on the modified Peregrine system. It is shown, that for long positive pulses, dispersive effects are not so important and nonlinear shallow water theory can be used. However, for periodic sinusoidal and bi-harmonic pulses of the same period, the dispersive effects result in significant wave transformation during its propagation but do not have a strong impact on its maximal run-up height. Overall, for maximum wave run-up height, we could not find a preference of dispersive model against the nondispersive one, and, therefore, suggest using nonlinear shallow water model for longwave run-up height estimation.

Keywords

Long wave run-up • Frequency dispersion • Nonlinear shallow water theory • Modified Peregrine system

A. Abdalazeez · I. Didenkulova (✉)
Department of Marine Systems, Tallinn University of Technology, Tallinn, Estonia
e-mail: irina.didenkulova@taltech.ee

I. Didenkulova
Nizhny Novgorod State Technical University n.a. R.E. Alekseev, Nizhny Novgorod, Russia

D. Dutykh
Univ. Grenoble Alpes, Univ. Savoie Mont Blanc, CNRS, LAMA, 73000 Chambéry, France

© Springer Nature Switzerland AG 2020
F. Fernandes et al. (eds.), *Advances in Natural Hazards and Hydrological Risks: Meeting the Challenge*, Advances in Science, Technology & Innovation,
https://doi.org/10.1007/978-3-030-34397-2_28

1 Introduction

There are several reasons why the nonlinear shallow water theory (NLSW) is favored for long wave run-up calculations as compared to dispersive wave models, often represented by Boussinesq-type approximations. First of all, wave run-up calculated using dispersive codes is prone to numerical instabilities, which make computations more sensitive to numerical parameters (Bellotti and Brocchini, 2002). Second, the Boussinesq terms in dispersive models tend to zero at the shoreline, so that dispersive equations simplify to NLSW in this region (Madsen et al. 1997).

Horrillo et al. (2006) studied dispersive effects during 2004 Indian Ocean tsunami propagation by comparing NLSW with the fully nonlinear Navier–Stokes equations (FNS). They came to conclusion that, NLSW is more suitable in hazard assessments (e.g. it has a very low computation cost and often it over-predicts the maximum wave run-up height, whereas this property is considered a safety factor). Moreover, Glimsdal et al. (2013) suggested that NLSW is more appropriate for warning purposes. However, dispersive effects become more important in trailing waves, whereas the leading waves can be well described by NLSW (Løvholt et al. 2012). Note that maximum wave is often not the first one, at least for tsunamis propagating over a long distance, see, for example, Candella et al. (2008).

Most of the mentioned studies were based on numerical results and were missing the fidelity control mechanism. Therefore, in this paper we take an advantage of available experimental data collection of different wave types (single pulses, sinusoidal waves, bi-harmonic waves, and frequency modulated wave trains) and simulate their run-up using two models: (i) NLSW and (ii) dispersive model of Boussinesq-type based on the modified Peregrine system (mPer).

The paper is organized as follows. In Sect. 2, we describe the available experimental dataset. The numerical models (NLSW and mPer) are briefly described in Sect. 3. In

Sect. 4, we compare numerical results of NLSW and mPer models with the experimental data. The main results are summarized in Sect. 5.

2 Experimental Data

The experiments were conducted in the Large Wave Flume (GWK), Hannover, Germany. The experimental set up consisted of a 251 m long section of constant depth 3.5 m and a plane beach with a slope angle 1:6 (fixed asphalt bed). There were from 16 to 18 wave gauges recording wave propagation along the flume. The wave run-up was measured by capacitance probe, which was supplemented by two regular video cameras. Waves were generated by the piston type wave maker, which showed to be efficient for long wave generation (Schimmels et al. 2016). The generated waves are listed in Table 1. Details of this experiment can be found in Didenkulova et al. (2013).

3 Numerical Set up

In the present work, two different numerical models are used: the nondispersive nonlinear shallow water model (NLSW) and dispersive Boussinesq-type model based on the modified Peregrine system (mPer). Both models were set not taking into account bottom friction and assumed the fluid being perfect and the flow to be incompressible and irrotational.

The model left boundary conditions corresponded to the experimental wave conditions in Table 1. On the right, we placed a sufficiently long plane beach to avoid any interactions with the right boundary.

More precisely, the bathymetry in numerical experiments was set up to reproduce the Large Wave Flume conditions:

$$h(x) = \begin{cases} h_0, & x \in [a, b] \\ h_0 - (x - b)\tan\alpha, & x \in [b, c] \end{cases}, \tag{1}$$

where h_0 = 3.5 m is the constant water depth, α is the bottom slope ($\tan\alpha$ = 1:6). The distance $x \in [a, b]$ is the cell interfaces, x_c is the centre of a cell, $[a, c]$ are left and right boundaries of the numerical flume (a = 0 m), and b = 251 m is the point where the slope starts. The distance $x \in [a, b]$ has been divided into number of cells $c_i = (X_{i-1/2}, X_{i+1/2})$, where X_i is the center of cell i, $X_i = 1/2(X_{i-1/2} + X_{i+1/2})$.

The numerical time steps can be calculated as $\Delta t = \left(\frac{t_f - t_i}{M}\right)$, where t_i is the initial time, t_f is the final time and M is the number of time steps.

3.1 Nonlinear Shallow Water (NLSW) Model

The 1D nonlinear shallow water equations are:

$$H_t + (Hu)_x = 0, \tag{2}$$

$$(Hu)_t + \left(Hu^2 + \frac{g}{2}H^2\right)_x = gHh_x, \tag{3}$$

where $H = h + \eta$ is the total water depth, $\eta(x, t)$ is the water elevation with respect to the still water level, $u(x, t)$ is the depth-averaged flow velocity, $h(x)$ is an unperturbed water depth described by Eq. (1), g is the gravitational acceleration, x is the coordinate directed onshore, and t is time. We use the finite volumes method. The numerical scheme is

Table 1 Generated waves and their run-up heights

Type of waves	Wave period (s)	Initial wave amplitude (m)	Experimental run-up (m)	NLSW run-up (m)	mPer run-up (m)
Positive pulse	20	0.10	0.259	0.268	0.254
Positive pulse	20	0.24	0.795	0.840	0.780
Sine wave	20	0.05	0.096	0.13	0.14
Sine wave	20	0.60	2.27	1.80	1.81
Bi-harmonic wave	20	0.12	0.79	0.89	0.85
Bi-harmonic wave	20	0.15	1.3	1.37	1.32
Wake-like train	20 → 10	0.10	0.46	0.60	0.51
Wake-like train	20 → 10	0.40	2.14	1.68	2.57

based on the second-order UNO2 reconstruction, for more details see (Dutykh et al. 2011).

3.2 Modified Peregrine (mPer) Model

The Boussinesq equations for long dispersive wave propagation, derived by Peregrine (1967), are:

$$\eta_t + ((h+\eta)u)_x = 0, \tag{4}$$

$$u_t + uu_x + g\eta_x - \frac{h}{2}(hu)_{xxt} + \frac{h^2}{6}(u_{xxt}) = 0. \tag{5}$$

This classical Peregrine system was modified by Durán et al. (2018) in order to recover the conservative form of equations.

Equation (4) of the mass conservation in new variables becomes:

$$H_t + Q_x = 0, \tag{6}$$

where Q is the horizontal momentum, H is the total water depth.

The momentum conservation equation obtained from Eq. (5) becomes (Durán et al. 2018):

$$\begin{aligned} &\left(1 + \frac{1}{3}H_x^2 - \frac{1}{6}HH_{xx}\right)Q_t - \frac{1}{3}H^2 Q_{xxt} \\ &- \frac{1}{3}HH_x Q_{xt} + \left(\frac{Q^2}{H} + \frac{g}{2}H^2\right)_x = gHh_x \end{aligned} \tag{7}$$

Equations (6) and (7) are called the modified Peregrine equations and are studied in detail in Durán et al. (2018).

4 Results

The two models described above have been used to reproduce the Large Wave Flume experiments listed in Table 1. The water surface elevation has been recorded by different wave gauges at different distances from the wave maker, including wave run-up. The comparison simulated by NLSW and mPer models against measured experiments. The maximum wave run-up heights are shown in Table 1. For periodic waves, the comparison is made for experimentally measured wave with the maximum run-up height. The deviations of computational run-up heights from the measured ones are shown in Table 2.

It can be noticed that for long positive pulses results of both NLSW and mPer models are in good agreement with experimental data (Fig. 1). In the weak amplitude case NLSW overestimates the experimental result by 3%, while mPer underestimates it by 2%. For the initial wave of higher amplitude, the discrepancy is also higher: 6% for NLSW and 2% for mPer (Table 2). Therefore, one can conclude that the dispersive effects are not so important to predict the run-up height for the class of long single waves of positive polarity and the NLSW can be used.

For dispersive wake-like trains and even for periodic waves (sine waves and bi-harmonic waves) the situation is quite different. Most of the results of both NLSW and mPer have an accuracy of 20–30% and do not really show the preference of one model against the other. Note, most of wave gauges located along the basin demonstrate the importance of dispersive effects, which are captured by mPer model and ignored by NLSW, but according to Table 2 they do not have principal impact on maximum run-up height.

Table 2 Deviations of computational run-up heights from the measured ones in %

Type of waves	Initial wave amplitude (m)	Experimental run-up (m)	NLSW (%)	mPer (%)
Positive pulse	0.10	0.259	3	−2
Positive pulse	0.24	0.795	6	−2
Sine wave	0.05	0.096	35	45
Sine wave	0.60	2.27	−21	−20
Bi-harmonic wave	0.12	0.79	13	8
Bi-harmonic wave	0.15	1.3	5	2
Wake-like train	0.10	0.46	30	11
Wake-like train	0.40	2.14	−22	20

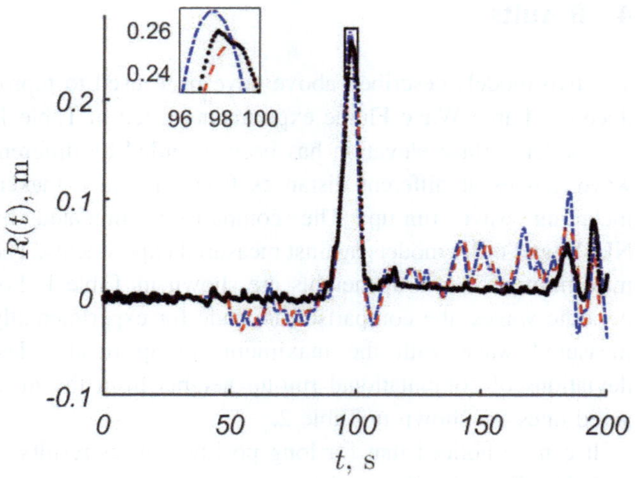

Fig. 1 Run-up height time-series for a positive pulse of $A = 0.1$ m, $T = 20$ s, calculated by mPer (red dashed line), NLSW (blue dash-dotted line) and measured experimentally (black dots)

This somehow supports the existing general opinion that for the maximum wave run-up estimation NLSW is enough.

5 Concluding Remarks

We used the nonlinear shallow water model (NLSW) and dispersive Boussinesq-type model based on the modified Peregrine system (mPer) to reproduce physical experiments of long wave propagation and run-up. The studied waves were of different shapes and amplitudes and included long single pulses of positive polarity, periodic sine and bi-harmonic waves and dispersive wake-like wave trains.

It is found, that in our wave collection, the single pulses had the best agreement with both models and dispersive effects, in this case, were negligible. For periodic waves and dispersive wave trains, the dispersive effects were important, and mPer model gave much better fit to the records of tide-gauges, located along the flume. However, for maximum run-up height this better capture of dispersive effects did not play a big role, and both models resulted in 20–30% deviation from the experimental data.

Acknowledgements Numerical simulation of wake-like wave trains was carried out with a financial support from Russian Science Foundation grant 16-17-00041, simulation of other wave types was supported by ETAG grant PUT1378. The experimental data were obtained within Hydralab IV Grant HyIV-FZK-03. Authors also thank the PHC PARROT project No 37456YM, which funded the authors' visits to France and Estonia and allowed this collaboration.

References

Bellotti G, Brocchini M (2002) On using Boussinesq-type equations near the shoreline: a note of caution. Ocean Eng 29(12):1569–1575

Candella RN, Rabinovich AB, Thomson RE (2008) The 2004 Sumatra tsunami as recorded on the Atlantic coast of South America. Adv Geosci 14:117–128

Didenkulova I, Denissenko P, Rodin A, Pelinovsky E (2013) Effect of asymmetry of incident wave on the maximum runup height. J Coastal Res 65:207–212

Durán A, Dutykh D, Mitsotakis D (2018) Peregrine's system revisited. In: Abcha N, Pelinovsky E, Mutabazi I (eds) Nonlinear waves and pattern dynamics. Springer, Cham, pp 3–43

Dutykh D, Katsaounis T, Mitsotakis D (2011) Finite volume schemes for dispersive wave propagation and runup. J Comput Phys 230 (8):3035–3061

Glimsdal S, Pedersen GK, Harbitz CB, Løvholt F (2013) Dispersion of tsunamis: does it really matter? Nat Hazards Earth Syst Sci 13:1507–1526

Horrillo J, Kowalik Z, Shigihara Y (2006) Wave dispersion study in the Indian Ocean-tsunami of December 26, 2004. Mar Geodesy 29 (3):149–166

Løvholt F, Pedersen G, Bazin S, Kühn D, Bredesen RE, Harbitz C (2012) Stochastic analysis of tsunami runup due to heterogeneous coseismic slip and dispersion. J Geophys Res Ocean 117:C03047

Madsen PA, Banijamali B, Schäffer HA, Sørensen OR (1997) Boussinesq type equations with high accuracy in dispersion and nonlinearity. In: Edge B (ed) 25th conference 1996, ASCE, vol 1. Coastal Engineering, Hørsholm, pp 95–108

Peregrine DH (1967) Long waves on a beach. J Fluid Mech 27: 815–827

Schimmels S, Sriram V, Didenkulova I (2016) Tsunami generation in a large scale experimental facility. Coast Eng 110:32–41

Analysis of Nonlinear Wave Parameters on Ofir Sandy Beach (NW Portugal)

Tiago Abreu, Paulo A. Silva, Paulo Baptista, Joaquim Pais-Barbosa, Sandra Fernández-Fernández, Caroline Ferreira, and João Matos

Abstract

The characterization of wave transformation processes in the nearshore is of paramount importance when it comes to assessing storm and flooding impacts, sediment transportation and deposition, harbors safety or design of coastal protective structures. This study analyzes nonlinear wave parameters on Ofir sandy beach. This beach is located along the northwest Portuguese coast which is a highly energetic coast exposed to waves generated far away in the Atlantic Ocean. Despite the existence of rocky outcrops in the nearshore and intertidal zones at the study site, reducing the wave energy that reaches the beach, the study site exhibits pronounced erosive processes. Field observations of six near-bottom pressure records collected at the intertidal zone help to characterize the evolution of wave nonlinearities which are directly associated with sediment transport mechanisms. Data results show that there is an interrelation between the characteristics of the waves and the local morphology. It is also possible to ascertain, more clearly, the level of asymmetry present in the waves propagated at different depths, contributing to a better understanding of the local morpho-hydrodynamics.

Keywords

Nonlinear waves • Sandy beach • Relative wave height • Skewness • Asymmetry

1 Introduction

It is recognized that an accurate description of the wave phenomena and the wave governing mechanisms are of paramount importance for a good understanding of the morphodynamics of the coastal zone. In particular, within the context of coastal erosion, there is the need to collect data with sufficiently long records to establish wave conditions for engineering purposes. This kind of monitoring helps to improve our knowledge of sediment transport and it is obviously a good contribution to an integrated coastal management strategy (e.g., Ferreira and Matias, 2013).

The highly energetic northwest coast of Portugal faces severe erosion and Ofir is one of the example areas under threat, facing serious problems (Lira et al. 2016). Despite the existence of rocky outcrops in the nearshore and intertidal zones, that reduce the wave energy reaching the beach, the study site exhibits pronounced erosive processes.

This study analyzes nonlinear wave parameters on Ofir sandy beach based on a field survey data, where local hydrodynamic conditions and beach topography were recorded. Local wave gauge measures of the sea surface were obtained through bottom-mounted pressure sensors. In total, six near-bottom pressure transducers were deployed at different depths, collecting data at the intertidal zone. The instruments help to characterize the evolution of nonlinearities associated with nearshore wave transformations, which is intrinsically correlated with sediment transport mechanisms (e.g., Sancho et al. 2011).

T. Abreu (✉)
Civil Engineering Department, School of Engineering (ISEP), Polytechnic of Porto, Porto, Portugal
e-mail: taa@isep.ipp.pt

T. Abreu · P. A. Silva · P. Baptista · J. Pais-Barbosa · S. Fernández-Fernández · C. Ferreira
CESAM-Centre for Environment and Marine Studies, University of Aveiro, Aveiro, Portugal

P. A. Silva · S. Fernández-Fernández · C. Ferreira
Department of Physics, University of Aveiro, Aveiro, Portugal

J. Pais-Barbosa
Universidade Lusófona do Porto, Porto, Portugal

P. Baptista
Department of Geosciences, University of Aveiro, Aveiro, Portugal

J. Matos
Department of Mathematics, School of Engineering (ISEP), Polytechnic of Porto, Porto, Portugal

© Springer Nature Switzerland AG 2020
F. Fernandes et al. (eds.), *Advances in Natural Hazards and Hydrological Risks: Meeting the Challenge*, Advances in Science, Technology & Innovation,
https://doi.org/10.1007/978-3-030-34397-2_29

2 Study Site

2.1 Ofir Beach

Ofir beach is located between Cávado river mouth (Esposende city), at north, and Apúlia beach, at south (Fig. 1). The area is composed by palaeozoic schists and quartzite outcrops with an approximately NW-SE orientation. The nearshore area is clearly influenced by these outcrops due to a large stand between 5 and 10 m isobaths. Furthermore, at Ofir beach, the isobath of 5 m is 800 m far away of shoreline. Predominant waves come from WNW to NNW and have a mean significant wave height of 1.5–2.5 m and a peak period of 8–11 s. The existence of rocky outcrops in large areas of the nearshore and intertidal zone at the study site influences the wave characteristics, reducing the wave energy reaching the beach. The tidal regime is semidiurnal. Beach sediments are composed by medium sand (d_{50} = 0.34 mm) with significant amount of mafic minerals representative of the rocky outcrops (Fernández-Fernández et al. 2016).

2.2 Field Survey

The field survey was conducted on October 4th, 2013, during a complete tidal cycle from 9 h 06 min to 21 h 29 min (local time) with a tidal range of 2.85 m. Two shore-normal arrays of pressure transducers (PT—Level TROLL 500 In Situ) were installed at approximately 0.10 m above the sand bed at the intertidal zone and small changes of around 0.02 m were observed at the end of the tidal cycle. The PT was attached to metallic structures and recorded near-bottom pressure time-series with a sampling frequency of 2 Hz. The PT horizontal positions (cross-shore coordinate x) and the vertical coordinate z referred to the high tide level as $z = 0$ are shown in Fig. 2. Both PT1 and PT4 were installed at the seaward-most locations and although in different shore-normal arrays, they practically possess the same cross-shore position.

The offshore wave conditions (significant wave height, H_s, wave peak period, T_p, and direction) were recorded at the Leixões wave buoy. This buoy is located approximately 28 km SW from the study site, at a depth of 83 m

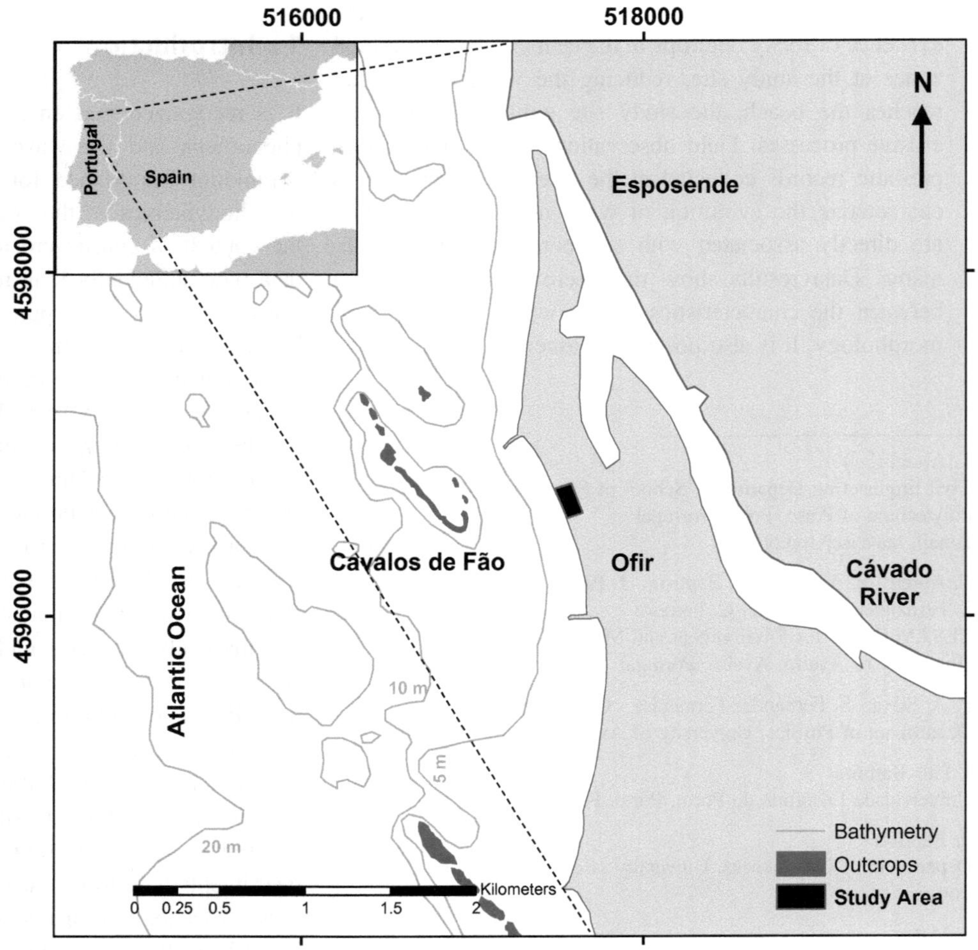

Fig. 1 Study area

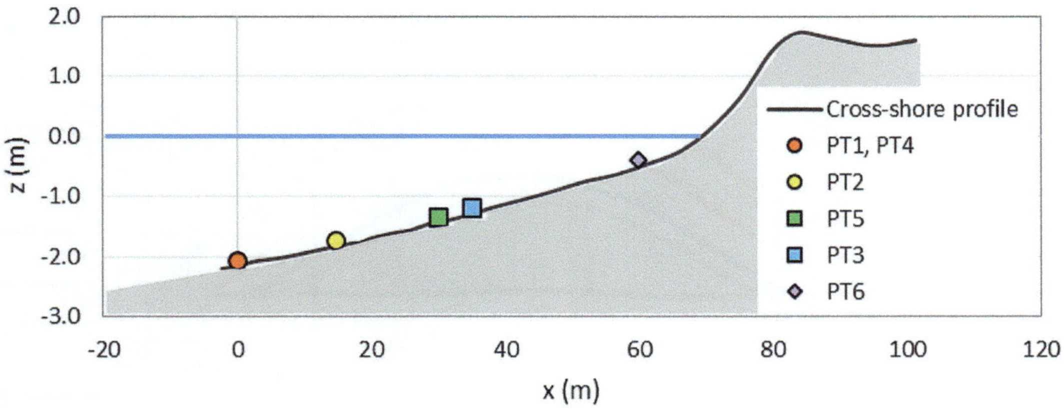

Fig. 2 Ofir beach cross-shore profile and pressure transducer (PT) locations

(41.31 N, 8.95 W) and is operated by the Portuguese Hydrographic Institute (IH). The offshore wave climate showed incident waves from NW, with a mean value of H_i of 1.5 m, during the survey period and an average T_p of 8.0 s.

The INSHORE (INtegrated System for High Operational REsolution in shore monitoring) system was used to obtain the topography of the beach (Baptista et al. 2011). Two surveys of the site were made, at both low tides of that day.

3 Results

All the information gathered with the PT when they were immersed, in the shoaling and surf zones, was used to characterize the wave data at the beach during this tidal cycle. The pressure data measured with the PT were corrected taking into account the atmospheric pressure values measured when the sensor was dry (at low tide). To compute the water depth, the distances of the sensor to the bed were also taken into account. Because of the attenuation of wave-induced pressure fluctuations with increasing depth and frequency, a pressure response correction was considered to compute the sea surface elevation.

The water depth at each instant was computed from the measured PT sea surface water level, by applying a Butterworth low-pass filter with a cut-off frequency of 0.05 Hz. The remaining signal at high frequencies consists of short gravity waves. Both records were divided into 20-min intervals from which the mean water depth, h, and both H_s and T_p were computed. The data corroborates the offshore T_p of 8.0 s but the mean H_s value at high tide is 0.8 m, falling to about half of the offshore value ($H_s \approx 1.5$ m). This difference reveals a strong attenuation of wave energy, which is expected to occur in Ofir Beach as a consequence of the natural outcrops (e.g., Fernández-Fernández et al. Silva 2016).

Figure 3 shows the water depth obtained for each pressure transducer (PT) and the corresponding significant wave height to water depth ratio ($\gamma = H_s/h$). As can be seen in Fig. 3a, the landward-most PT (PT6) was located above the other devices and recorded mainly data in the swash zone. Therefore, it was cut-off from the PT records and it is not considered in the present analysis. For the other PT, the comparison of both panels in Fig. 3, allows to estimate at which time the data should not be considered in the analysis. For example, the peak values of water depth ratio for PT1 or PT2 (see Fig. 3b) indicate that the shoaling and surf zone periods is between 11 h 25 min and 19 h 31 min. Immediately before or after, the devices are in the swash zone or the tide level did not reach the instruments.

The relative wave height γ presented in Fig. 3b is one of the most important parameters in the nearshore. Indeed, most of the computational models rely on linear wave theory, aiming to predict surf zone hydrodynamics. Inside the surf zone, one expects a decrease in wave height, to a point where it can be described as a linear function of the local depth (e.g., Poate et al. 2018). Upper bound values of γ were extensively studied, both at breaking (the so-called breaker index γ_b) and non-breaking waves (γ) and its limits are of obvious significance for engineering purposes. The central part of Fig. 3b, corresponding to the high tide, shows that the values of γ increase for decreasing depths from about 0.38 (PT1) to 0.68 (PT3). Similar reasoning can be observed along time when the highest γ values are due to broken waves that start to reach the devices and its values reduce when the water depth increases. This is consonant with breaking characteristics since the wave propagation phenomenon exerts its influence on the relative wave heights γ inside the surf zone (Masselink, 1993).

Figure 4 shows two other parameters calculated from the acquired sea surface elevation, η, that measure the wave shape transformation as waves shoal and break into the nearshore for PT1 and PT4. These parameters are the

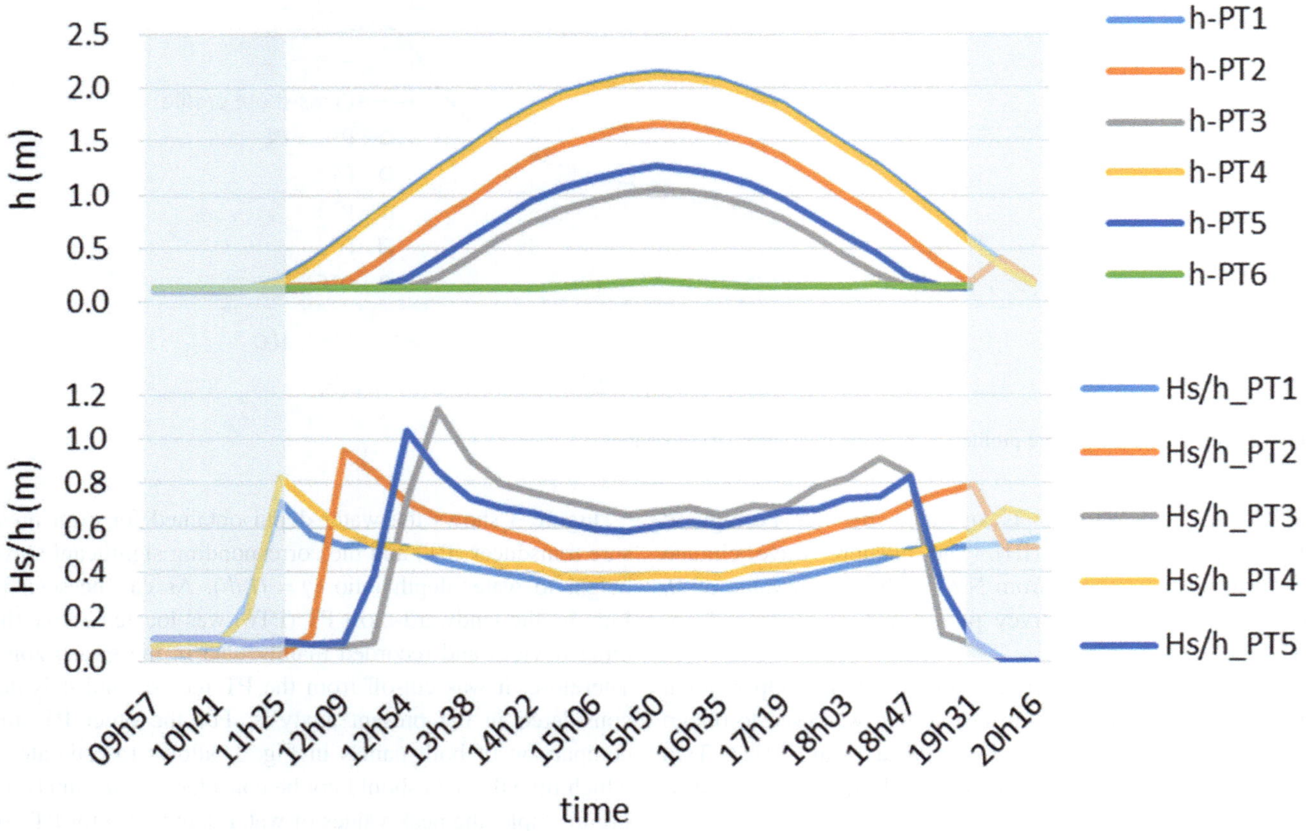

Fig. 3 **a** Depth obtained for each pressure transducer (PT) location and **b** corresponding significant wave height to water depth ratio

skewness coefficient ($R = \eta_{max}/(\eta_{max} - \eta_{min})$) and the "acceleration skewness parameter" ($\alpha = 2T_{pc}/T$), where T is the wave period η_{max} and η_{min} are the η values at the crest and wave trough, respectively, and T_{pc} is the time interval measured from the zero up-cross point to η_{max}. Respectively, these parameters are linked with the so-called skewness *Sk* (which is a measure of horizontal wave shape mutation, i.e., crest-trough differences in a wave) and asymmetry *As* (which is a measure of vertical wave shape transformation, i.e., left-right differences in a wave).

Looking to the variation of R and α for PT1 and PT4, a better insight on the phenomena can be provided. For symmetric oscillatory motions, R equals 0.5, whereas when the magnitude of velocity at the crest is larger than that at the trough one obtains $R > 0.5$. It is clearly seen that the value of R increases with depth reduction, h.

The values of α are also different than 0.5, which is the representative value of a linear (sinusoidal) motion. The observed differences mean that the depth reduction is associated with nonlinear transformations, the wave is pitching forward (sawtooth shape) and the acceleration skewness is playing an important role in sediment transport (Sancho et al. 2011).

4 Concluding Remarks

This study analyzes nonlinear wave parameters on Ofir sandy beach, assessing systematic changes in the wave shape as they shoal and break into the nearshore. The data was collected during a complete tidal cycle at the intertidal zone during a field survey, using six near-bottom pressure transducers. The instruments help to characterize the evolution of nonlinearities associated with nearshore wave transformations, which is intrinsically correlated with sediment transport mechanisms.

The analysis of the relative wave height discloses that its values increase for decreasing depths, revealing an interrelation between the characteristics of the waves and the local morphology. It is also possible to ascertain, more clearly, the

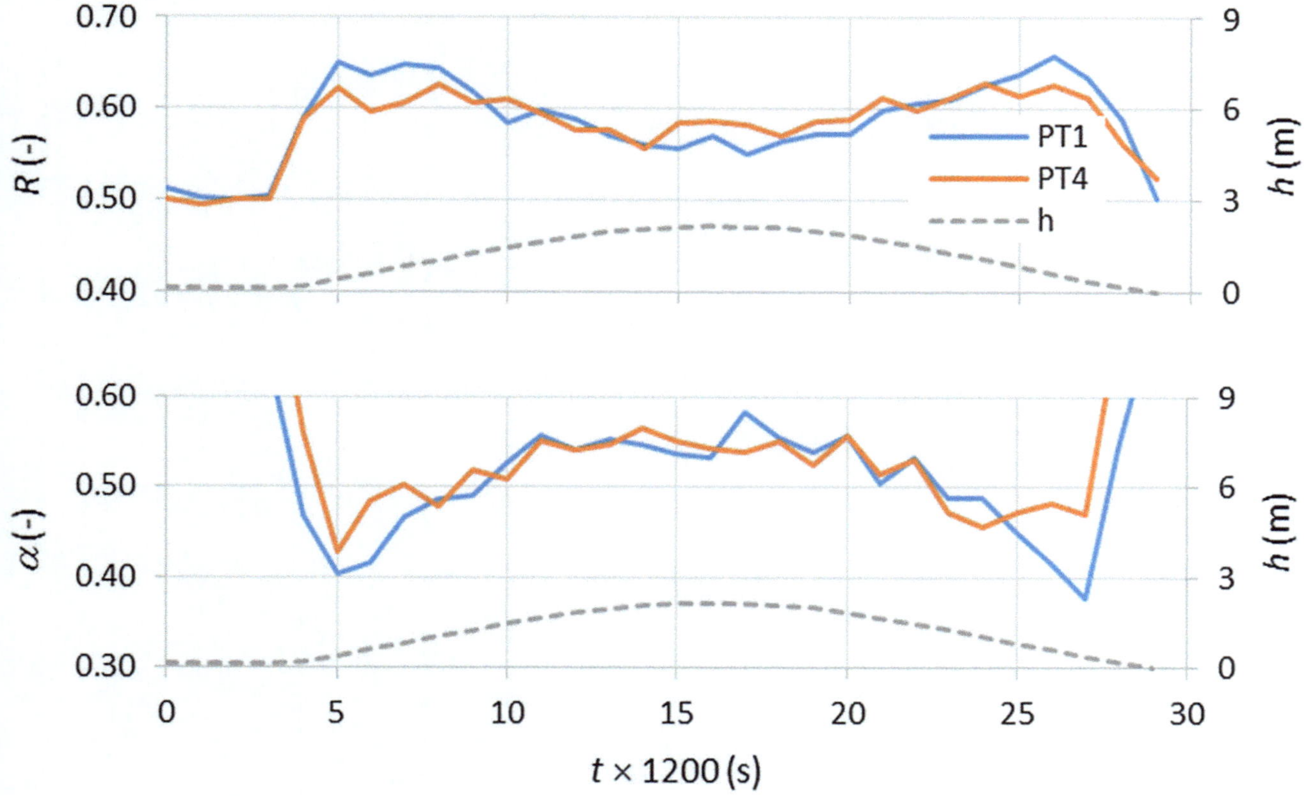

Fig. 4 **a** Skewness coefficient for PT1 and PT4 and **b** "acceleration skewness parameter" for PT1 and PT4

level of skewness and asymmetry present in the waves propagated at different depths, contributing to a better understanding of the local morpho-hydrodynamics. The results are analyzed for the complete tidal cycle highlighting an increase of the nonlinearities with the reduction of depth.

Acknowledgements Thanks are due to the financial support to CESAM (UID/AMB/50017/2019), to FCT/MEC through national funds, and the co-funding by the FEDER, within the PT2020 Partnership Agreement and Compete 2020. This work is a contribution to project SANDTRACK (POCI-01-0145-FEDER-031779) funded by FEDER, through COMPETE2020 - Programa Operacional Competitividade e Internacionalização (POCI), and by national funds (OE), through FCT/MCTES.

References

Baptista P, Cunha T, Bernardes C (2011) The validation analysis of the INSHORE system: a precise and efficient coastal survey system. Environ Monit Assess 179(1–4):589–604

Fernández-Fernández S, Baptista P, Martins V, Silva, PA, Abreu T, Pais-Barbosa J, Bernardes C, Miranda P, Rocha M, Santos F, Bernabeu A, Rey D (2016) Longshore transport estimation on Ofir beach (NW Portugal): sand tracers experiment. J Waterw Port Coast Ocean Eng 142(2), 04015017:1-14

Ferreira O, Matias A (2013) Coastal erosion and protection in Europe, 1st edn. Pranzini E, Williams A, London

Lira CP, Silva AN, Taborda R, Andrade CF (2016) Coastline evolution of Portuguese low-lying sandy coast in the last 50 years: an integrated approach. Earth Syst Sci Data 8:265–278

Masselink G (1993) Attenuation coefficients on natural beaches. In: 11th Australasian conference on coastal and ocean engineering, 105–111

Poate T, Masselink G, Austin MJ, Dickson M, McCall R (2018) The role of bed roughness in wave transformation across sloping rock shore platforms. J Geophys Res Earth Surf 123(1):97–123

Sancho F, Abreu T, D'Alessandro F, Tomasicchio GR, Silva PA (2011) Surf hydrodynamics in front of collapsing coastal dunes. J Coast Res SI 64(1):144–148

Morphodynamics of an Embayed Beach in Majorca Island

Tiago Abreu, Benjamín Parreño-Mas, and José Pinto-Faria

Abstract

Cala Millor Beach is located in Mallorca Island and is a small sandy beach, embedded in a rocky bay. It remained generally in a state of equilibrium but, in the 60s, the urbanistic area began to develop towards the beach, interfering with the natural sediment transport of the bay. The seabed is covered by native vegetation known as Posidonia Oceanica, playing an important role in the morphodynamic system. This vegetation is responsible for the lamination of the waves through energy dissipation. To characterize the local dynamics, the SMC program (System of Coastal Modelling) is employed to study sediment transport trends and to evaluate morphological evolutions in the short term. The results seem to support the observed problem, resulting from an imbalance of sediment transport due to the drift from north to south. Moreover, the results support the idea that the morphological variations can be significantly attenuated in the presence of Posidonia Oceanica. This is of great importance for coastal management since the influence of the vegetation can be assessed, improving our knowledge on the coastal morphology and enabling to identify ways to protect the beach.

Keywords

Coastal management • Sediment transport • Hydrodynamic • Morphodynamics • Vegetation effects

T. Abreu (✉) · J. Pinto-Faria
Civil Engineering Department, School of Engineering (ISEP), Polytechnic of Porto, Porto, Portugal
e-mail: taa@isep.ipp.pt

T. Abreu
CESAM-Centre for Environment and Marine Studies, University of Aveiro, Aveiro, Portugal

B. Parreño-Mas
Universidad Politecnica de Valencia, Valencia, Spain

© Springer Nature Switzerland AG 2020
F. Fernandes et al. (eds.), *Advances in Natural Hazards and Hydrological Risks: Meeting the Challenge*, Advances in Science, Technology & Innovation,
https://doi.org/10.1007/978-3-030-34397-2_30

1 Introduction

Cala Millor Beach is an urban beach located in Majorca Island, which is one of the Balearic Islands of an archipelago placed in the western Mediterranean Sea. The beach presents a high occupation during the summer months. The economy of the area depends mostly on maintenance, in terms of comfort and quality, of both water and rest areas for this type of sandy beaches. The beach is embedded in a rocky bay and possesses a fine sandy layer, resting on a rocky substrate that continuously emerges and prevents recreational use of the beach for long periods. Sometimes that also takes place during the bathing season, which consequently causes damages to the tourism sector and business-related services.

In the past decades, the urbanistic area began to develop towards the beach, destroying practically the entire pre-existing dune system. This modified the natural equilibrium beach state, making the beach more reflective and vulnerable than before. Some solutions could be adopted to solve this problem, but it is important to bear in mind that, close to the shore, the sea bed is covered by native vegetation known as Posidonia Oceanica (Abadiea et al. 2018). It is recognized that this vegetation has an important role in the system, but few studies were done to evaluate its influence on the coastal morphology (Guillén et al. 2013). This kind of vegetation is responsible for the lamination of the waves through energy dissipation and different models that deal with this phenomenon can be found in the literature (e.g., Kobayashi et al. 1993; Lima et al. 2006; Sánchez-González et al. 2011).

In this work, it is intended to study the morphodynamic of the beach, considering representative wave regimes for that location. For the simulations, the SMC program (System of Coastal Modelling) was used, which integrates a series of numerical models and allows to model the study site. The results of this work enable a better understanding of Cala Millor beach morphology, taking into consideration the wave climate and, also, the roughness effects caused by

Posidonia Oceanica. This is of great importance for coastal management since the influence of the vegetation could be assessed, improving the knowledge on the coastal morphology and enabling to identify forms of intervening on that beach.

2 Study Site

2.1 Cala Millor Beach

The municipality of Son Servera is located in the eastern part of the island of Mallorca, belonging to the largest island in the Balearic Islands (see Fig. 1). Son Servera's economy was based primarily on rich agriculture, livestock and fishing. However, nowadays, the tourism is the actual main source of income in this municipality and Son Servera has become one of the tourist centers of the island. Besides tourism, other different activities are rare in Son Servera.

The Cala Millor Beach is also called "Arenal de Son Servera" and it is located in the south bay area of Son Servera. The beach is embedded in 1600 m length and 35 m wide, being orientated NNE-SSW. The beach is composed of two distinct layers: a first sand stratum that lies on a second rock stratum. The sand essentially consists of bioclastic sand. The coarse fraction does not exist and the average of the particle size is uniform in all the studied samples, with a median grain size $d_{50} = 0.33 \pm 0.3$ mm (Gómez-Pujol et al. 2011).

2.2 Hydrodynamics

To characterize the hydrodynamics affecting Cala Millor it is necessary to know the local wave climate. For that purpose, the data obtained from the Capdepera buoy was analyzed. The buoy is located at a depth of 48 m (3.49° E, 39.65° N). Over the past decade, the buoy registered the beach is mainly affected by waves coming from the first and second quadrants, where swells from the first quadrant possess a higher frequency and present an average direction of 18° (N18E). The second quadrant presents lower frequencies with an average direction of approximately 117° (S63E).

From the buoy data, it is also possible to analyze the fluctuations of the wave climate over the past three decades. Figure 2 shows the annual/seasonal variability in terms of significant wave height (H_s) and wave peak period (T_p). Some usual variability associated with the Mediterranean climate (defined by hot, dry summers and rainy winters). The figure presents the average values as well as their fluctuations regarding the standard deviation around the average. During summer months (Jun–Aug.), the wave height and the peak period are smaller in comparison to colder months (Dec.–Feb.). On average, we can adopt a value $H_s \approx 1.0$ m and $T_p \approx 6.2$ s.

3 Results

To model the study site regarding the hydrodynamics and sediment transport, the SMC program was used. This program was developed by the *Grupo de Ingeniería Oceanográfica y de Costas* (GIOC) of Cantabria University. The program includes a series of numerical models, allowing the hydro-morphodynamic modelling of the study site. BACO model is used to incorporate the local bathymetry.

In this model, it is possible to couple different conditions of sea waves and to simulate its spread on the bathymetry (OLUCA model). It is also possible to combine these results with MOPLA model, enabling morphodynamic computations.

For the simulation, one selected the data from Capdepera buoy mentioned in the previous section. Because it is recognized that the erosion is directly associated to the wave height, two situations were chosen for winter months where the biggest wave heights were recorded (Fig. 2—February). The monthly average value and such value aggravated by the standard deviation were considered. The corresponding peak period was also taken into account. Since the predominant propagation comes from NNE and SSE, both directions have been considered for the simulations.

Figure 3 shows sediment transport estimates obtained by the SMC program for a sandy bottom. The difference between the two panels lies on the wave direction related to the two dominant directions registered for the 1st and 2nd quadrants. It is noted that these results do not include effects associated with the vegetation seabed or tidal variations. The comparison reveals a clear imbalance of the sediment transport, reflecting the differences whenever the waves are originated in the first or second quadrant. This difference is evidenced by the magnitude of sediment transport in both situations, resulting in a pronounced transport to the south. This imbalance between north and south can lead to an annual net loss of sediments in the bay, resulting in the disappearance of the sand substrates of the southern dry beach area.

4 Discussion

To address the roughness effects associated to Posidonia Oceanica, other runs were performed with SMC. The area covered by the seagrass is extended from 5 to 30 m depth, reaching in some areas up to 35 m. The computational meshes were adapted to easily implement and meet the spatial arrangement of the seabed vegetation. Therefore, the

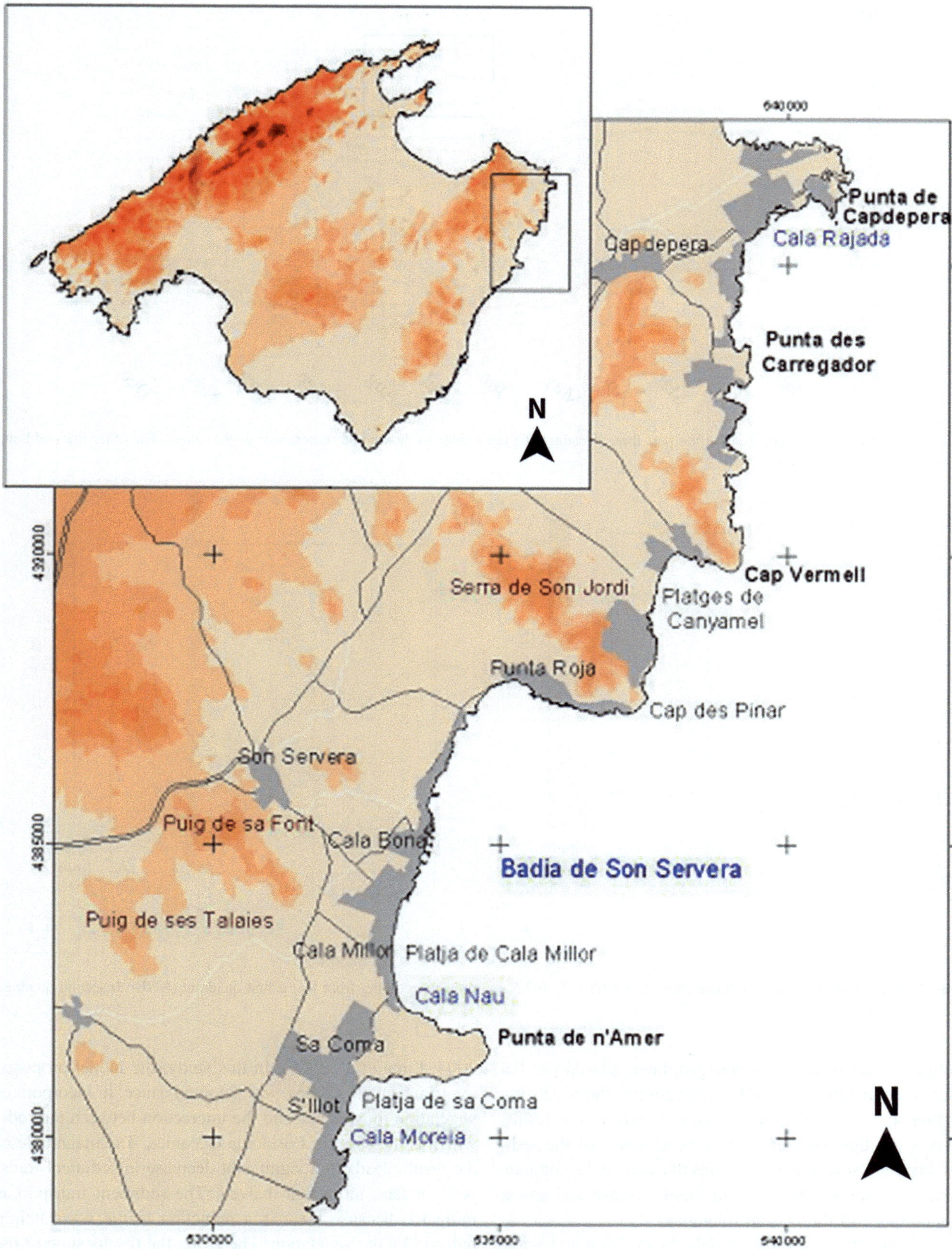

Fig. 1 Location of Son Servera's bay in Majorca Island

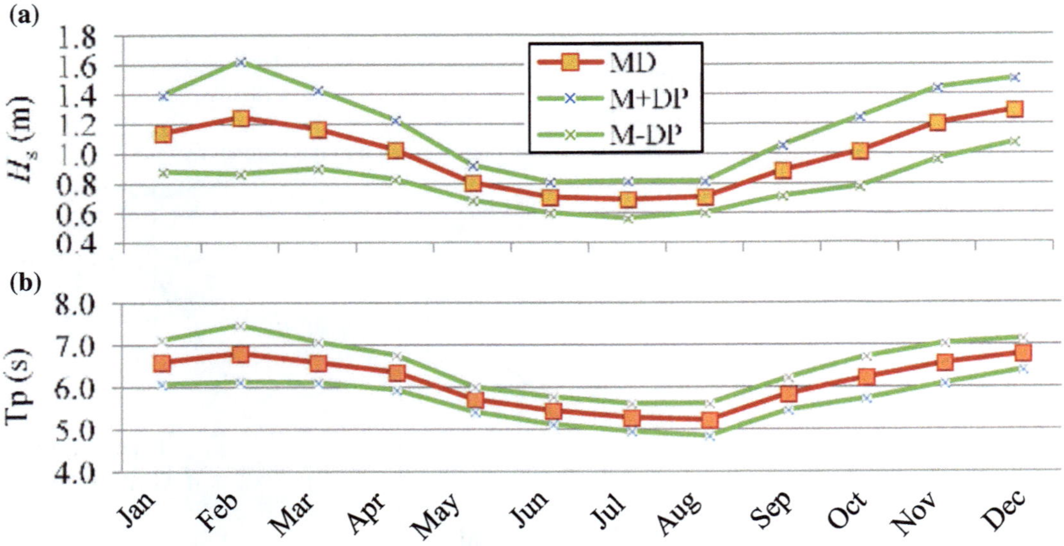

Fig. 2 Capdepera buoy: H_s and T_p over the past three decades. The red curves in the middle represent average values. The other lines add and subtract the standard deviation

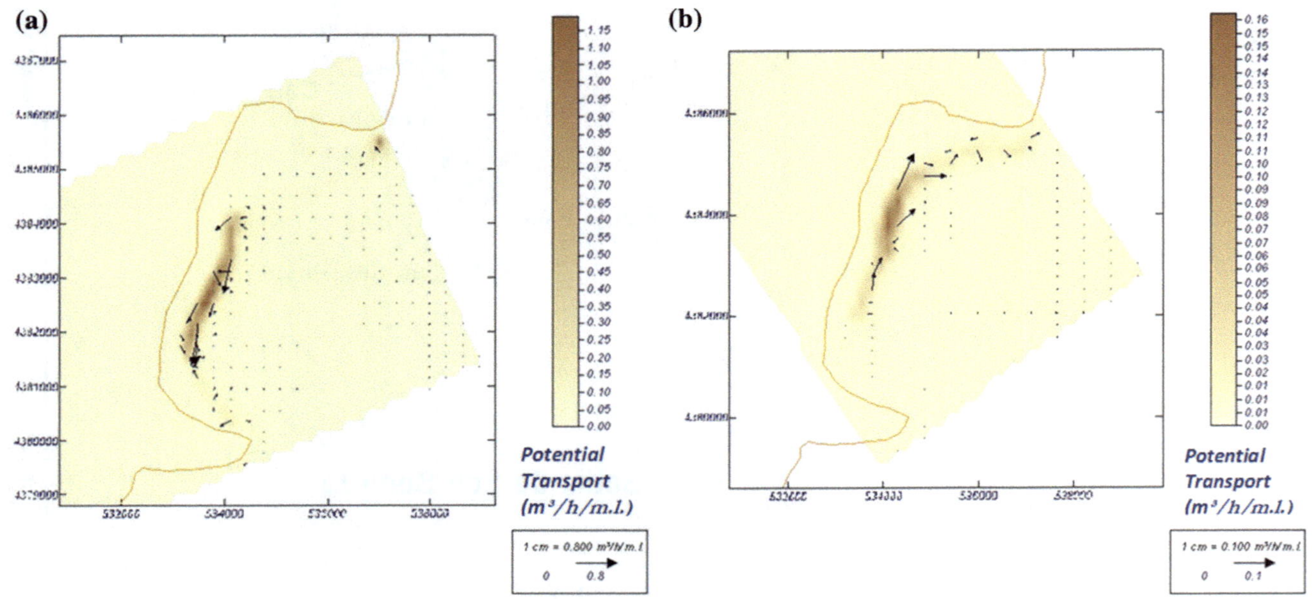

Fig. 3 Sediment transport values obtained for H_s = 1.6 m, T_p = 7.5 s with waves coming from the: **a** first quadrant (NNE); **b** second quadrant (SSE)

direction of propagation was changed, being placed parallel to the bathymetric lines. One recognizes these stirring characteristics are not dominant and, therefore, the results derived from this work are indicative in terms of the sediment budget. Nevertheless, the results enable to compare sediment transport trends in a qualitative sense and assess the importance of the bottom roughness.

In the literature, several models to evaluate the wave attenuation concerning vegetation effects on the seabed can be found (e.g., Dalrymple et al. 1984; Mendez and Losada 2004; Lima et al. 2006). In this study, the model proposed by Lima et al. (2006) was adopted since it incorporates vegetation movements and the interaction between the rods, which is the case for Posidonia Oceanica. The magnitude of the results leads to a significant decrease in sediment transport, in fact, more than halved. The sediment transport is mitigated because there is a reduction in the wave height induced by the vegetation. Therefore, the results support the idea that the morphological changes can be significantly attenuated in the presence of Posidonia Oceanica.

One stress that the model response is not taking into consideration, vertical dykes built on the beach in recent decades, as well as existing rock outcrops on the South Beach area. In addition, it is important to consider different scenarios with rising sea levels, incorporation of tides and storm surges. From a climate change perspective, Posidonia Oceanica can be severely affected and its effects should be addressed in the future (e.g., Pergent et al. 2016).

5 Concluding Remarks

Cala Millor Beach suffers a persistent loss of sediments since its natural balance was changed due to anthropic activities that took place on the Spanish coast in the 60s. The observed erosion turns out to be a serious problem for the development of this beach since the main incomes result from the tourist sector. It is necessary to correct the problem, or at least mitigate it, by means of sustainable and economic solutions.

In order to study the local dynamics of this beach the SMC program is employed. The model is used to study sediment transport trends and its evolution in the short term. The results seem to support an observed problem that results from an imbalance of sediment transport due to the drift from north to south during storm situations. Local conditions suggest a greater loss of sediments to the south, which can significantly affect the future use of the beach. For coastal management purposes, this kind of studies and tools are very important, essential for scenario-based planning and helping to propose alternatives to protect and mitigate sediment losses. The solutions require a deeper knowledge accounting for technical, functional, economic and environmental aspects.

In addition, it is recommended to undertake protective measures concerning the conservation of the Posidonia Oceanica meadow because it presents a clear positive effect on the mitigation of the energy reaching the beach.

Acknowledgements Thanks are due for the financial support to CESAM (UID/AMB/50017/2019), to FCT/MEC through national funds, and the co-funding by the FEDER, within the PT2020 Partnership Agreement and Compete 2020. This work is a contribution to project SANDTRACK (POCI-01-0145-FEDER-031779) funded by FEDER, through COMPETE2020—Programa Operacional Competitividade e Internacionalização (POCI), and by national funds (OE), through FCT/MCTES.

References

Abadiea A, Pace M, Gobert S, Borg JA (2018) Seascape ecology in Posidonia oceanica seagrass meadows: linking structure and ecological processes for management. Ecol Ind 87:1–13

Dalrymple RA, Kirby JT, Hwang PA (1984) Wave diffraction due to areas of energy dissipation. J Waterw Port Coast Ocean Eng 110 (1):67–79

Gómez-Pujol L, Orfila A, Alvarez-Ellacuria A, Tintoré J (2011) Controls on sediment dynamics and medium-term morphological change in a barred microtidal beach (Cala Millor, Mallorca, Western Mediterranean). Geomorphology 132:87–98

Guillén J, Lizaso J, Jiménez S, Martínez J, Codina A, Montero M, Triviño A, Soler G, Zubcoff J (2013) Evolution of Posidonia oceanica seagrass meadows and its implications for management. J Sea Res 83:65–71

Kobayashi N, Raichle AW, Asano T (1993) Wave attenuation by vegetation. J Waterw Port Coast Ocean Eng 119(1):30–48

Lima SF, Neves CF, Rosauro N (2006) Damping of gravity waves by fields of flexible vegetation. In: Proceedings of the 30th international conference on coastal engineering. World Scientific, USA, pp 491–503

Mendez F, Losada IJ (2004) An empirical model to estimate the propagation of random breaking and nonbreaking waves over vegetation fields. Coast Eng 51(2):103–118

Pergent G, Pergent-Martini C, Bein A, Dedeken M, Oberti P, Orsini A, Santucci JF, Short F (2016) Dynamic of Posidonia oceanica seagrass meadows in the northwestern mediterranean: could climate change be to blame? CR Biol 338:484–493

Sánchez-González JF, Sánchez-Rojas V, Memos CM (2011) Wave attenuation due to Posidonia oceanica meadows. J Hydraul Res 49 (4):503–514

Dike Vulnerability Due to Sea-Level Rise (Western Canada)

Otavio Sayão and Guy Félio

Abstract

This paper discusses the vulnerability of sea dikes in Western Canada, including examples of existing dike designs in British Columbia (BC). A dike means an embankment, revetment wall, or other coastal structures that protect upland areas and prevent the flooding of the land. There are more than 200 regulated dikes in BC, with a total length of over 1100 km, protecting 160,000 ha of valuable land. The BC Ministry of Environment guidelines for the design of sea dikes to protect low lying lands are used to define the upgraded sea dikes flood construction levels (FCL) that are now necessary due to climate change and sea-level rise (SLR). Thus, dike crest elevations will need to rise so that they continue to exert their protective function in the future, towards a lifetime up to the year 2100 and beyond. The vulnerability of sea dikes is evaluated using an example dike for the application of the PIEVC (Public Infrastructure Engineering Vulnerability Committee) Protocol method, which is used to identify the sea dike components that are especially vulnerable to climate- and weather-related impacts and to evaluate the risks to these infrastructure elements, as well as to propose typical mitigation measures.

Keywords

Sea dikes • Coastal flooding • Sea-level rise • Climate change • PIEVC protocol

1 Introduction

It is estimated that about 40% of the world's population lives within 100 km of coastline. This coastal population and the coastal infrastructure it relies on for economic viability is highly vulnerable in the face of global sea-level rise projections. This vulnerability will be exacerbated by projected increases in the frequency of the most intense tropical storms. Sea level rise along with more intense storm events could mean more frequent damage to sea dikes and increased coastal flooding.

This paper will discuss the vulnerability to sea dikes in Western Canada, including examples of existing dike designs in British Columbia (BC). A dike means an embankment, revetment wall, fill, and other coastal structures that together with pumps, culverts, canals, and/or ditches may drain overtopping waters from upland areas to prevent the flooding of the land. There are more than 200 regulated dikes in BC, with a total length of over 1100 km, protecting 160,000 ha of valuable land.

The BC Ministry of Environment provides guidelines for the design of sea dikes to protect low lying lands that are exposed to coastal flood hazards arising from their exposure to the sea and to expected sea-level rise due to climate change (Province of BC 2018). These guidelines are used to define the upgraded sea dikes flood construction levels (FCL) that are now necessary to raise dike elevations so that they continue to exert their protective function in the future, towards a lifetime up to the year 2100. In some cases, the crest elevation of dikes needs to be raised by about 2.5–3 m to be able to protect lands and residences in its lee.

The vulnerability of sea dikes is evaluated in this paper using an example dike for the application of the PIEVC (Public Infrastructure Engineering Vulnerability Committee) Protocol method (Engineers Canada 2016), which was developed by Engineers Canada and since 2007 has been applied to over 50 infrastructure risks assessments in Canada and in other international sites. The PIEVC method is used

O. Sayão (✉)
Acqua Engineering Inc., Vancouver, Canada
e-mail: osayao@gmail.com

G. Félio
Stantec Consulting Ltd., 400-1331 Clyde Ave, Ottawa, ON K2C3G4, Canada

© Springer Nature Switzerland AG 2020

F. Fernandes et al. (eds.), *Advances in Natural Hazards and Hydrological Risks: Meeting the Challenge*, Advances in Science, Technology & Innovation,
https://doi.org/10.1007/978-3-030-34397-2_31

to assess the vulnerability of built assets (infrastructure, buildings, facilities) to the impacts of climate change. In this paper, the PIEVC protocol is used to identify the sea dike components that are especially vulnerable to climate- and weather-related impacts and to evaluate the risks to these infrastructure elements, as well as to propose typical mitigation measures.

2 Dike Design Guidelines for Sea Level Rise

The original BC provincial document (Province of BC 2018) dated May 2004 quotes: "new site-specific studies containing professional evaluation, advice and recommendations including mapping, may be required where the risk to life and property is high, where advice is required to meet provincial flood hazard management guidelines or where modified or new protective works are proposed". Also, the BC guideline document specifies "If the training works, when constructed, will protect multiple properties of more than one person, then an ongoing operation and maintenance program and access to structures must be assured by the local government. Works are to be designed by a professional engineer" (a person who is registered or licensed under the provisions of the Engineers and Geoscientists Act of BC).

As shown in Fig. 1, it is recommended that a 1.0 m sea-level rise (SLR) be used by the year 2100, for all coastal defense (flood protection) projects with a design life of 50–100 years, while recognizing the uncertainties of all SLR predictions.

Province of BC (2018) recommendations are in line with the Intergovernmental Panel on Climate Change (IPCC) AR5 Summary for Policymakers (IPCC 2013), which reports that global mean sea-level rise (SLR) for RCP8.5 scenario by the year 2100 is 0.52–0.98 m. For most locations, RCP8.5, which is the scenario giving the largest amount of global SLR, provides the largest projection of relative sea-level rise. Other recent USA guidelines predict similar global SLR 2100 values, 3.1 ft. (0.95 m) to 3.4 ft. (1 m), see Miller et al. (2018), California Ocean Protection Council (2018).

Figure 2 gives an example of a typical section of an existing dike design protecting reclaimed land, today. To reflect future design including the recommended SLR allowance, a new engineered section is necessary to resist the 50-year and 100-year return period extreme high-water levels and overtopping due to waves.

This paper develops an example of adaptation of existing dike protection structures to account for future SLR, following guidance defined in Province of BC (2018).

2.1 Flood Construction Level, Typical Results

Table 1 summarizes the steps for determining the Flood Construction Level (FCL) for coastal lands exposed to potential flood hazard risks and future SLR due to climate change, following the Province of BC 2018 guidelines (Province of BC 2018) and other reviewed literature sources, both in BC and worldwide. Please note that FCL means the recommended elevation for the top of dike assuming occupied structures in the lee of the protection are outside of the recommended setback distance (Province of BC 2018) and the wave overtopping is estimated such that discharge threshold values are not to be exceeded (EurOtop 2018).

Using the estimates of FCL (Table 1) upgraded dike protection for the future years may be developed, which typically could look like the cross-section of Fig. 3.

It may be necessary to elevate the fill at the back area of the dike (Fig. 3), and to further design the crest of the revetment with a wave wall in required locations, to add geotextiles or drainage canals at the lee of the dike, etc.; however, these detailed engineering design features are not considered in this paper.

Fig. 1 Recommended global SLR curve for planning and design (Province of BC 2018)

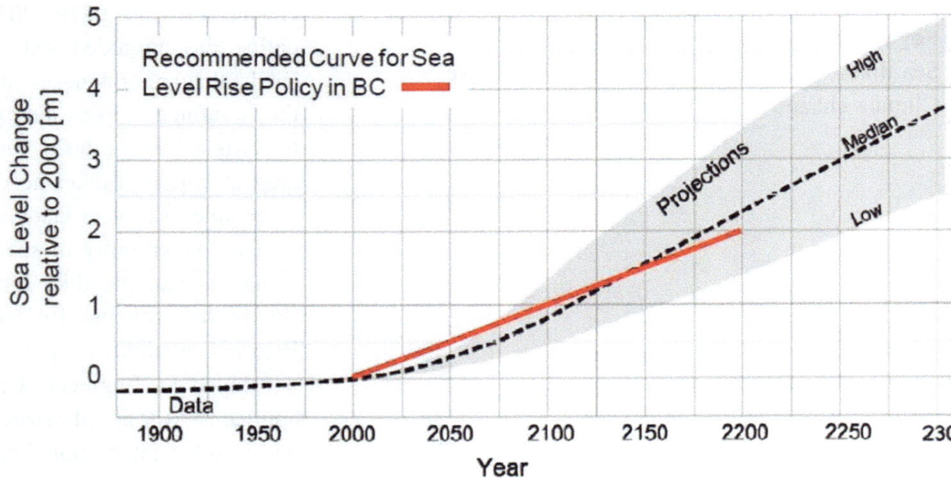

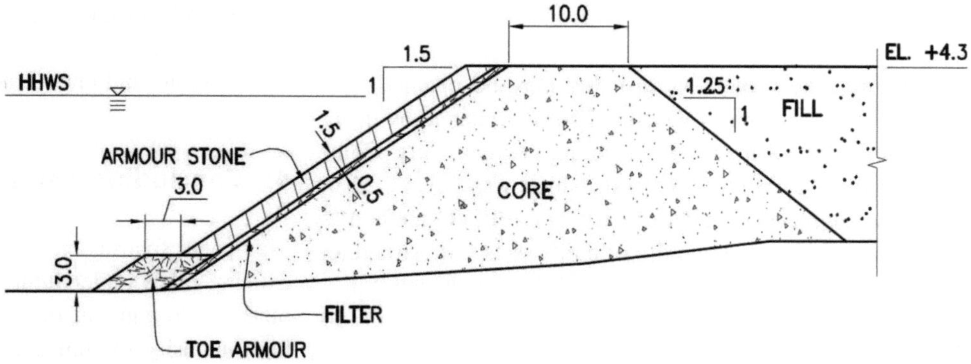

Fig. 2 Typical dike protection section, today (all elevations and dimensions are in meters)

Table 1 Flood construction level estimate for a rubblemound dike

2018 Ministry guidelines method (Province of BC 2018)	Estimates/review
2100 future sea-level rise allowance	0.77 m[a]
Regional adjustment (uplift/subsidence)	−0.23 m[b]
High tide (HHWS)	4.8 m CD[c]
Storm surge allowance	1.3 m[d]
Wave effect allowance	1.7 m[e]
75-year post-construction settlement	−0.28 m[f]
Flood construction reference plane, CD	9.08 m
Freeboard allowance	0.6 m[g]
Flood construction level (FCL)	9.68 m CD[h]

[a]From James et al. (2014), Fig. 2 for Vancouver and Surrey, range from 0.52 to 1.2 m. For Vancouver, BC, the uplift rate is 0 (James et al. 2014). For 2100 SLR for Vancouver, Table C3 estimates from James et al. (2014) was used (Sea-level projections at 2081–2100 relative to 1986–2005 for RCP8.5, with a 95% confidence range yields a value of 0.77 m)
[b]From rates estimate given in the local literature with a 1/100-year design wave event, for 90 years (2020–2100) yield a 0.23 m subsidence
[c]Higher high-water spring tide elevation, Canadian Hydrographic Service tide information
[d]Used storm surge allowance based on local literature, where storm surge for the Georgia Strait (BC) is reported as 1.3 m, with an estimated return period of 1/200 year
[e]Using an expression of about 0.5*wave runup estimates on the future shoreline (rubblemound sloped revetment), the wave effect allowance is approximately 1.7 m
[f]Value based on static settlement analysis based on local literature
[g]As the structure is subjected to wave overtopping, the higher freeboard value (0.6 m) was used
[h]*CD* Chart datum

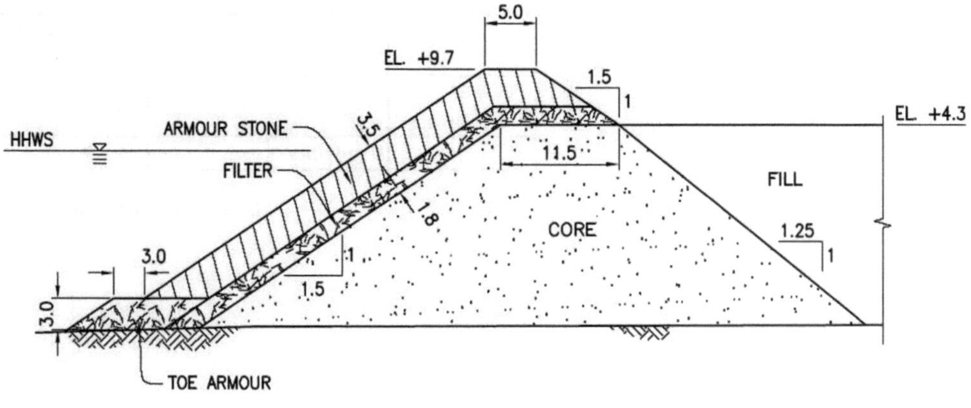

Fig. 3 Typical dike protection section, future (all elevations and dimensions are in meters)

3 The PIEVC Protocol

The future dike protection structure may be planned as a resilient infrastructure to withstand climate change by using the Public Infrastructure Engineering Vulnerability Committee—PIEVC Protocol method, which was developed in 2005 by Engineers Canada (2016). This PIEVC method is used to assess the vulnerability of built assets (infrastructure, buildings, facilities) to the impacts of climate change. Since 2007 the PIEVC Protocol has been applied to over 50 infrastructure risk assessments in Canada and in more than five international sites.

This paper will apply the PIEVC Protocol method to identify the dike infrastructure components that are especially vulnerable to climate- and weather-related impacts and to evaluate the risks with respect to SLR, for future years beyond 2050.

Figure 4 presents an illustration of the application of the PIEVC Protocol method to a general infrastructure component, similar to the approach used in this paper for a dike protection section.

The PIEVC approach is well-suited for this application since it is not restricted to any type of infrastructure and is specifically designed for climate change risks considering the potential impacts on the assets for the climate parameters that are most relevant to the structure under investigation. The PIEVC Protocol is consistent with the IPCC (2013) guidelines. The process is also well suited to community consultations and engagement.

4 Concluding Remarks

This paper focuses on the potential of wave overtopping of existing dikes under future sea-level rise (SLR) along with other metocean factors. Although it is recognized that high-intensity rainfall events may occur concurrently with the storm wave overtopping and reduce the design capacity of the drainage system behind the dike, this was not included in the present evaluation.

The PIEVC approach helped on the selection of climate parameters and infrastructure thresholds that are commonly related to dike protection and wave overtopping. Some events that were considered included a combined event of high tides, high winds and future SLR, which yielded the highest risks; and a combined spring freshet from a river delta and an extremely high-water level including storm surge, which was the runner up event for higher risks. These two events were part of the PIEVC risk assessment and the principal infrastructure affected included the dikes, the foreshore ecosystem, and pump station structures internal to the dike protected shoreline.

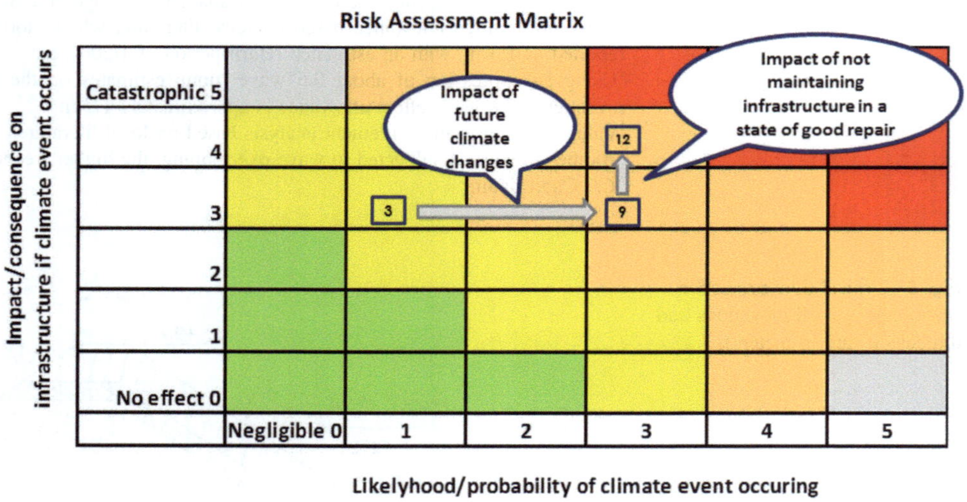

Fig. 4 Typical PIEVC risk assessment matrix for application to a dike protection section

References

California Ocean Protection Council (2018) State of California sea-level rise guidance 2018 update. Ocean protection council report, in coordination with the California Natural Resources Agency, the Governor's Office of Planning and Research, and the California Energy Commission

Engineers Canada (2016) PIEVC engineering protocol for infrastructure vulnerability assessment and adaptation to a changing climate. Principles and guidelines, version PG-10.1, dated June 2016

EurOtop (2018) Manual on wave overtopping of sea defences and related structures. An overtopping manual largely based on European research, but for worldwide application. In: Van der Meer JW, Allsop NWH, Bruce T, De Rouck J, Kortenhaus A, Pullen T, Schüttrumpf H, Troch P, Zanuttigh B (eds) European overtopping manual

IPCC (2013) Summary for policymakers. In: Stocker TF, Qin D, Plattner G-K, Tignor M, Allen SK, Boschung J, Nauels A, Xia Y, Bex V, Midgley PM (eds) Climate change 2013: the physical science basis. Contribution of working group I to the fifth assessment report of the intergovernmental panel on climate change. Cambridge University Press, Cambridge, United Kingdom and New York

James TS, Henton JA, Leonard LJ, Darlington A, Forbes DL, Craymer M (2014) Relative sea level projections in Canada and the adjacent Mainland United States. Geological Survey of Canada, Open File 7737, 72p

Miller IM, Morgan H, Mauger G, Newton T, Weldon R, Schmidt D, Welch M, Grossman E (2018) Projected sea level rise for Washington State—a 2018 assessment. A collaboration of Washington Sea Grant, University of Washington Climate Impacts Group, Oregon State University, University of Washington, and US Geological Survey. Prepared for the Washington Coastal Resilience Project

Province of BC (2018) Flood hazard area land use management guidelines; originally dated May 2004 by the Ministry of water, land and air protection of British Columbia and amended by the Ministry of forests, lands, natural resource operations and rural development in January 18, 2018

Iber+ : A New Code to Analyze Dam-Break Floods

José González-Cao, Orlando García-Feal, Diego Fernández-Nóvoa, and Moncho Gómez-Gesteira

Abstract

Dams provide a wide range of benefits but they have risks associated with their potential fails, e.g. dam-breaks. Floods derived by this type of fails usually have dramatic consequences in terms of economic losses and casualties. Therefore, the ability to predict the extent and the velocity of propagation of this type of floods is crucial. In this work the numerical code Iber+ is used to reproduce the flood derived by the break of the Malpasset dam. The accuracy of the numerical results along with the high computational efficiency of the code makes Iber+ a suitable tool for forecasting of flood events even in early-warning systems.

Keywords

Flood • Dam-break • Numerical simulation • Iber+ • Early-warning systems

1 Introduction

Dam-breaks floods are one of the most dangerous natural disasters that can affect human activity since they are characterized by a sudden release of a high volume of water stored in the impoundment of the dam. These events are registered from the ancient antiquity (break of the Great Dam of Marib in the eighth century BC, Yemen) until present (break of the dam of Brumadinho on January 25th, 2019, Brasil) and can increase under a scenario of climate change where extreme rain events will be more frequent. Therefore, the development of accurate and computationally efficient numerical tools to predict the effects of these events is an essential task of scientific community.

J. González-Cao (✉) · O. García-Feal · D. Fernández-Nóvoa · M. Gómez-Gesteira
Environmental Physics Laboratory, Universidad de Vigo, Orense, Spain
e-mail: jgcao@uvigo.es

In this work, the well-known case of the Malpasset dam-break on December 2nd, 1959 is used to analyse the accuracy and the efficiency of the numerical code Iber+ (García-Feal et al. 2018) to reproduce dam-break floods.

2 Materials and Methods

2.1 Numerical Code: Iber+

Iber+ is a new implementation of the hydraulic module of the numerical code Iber (Bladé et al. 2014) that solves the 2D Shallow Water Equations (Eqs. 1–3) by means of finite volume schemes.

$$\frac{\partial h}{\partial t} + \frac{\partial hU_x}{\partial x} + \frac{\partial hU_y}{\partial y} = r - i \quad (1)$$

$$\frac{\partial hU_x}{\partial t} + \frac{\partial}{\partial x}\left(hU_x^2 + g\frac{h^2}{2}\right) + \frac{\partial}{\partial y}(hU_xU_y) = -gh\frac{\partial Z_b}{\partial x} - \frac{\tau_{b,x}}{\rho} + \Psi_x(v_t, h, U_x) \quad (2)$$

$$\frac{\partial hU_y}{\partial t} + \frac{\partial}{\partial y}\left(hU_y^2 + g\frac{h^2}{2}\right) + \frac{\partial}{\partial x}(hU_xU_y) = -gh\frac{\partial Z_b}{\partial y} - \frac{\tau_{b,y}}{\rho} + \Psi_y(v_t, h, U_x) \quad (3)$$

In Eqs. (1–3) h refers to the water depth, U_x and U_y are the averaged horizontal velocities, g is the acceleration of the gravity, ρ is the density of the water, Z_b is the bed elevation, τ_b is the river-bed friction, υ_t is the turbulent viscosity, r is the rainfall intensity and i is the infiltration rate. Ψ_x and Ψ_y are functions on υ_t, h, U_x and U_y.

The river-bed friction is computed using the Manning's equations (Eq. 4)

$$\tau_{b,x} = \rho gh\frac{n^2U_x|U|^2}{h^{4/3}}, \quad \tau_{b,y} = \rho gh\frac{n^2U_y|U|^2}{h^{4/3}} \quad (4)$$

© Springer Nature Switzerland AG 2020
F. Fernandes et al. (eds.), *Advances in Natural Hazards and Hydrological Risks: Meeting the Challenge*, Advances in Science, Technology & Innovation,
https://doi.org/10.1007/978-3-030-34397-2_32

where n is the Manning's coefficient.

The numerical code Iber+ is implemented using C++ and it is parallelized for shared memory systems with OpenMP that allows computations using several CPUs. Iber+ also provides an Nvidia CUDA implementation for execution in GPUs.

2.2 Application Case: Malpasset Dam-Break

Iber+ was applied to a real dam-break case: The Malpasset dam-break. The Malpasset dam was located in the Reyran river basin, 12 km north of the city of Fréjus (southern France). It was built in 1954 for water supply and irrigation purposes. It was 66.5 m high with a maximum capacity of 55 Hm^3. Figure 1 shows the location and a detailed image of the dam.

The rainfall event registered from November 19th to December 2nd 1959 increased the water level to the dam crest (Valiani et al. 2002). The dam broke explosively at 21:14 on December 2nd, 1959. Due to its geometry the break was extremely rapid and violent. The resulting wave destroyed roads, bridges and caused more than 400 casualties. The water arrives at Fréjous 20 min after the break and reached the sea 16 min later. The reader can find more information about this event in Zhao (2012), Shi (2006) and Valiani et al. (2002).

In order to analyze this event, a physical model was built in 1964 by the LNHE/EDF (Laboratoire National d'Hydraulique et Environnement de l'EDF). Values of the maximum water elevation collected by the local police at different points on both banks of the river and the times of electricity cutting off of three different electric transformers were used to characterize the physical model. In addition to this field data, fourteen water depth gauges were defined in the physical model. Only the data of the last 8 depth gauges are used in this work. The location of the gauges of the physical model (S6–S14) transformed to the real coordinates of the case, electric transformers (A, B and C), control points (P1–P17) and the topography of the area of study is shown in Fig. 2.

The numerical domain was discretized using 67,319 triangular elements. A physical time of 90 min was simulated with a sampling frequency of 1 Hz. The numerical scheme used for the simulations was Roe 2nd order and the CFL condition was defined equal to 0.45. The value of the wet-dry limit was equal to 0.01 m. A constant value of the Manning's coefficient for the entire domain was computed in a calibration stage starting from the arrival times obtained in the physical model (S-gauges). The arrival times are defined as the first instant when the water depth exceeds the value of the wet-dry limit. An initial water depth equal to 100 m was defined in the impoundment. Unless specifically specified otherwise, the simulations were carried out using GPUs.

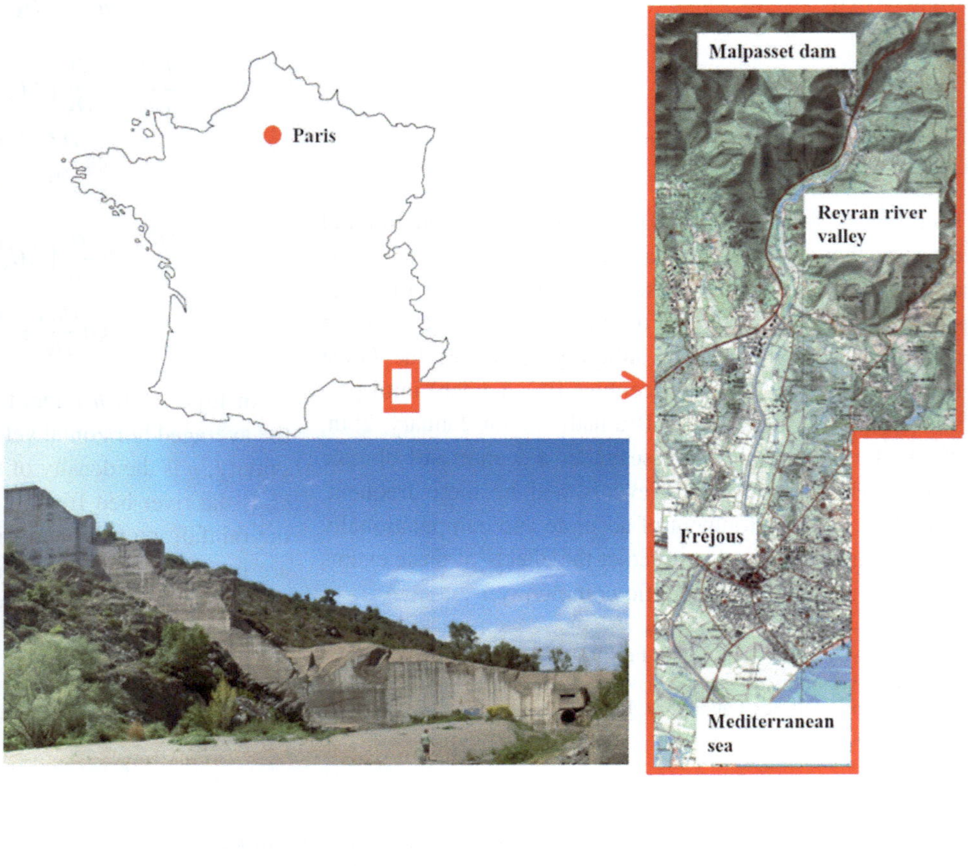

Fig. 1 Location of the Malpasset dam and an actual image of the wreckage of the dam

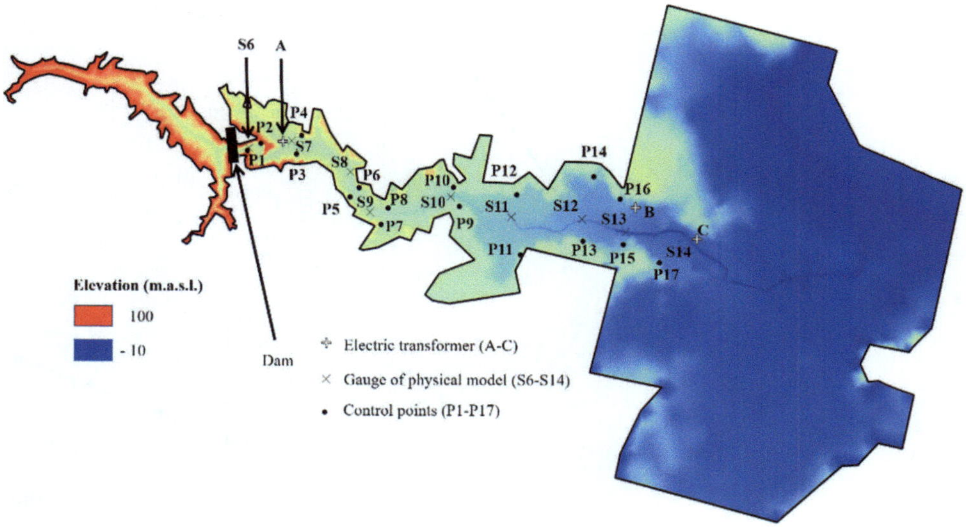

Fig. 2 Location of the dam, the electric transformers (A, B and C), gauges of the physical model (S6–S14) and control points (P1–P17). The topography of the case is also depicted

The metric used to assess the accuracy of the numerical results is the average quadratic relative error, ε_r (Eq. 5).

$$\varepsilon_r = \frac{\sqrt{\sum_{i=1}^{N}\left[\frac{y_r(i)-y_M(i)}{y_r(i)}\right]^2}}{N} \tag{5}$$

where $y_r(i)$ refers to the *i*-reference data (physical model or field data), $y_M(i)$ refers to the *i*-numerical value and *N* refers to the total number of points.

3 Results and Discussion

3.1 Calibration of the Manning's Coefficient

In this first stage, a series of numerical simulations using different values of the Manning's coefficient (*n*) were carried out to obtain the value of the Manning's coefficient that best fits the arrival times obtained at S-gauges of the physical model (n_{opt}). Left panel of Fig. 3 shows that the value of n_{opt} is 0.0315 $m^{-1/3}$ s ($\varepsilon_r \sim 1.3\%$) provides the best fit. Figure 3 (right panel) shows that the arrival time at the different gauges calculated for that particular n is in good agreement with the values provided by the physical model. This value will be considered in the rest of the study.

3.2 Numerical Results

In this subsection, the numerical maximum water elevations obtained at S-Gauges and P-Points using a value of *n* equal to n_{opt} along with the values obtained in the physical model and in the real event are depicted. The times of cutting off of the electric transformers will be compared with the arrival and maximum water elevation times obtained in the numerical simulations. The runtime using Iber+ is also shown.

Figure 4 shows the maximum water elevation obtained with Iber+ at P-Points (left panel) and S-Gauges (right panel) along with the field and the physical model data, respectively. The value of the quadratic average relative error, in terms of maximum water elevation, at P-Points is ~3%. In the case of the maximum water elevations at S-Gauges the quadratic average relative error is less than 1%. Figure 4 also shows the numerical values of maximum water elevation at P-Point versus the field data (detail in left panel) and the numerical values at S-Gauge versus the physical model data (detail in right panel). The equations of the regression lines are also shown. The differences between the numerical values and the reference ones are negligible. The values of R^2 (coefficient of determination) of the regression line corresponding to field data (P-Gauge) and to the physical model (S-Gauge) are equal to 0.98 and 0.99, respectively.

Table 1 shows the numerical arrival and maximum water elevation times at the electric transformers along with the field data corresponding to the time of the power cut off for every electric transformer. In all cases, the instant of the cut off lies within the interval defined by the arrival and the maximum water elevation time. Except for the first transformer (A) the delay between the arrival time and the maximum water elevation time is ~10 min. This delay is a good indicator of the uncertainty to define a "numerical time of cutting off" similar to the time recorded during the real event.

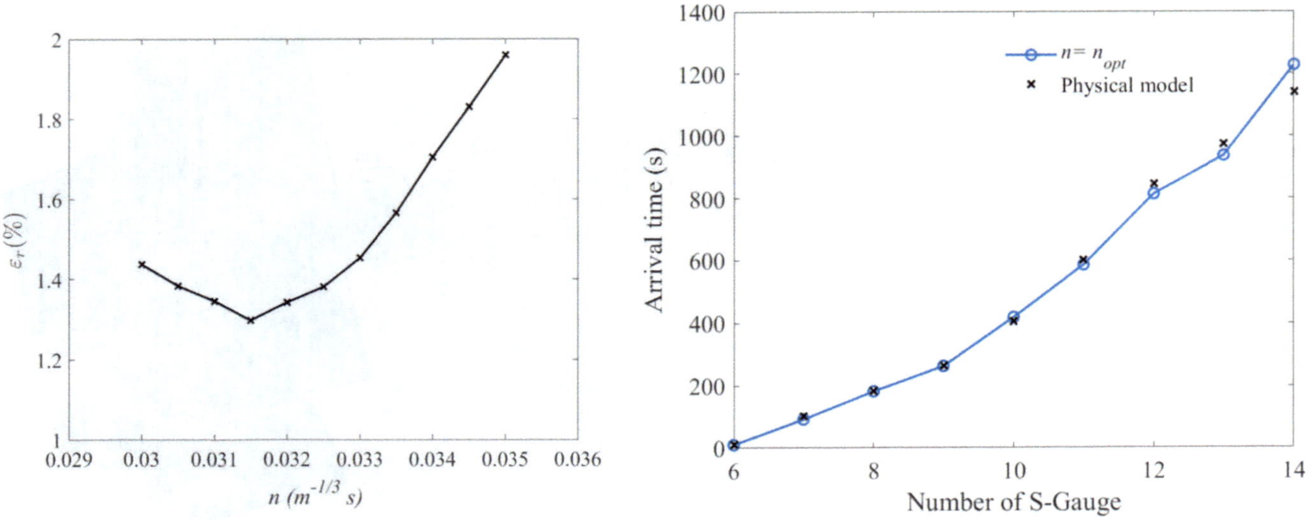

Fig. 3 (Left panel) Values of ε_r as a function of the Manning's coefficient (n) and (right panel) arrival times of the physical model (black crosses) along with the arrival times obtained in the numerical simulations using n_{opt} (blue circle and line)

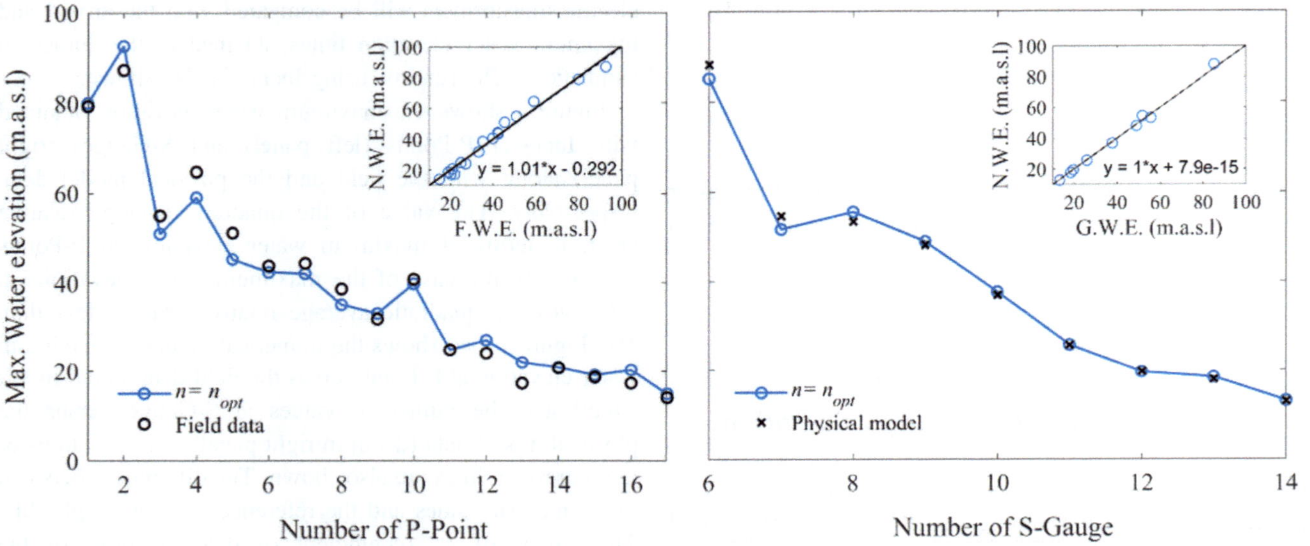

Fig. 4 Numerical maximum water elevation (blue line and circle) versus field data (black circle—left panel) and physical model data (black cross—right panel). The details in left and right panels show the numerical values of the maximum water elevation (N.W.E.) versus the reference values (F.W.E. or G.W.E.)

Table 1 Arrival time and maximum water elevation time to electric transformers and cut off time (in minutes)

Transformer	Arrival time	Max. water elevation time	Time of cutting off
A	1.6	8.4	1.7
B	17.8	28.5	20.7
C	20.2	29.8	23.7

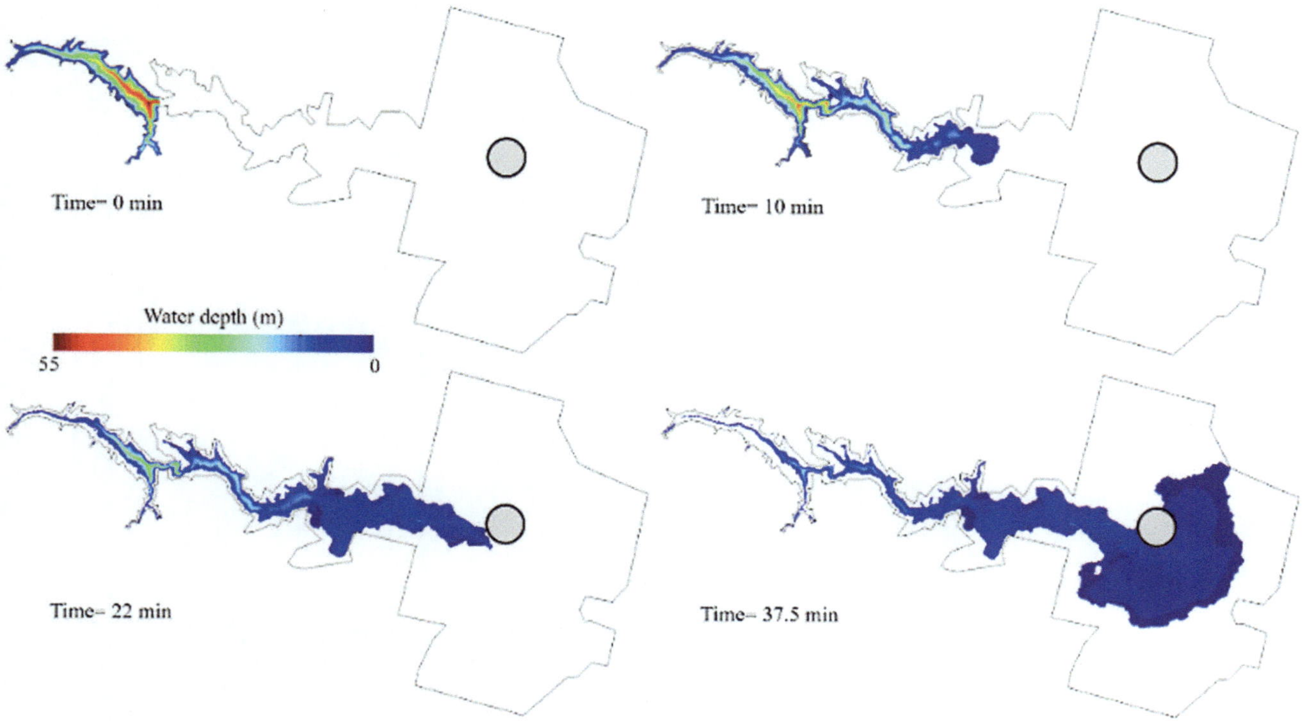

Fig. 5 Different instants of the numerical water depth. Shaded circle refers to Fréjous

Figure 5 shows different instants of the propagation of the dam-break wave. The water takes 22 min to arrive near Fréjous and 37.5 min to arrive at the sea in good agreement with field observations (Valiani et al. 2002).

The total time needed to run the case was 2.3 min. This high computational efficiency shows the capacity of Iber+ to extend its application to early-warning systems.

4 Concluding Remarks

In this work, the numerical code Iber+ was applied to a well-known real case of dam-break flood. The high accuracy of the results ($\varepsilon_r \leq 3\%$) and the small runtime (2.3 min) to simulate the event makes Iber+ a fast and accurate tool suitable to be used in real-time forecasting systems.

Acknowledgements This work was partially supported by Water JPI-WaterWorks (project: IMDROFLOOD—PCIN-2015-243), by INTERREG-POCTEP (project: RISC_ML—Code: 0034_RISC_ML_6_E) and by Xunta de Galicia (Project ED431C 2017/64-GRC).

References

Bladé E, Cea L, Corestein G, Escolano E, Puertas J, Vázquez-Cendón E, Dolz J, Coll A (2014) Iber: Herramienta de simulación numérica del flujo en ríos. Rev. Int Métodos Numéricos para Cálculo y Diseño en Ing 30:1–10

García-Feal O, González-Cao J, Gómez-Gesteira M, Cea L, Domínguez JM, Formella A (2018) An accelerated tool for flood modelling based on Iber. Water 10(10)

Shi Y (2006) Résolution numérique des équations de Saint-Venant par la technique de projection en utilisant une méthode des volumes finis dans un maillage non structuré. L'uniersité de Caen, Basse-Normandie, France (Ph.D. thesis)

Valiani A, Caleffi V, Zanni A (2002) Case study: Malpasset dam-break simulation using a two-dimensional finite volume method. J Hydraul Eng 128(5):460–472

Zhao J (2012) Development of a fast SPH model for non linear shallow water flows: application to coastal flooding and dam breaking. École Centrale de Nantes, France (Ph.D. thesis)

Rivers' Confluence Morphological Modeling Using SRH-2D

Eman AlQasimi and Tew-Fik Mahdi

Abstract

Efficient river morphology modeling is needed for hydraulics engineers. Several one-dimensional software supporting sediment transports are available and widely used, regardless of their limitations. However, two-dimensional sediment transport is still not widely used in practice for several reasons. Besides the high simulation times, the efficiency of such tools is not widely accepted. This paper tests the ability of the US Bureau of Reclamation two-dimensional sediment transport software, SRH-2D, to simulate the morphological evolution of two rivers' confluence. Even if no quantitative data are available to judge the results' quality, SRH-2D is able to capture the trend of the observed morphological evolution for this complex case.

Keywords

Rivers' confluence • 2D numerical simulation • SRH-2D • Morphological evolution

1 Introduction

In river engineering, efficient morphological modeling tools are needed to predict river evolution after a dam construction of failure, for instance. Several one-dimensional sediment transport softwares have been developed during the last decades. For example, GSTARS (Yang and Simoes 2002) and MHYSER (Mahdi 2009) use the flow quasi-steadiness hypothesis to address erosion/deposition problems, while CONCEPTS (Langendoen 2000), SRH-1D (Huang and Greimann 2012), MIKE11 (DHI 2009), CCHE1D (Vieira and Wu 2002) or ISIS (Wallingford 2001) solves the St. Venant and sediment transport equations. More recently, AlQasimi and Mahdi (2018a) developed UMHYSER-1D, Unsteady Model for HYdraulics of SEdiments in Rivers-One Dimensional, to predict river morphology changes and successfully applied it to the 1996 river Ha!Ha! flooding (AlQasimi and Mahdi 2018b) and to address the Aux Sables River flood mitigation solutions (AlQasimi et al. 2018). For two-dimensional river morphology modeling, even if few software exists, their use is not widely spread within the practicing engineers' community.

In this work, two rivers' confluence undergoing flow regime changes will be modeled using the USBR two-dimensional software SRH-2D (Lai 2008). This case might be used to test the capabilities of other two-dimensional models or to address the performances of existing one-dimensional software.

2 Materials and Methods

2.1 Two-Dimensional Software: SRH-2D

SRH-2D solves the shallow water equations and the convection-diffusion equation over a mobile bed and Boussinesq equations are used to compute the turbulence stresses (Lai 2008, 2010; Greimann et al. 2008).

SRH-2D proposes two turbulence models: *k*-epsilon and depth-averaged parabolic models. SRH-2D uses a wetting–drying front limit of 0.001 m. An implicit scheme is used to solve the finite-volume numerical method.

Pre-processing and post-processing of the model is executed in SMS (AQUAVEO 2019), a modeling software presented as a graphical user interface and analysis tool that holds all of the SRH-2D functionalities.

E. AlQasimi · T.-F. Mahdi (✉)
Hydro-environmental Laboratory, Polytechnique Montreal, Montreal, Canada
e-mail: tewfik.mahdi@polymtl.ca

© Springer Nature Switzerland AG 2020
F. Fernandes et al. (eds.), *Advances in Natural Hazards and Hydrological Risks: Meeting the Challenge*, Advances in Science, Technology & Innovation,
https://doi.org/10.1007/978-3-030-34397-2_33

2.2 Application Case: Petite Manouane and Manouane Rivers' Confluence

The study case is the confluence of the two rivers, Petite Manouane and Manouane, just upstream the Duhamen lake, located in the Quebec province of Canada (Fig. 1). The discharges are 250 m^3/s and 40 m^3/s for the Petite Manouane and Manouane rivers, respectively. The lake level is constant at the elevation 252.45 m. The bed material is composed mainly of uniform sand, having a median size of 3 mm. The Manouane river carries no sediment while the Petite Manouane river sediment discharges is not limited. The sediment transport equation of Engelund and Hanzen is used for this study. The domain is 4800 m long from the Petite Manouane River upstream boundary, and 3800 m long from the Manouane river upstream boundary.

For the two-dimensional simulation the numerical domain is discretized using 42,932 triangular elements. A constant value of the Manning's coefficient of 0.018, for the entire domain, is adopted according to Strickler formula (Yang 2003), and to guarantee numerical stability, the time step is reduced to 2 s. For the downstream boundary condition, a constant water level is maintained at 252.45 m.

3 Results and Discussion

Figure 2 shows the sand deposition/erosion evolution for the first 16 days.

No qualitative data are available for validation but, deposition has been observed to increase during the last years, when the Manouane river discharge started to decrease, at the same places as simulated.

The deposition and erosion zones correspond to the zones of low and high velocities, respectively. At the junction of the Petite Manouane and the Manouane rivers, the velocities' values are low, allowing sediments' deposition, as shown in Fig. 3. Note that erosion occurs in the Petite Manouane river, where the velocities are higher (small cross-section and higher discharge).

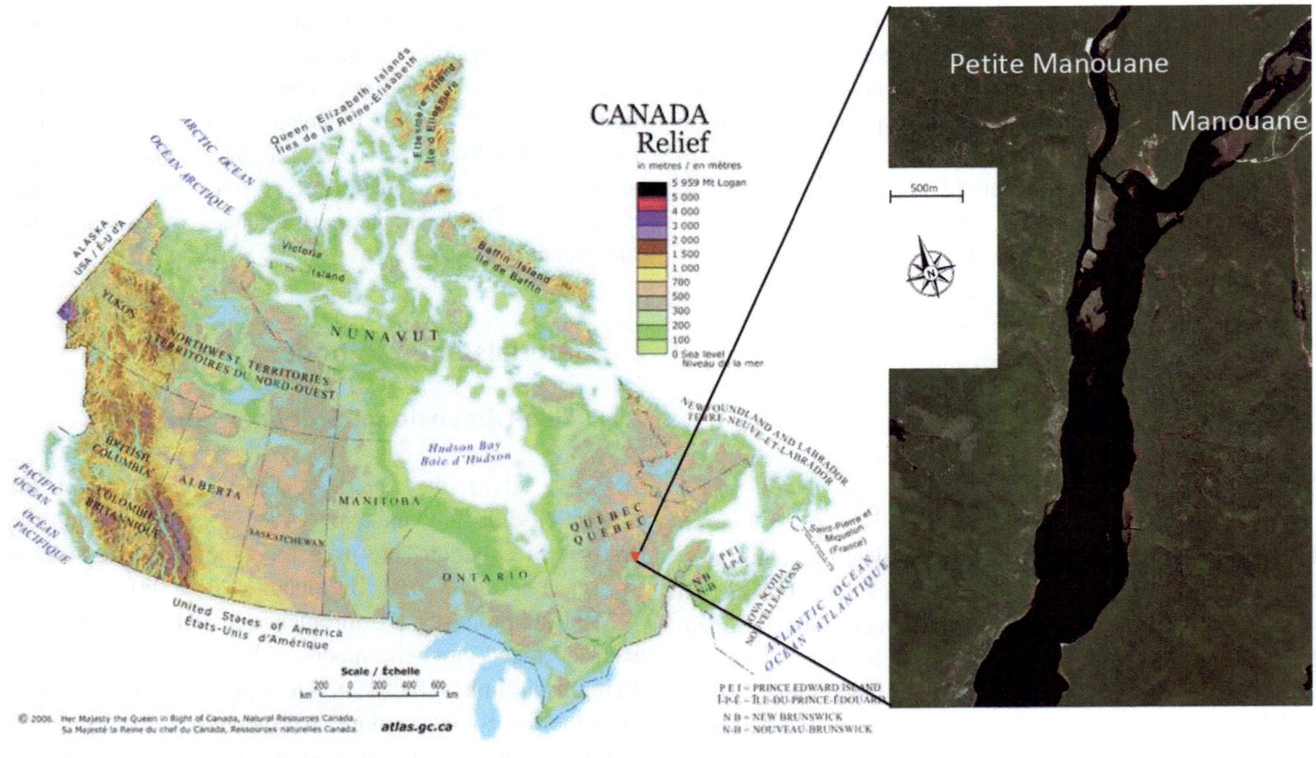

Fig. 1 Location of the rivers' confluence

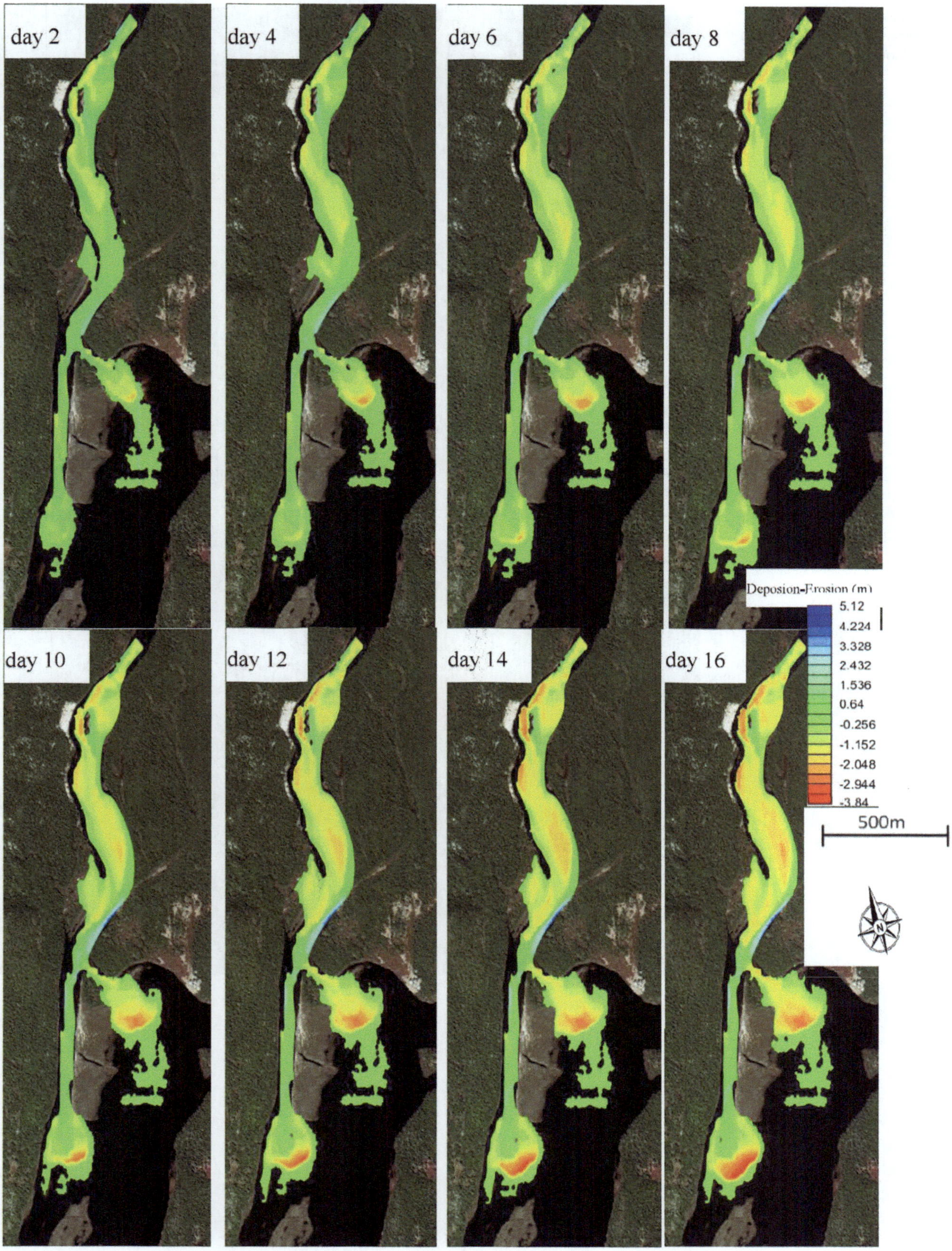

Fig. 2 Erosion/deposition evolution during the first 16 days (deposition values are negative)

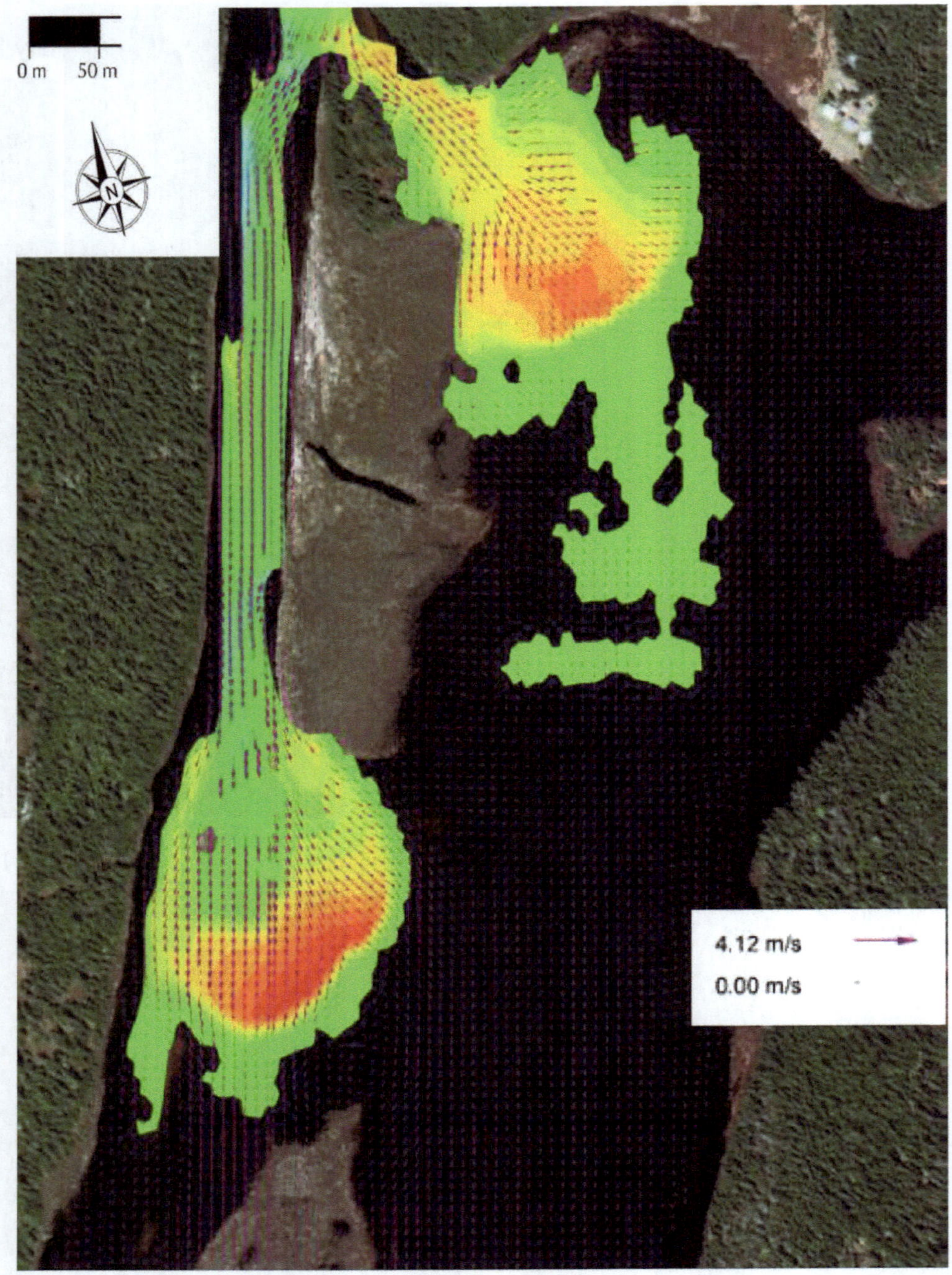

Fig. 3 Velocity field at the confluence. Deposition occurs in zones of low flow velocities

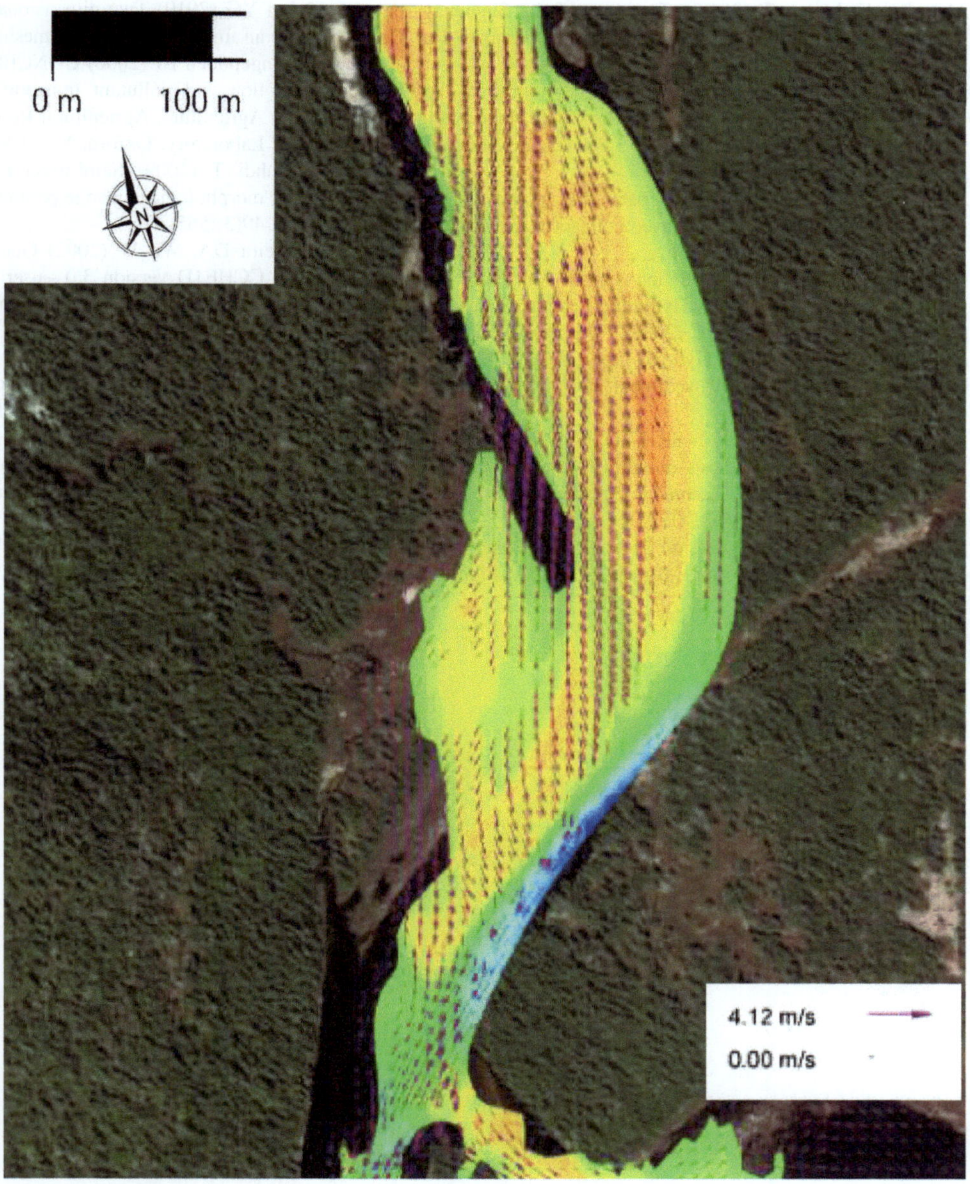

Fig. 4 High erosion occurring in zones of high flow velocities on the Petite Manouane river

Figure 4 shows that its last bend, before the confluence, experiments huge erosion on the left side (looking downstream), after the deposition increases on the right side. The maximum values of deposition and erosion are about 4 m to 5 m, respectively

4 Concluding Remarks

In this work, the numerical software SRH-2D software was used to simulate morphological evolution of the confluence of the Petite Manouane and Manouane rivers, located in the Quebec province, Canada. The results capture the trend of deposition/erosion observed in the field. The deposition and erosion zones correspond to the zones of low and high flow velocities. Future work will use these results to test the capabilities and limitations of the software UMHYSER-1D.

Acknowledgements This research was partially supported by a Canadian National Science and Engineering Research Council (NSERC) Discovery Grant for the corresponding author, application No: RGPIN-2016-06413.

References

AlQasimi E, Mahdi T (2018a) Unsteady model for the hydraulics of sediments in rivers. In: 26th annual conference of the computational fluid dynamics society of Canada. 10–12 June, Winnipeg, MB, Canada

AlQasimi E, Mahdi T (2018b) Flooding of the Saguenay region in 1996: part 1—modeling river Ha! Ha! flooding. J Nat Hazards. https://doi.org/10.1007/s11069-018-3443-4

AlQasimi E, Pelletier P, Mahdi T (2018) Flooding of the Saguenay region in 1996. part 2: Aux Sables river flood mitigation and environmental impact assessment. J Nat Hazards. https://doi.org/10.1007/s11069-018-3444-3

AQUAVEO (2019) SMS 13.0 beta—the complete surface-water solution. https://www.aquaveo.com/software/sms-surface-water-modeling-system-introduction

DHI (2009) MIKE 11: a modeling system for rivers and channels. Reference manual. Danish Hydraulic Institute, Denmark

Greimann B, Lai Y, Huang JC (2008) Two-dimensional total sediment load model equations. J Hydraul Eng 134(8):1142–1146

HR Wallingford (2001) ISIS sediment 2001-user manual, report, HR Wallingford, U.K.

Huang J, Greimann B (2012) SRH-1D 3.0 User's manual (sedimentation and river hydraulics—one dimension, version 3.0). Technical report SRH-2013-11, Technical Service Center, US Bureau of Reclamation, Denver, CO, USA

Lai YG (2008) SRH-2D version 2: theory and user's manual. U.S. Department of the interior—Bureau of Reclamation, Denver, CO, USA

Lai YG (2010) Two-dimensional depth-averaged flow modeling with an unstructured hybrid mesh. J Hydraul Eng 136(1):12–23

Langendoen EJ (2000) CONCEPTS—CONservational channel evolution and pollutant transport system report. U.S. Department of Agriculture, Agricultural Research Service, National Sedimentation Laboratory, Oxford, MS, USA

Mahdi T (2009) Semi-two-dimensional numerical model for river morphological change prediction: theory and concepts. Nat Hazards 49(3):565–576

Vieira DA, Wu W (2002) One-dimensional channel network model CCHE1D version 3.0—user's manual, report. National Center for Computing Hydroscience and Engineering, University of Missisipi, Oxford, MS, USA

Yang CT, Simoes F (2002) User's manual for GSTARS 3.0 (Generalized stream tube model for alluvial river simulation version 3.0). U.S. Bureau of Reclamation, Technical Service Center, Denver, CO, USA

Yang CT (2003) Sediment transport: theory and practice. Krieger Publishing Company, Malabar

Evaluation of the Hydraulic Design of Culverts Under a Climate Change Scenario: A Preliminary Analysis of Road Case Studies in Southern Portugal

Gonçalo Bastos, Jorge Matos, and José Neves

Abstract

The present work aims to contribute to the study of current capacity of culverts associated to transportation infrastructures, under a climate change scenario. The records of annual maximum daily precipitation were analysed for two meteorological stations, allowing to estimate such precipitation for both stations, namely through the adjustment of a statistical law for a return period of 100 years. Design precipitations were then determined through a methodology that makes use of annual maximum daily precipitation estimation. Through hydrological and hydraulic methodologies frequently used in hydraulic and drainage projects in Portugal, eight culverts were evaluated. The hydraulic analysis led to the conclusion that three out of eight culverts had an insufficient flow rate capacity. According to those results, it is not possible to conclude that climate change is directly influencing the adequacy of such culverts, in terms of flow rate capacity. Nevertheless, it seems reasonable to expect that a considerable number of culverts may no longer be suitable in terms of flow rate capacity, resulting in a potential high number of critical points in the Portuguese road network, jeopardizing the safety of the transportation infrastructure and its users.

Keywords

Climate change • Annual maximum daily precipitation • Design flow • Culvert • Hydraulic design • Flow rate capacity

G. Bastos (✉)
Instituto Superior Técnico, Universidade de Lisboa, Lisbon, Portugal
e-mail: bastos.g.goncalo@gmail.com

J. Matos · J. Neves
Department of Civil Engineering, Architecture and Georesources, CERIS, Instituto Superior Técnico, Universidade de Lisboa, Lisbon, Portugal

1 Introduction

Climate change has been identified as one of the main challenges that modern societies will face in the next few decades and remains as the main environmental challenge nowadays (Nordhaus 2018).

Scientific interest about intense precipitation events and the occurrence of floods has been growing in the last few decades, namely because it brings consequences to water resources infrastructures. In fact, one of the major concerns about climate change is that precipitation records have been changing due to global warming, which may lead to the occurrence of extreme events and its frequency increase (IPCC 2013). In turn, the increase of extreme precipitations events may possibly lead to increased flows in natural watercourses (see Fig. 1).

The present study focuses on the analysis of road culverts. The main problem related to culverts is associated to insufficient flow rate capacity. Thus, the estimation of the design flow is a critical process in culvert design. This study arises from the need to assess if the high number of culverts built in the past decades in Portugal still have a suitable flow rate capacity under a climate change scenario. In fact, facing the records of global climate change, it is valid to investigate, according to Portugal precipitation records, if the methodologies used in drainage projects lead to new design flows, which may not be accommodated by some of the existing culverts. Therefore, the adequate hydraulic operation of those culverts may be at risk.

2 Case Studies Description and Methodology

For the evaluation of culverts under a climate change scenario, eight case studies were selected from the Portuguese road network (see Table 1). Five case studies are in Beja district (built in the 1920s) and the other three are in Évora

© Springer Nature Switzerland AG 2020
F. Fernandes et al. (eds.), *Advances in Natural Hazards and Hydrological Risks: Meeting the Challenge*, Advances in Science, Technology & Innovation,
https://doi.org/10.1007/978-3-030-34397-2_34

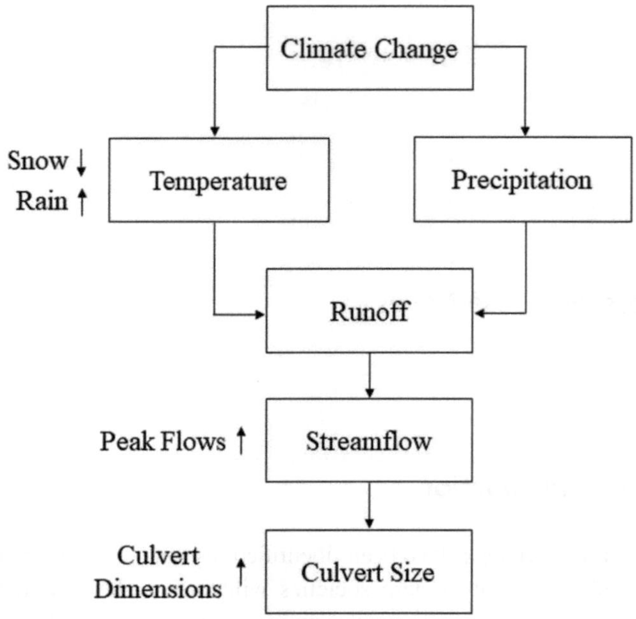

Fig. 1 Causal relationships between climate change and hydraulic culvert design. Adapted from Wilhere et al. (2017)

district (built in the 1950s). Two case studies are presented in Fig. 2. For both districts, located in southern Portugal, two meteorological stations were selected to obtain the precipitation records, based on the adequacy of their locations. For Beja district, Relíquias station was selected. Its annual maximum daily precipitation records suggest a slight decreasing tendency since the last five to six decades. For Évora district, Pavia station was selected, where an increasing tendency in the annual maximum daily precipitation is noticeable between about 1930 and 1970, followed by a decreasing tendency since then.

For the hydrological and hydraulic analyses of the case studies, a methodology divided into three main steps was followed.

The first step comprised the estimation of the design precipitation, P_t, based on the methodology developed by Portela (2006). This procedure combines annual maximum daily precipitation records with mainland Portuguese rainfall characteristics associated to the parameters of the IDF curves developed by Brandão et al. (2001). Annual maximum daily precipitation, P_{24}, was estimated for both meteorological stations using the Log-Normal statistical law for a return period of 100 years.

The design flow, Q, was calculated in the second step. Based on the watershed characteristics associated to each case study, it was possible to estimate the concentration-time, t_c, for each one. Design rainfall intensity, I, was then determined through the relation between P_t and t_c, assuming that the rainfall duration was equal to t_c. Through hydrological methods frequently used in drainage projects in Portugal, such as the Rational method and the Soil Conservation Service (SCS) method, it was possible do estimate the design flow for each case study.

Finally, the third step consisted of the hydraulic analysis. This analysis investigates if the culverts have enough flow rate capacity for the new design flows by using a methodology developed by U.S. Department of Transportation (Martins 2000; Ramos 2005; Schall et al. 2012). The flow rate capacity was evaluated by comparing the culvert height, D, with the critical depth, h_c. In case the culvert has, in fact, a flow rate capacity, then the inlet headwater, H_w, is calculated assuming an inlet flow control and then an outlet flow control. The higher one will define the control section of the flow. Note that, as a safety criterion, it was assumed that $H_w/D \leq 1.35$, in a centenary flood situation.

3 Results

All case studies were analysed through the previously mentioned methodology. The main results (see Table 2) showed that three out of eight culverts did not have a suitable flow rate capacity for the estimated design flows. On the other hand, the remaining five culverts had an adequate flow rate capacity, although two of them with an outlet control section.

Table 1 Case studies characteristics

District location	Culvert designation	NM[a]	Type	Dimensions[b]	Length (m)	Slope (%)
Beja	B-1	2	Rectangular	1.76 m × 1.54 m	19.15	1.00
Beja	B-2	1	Rectangular	0.63 m × 0.56 m	8.93	1.00
Beja	B-3	2	Rectangular	1.60 m × 2.00 m	14.78	1.00
Beja	B-4	1	Rectangular	2.60 m × 2.70 m	18.00	0.50
Beja	B-5	2	Rectangular	3.30 m × 3.40 m	22.71	1.20
Évora	E-1	1	Circular	1.20 m	28.15	1.00
Évora	E-2	1	Circular	1.00 m	17.05	0.50
Évora	E-3	1	Rectangular	1.60 m × 2.00 m	25.50	1.00

[a]Number of modules; [b]it refers to the diameter, for the circular cross-section

Fig. 2 Case studies: **a** B-1 (inlet); **b** E-1 (outlet)

(a) (b)

Table 2 Hydrological and hydraulic results

Culvert designation	P_{24} (mm)	P_t (mm)	Q (m³/s)	h_c (m)	Flow rate capacity	H_w/D (–)	Control section
B-1	112.45	53.83	29.21	1.91	No	–	–
B-2	112.45	31.65	2.58	1.20	No	–	–
B-3	112.45	63.04	28.19	1.99	Yes	1.59	Outlet
B-4	112.45	61.53	20.27	1.84	Yes	1.03	Inlet
B-5	112.45	68.60	44.74	1.67	Yes	0.74	Inlet
E-1	101.59	37.71	8.32	1.57	No	–	–
E-2	101.59	30.16	2.84	0.96	Yes	1.93	Outlet
E-3	101.59	34.54	6.71	1.05	Yes	0.79	Inlet

Case studies B-1, B-2 and E-1 did not have enough flow rate capacity for the new design flows. In fact, for these culverts, the critical depth was higher than the section height, leading to an absence of flow control section. On the other hand, case studies B-3 and E-2 had a sufficient flow rate capacity for the estimated design flows. However, those culverts would experience an outlet flow control section which may lead to subcritical flows inside the culverts or even to a submerged outlet. Also, culverts B-3 and E-2 did not satisfy the maximum upstream headwater. In fact, the relation H_m/D was larger than 1.35, which may put at risk the landfill and transportation infrastructure safety. The other three case studies, B-4, B-5 and E-2, did not show any problems related to flow rate capacity or upstream maximum headwater. Hence, they may be considered appropriate to accommodate the new design flows.

4 Discussion

The results obtained by the hydraulic analysis are related to a sample of eight culverts distributed in Beja and Évora districts, in southern Portugal.

According to those results, three out of eight culverts (two in Beja district and another one in Évora district) did not have a satisfactory flow rate capacity for the new design flows, representing almost 40% of the analysed sample. In contrast to what could possibly be expected, the number of problematic culverts was higher in Beja district, where the annual maximum daily precipitation records suggested a more stable decreasing tendency over various decades. In turn, only one culvert out of three seemed to show an insufficient flow rate capacity for Évora district. For all these

Table 3 Suggested solutions for culverts with inadequate flow rate capacity

Culvert designation	Type	Dimensions[a]	Length (m)	Slope (%)	H_w/D (–)	Control section
B-1	Rectangular	2.75 m × 2.75 m	19.15	0.55	1.24	Inlet
B-2	Circular	1.20 m	8.93	1.00	1.10	Inlet
E-1	Rectangular	1.75 m × 1.75 m	28.15	1.00	1.14	Inlet

[a]It refers to the diameter, for the circular cross-section

three culverts, the recommended solution is to replace the culvert by another one with suitable flow rate capacity (see Table 3).

All the other five culverts had enough flow rate capacity, although two of them with an outlet control section, which is not advisable for an adequate hydraulic functioning. A possible solution for an outlet controlled flow would be to readjust the longitudinal profile of natural watercourse downstream of the culvert, or even the installation of a ditch.

For every culvert, it is advisable to include an outlet energy dissipation structure such as a rockfill bed. Such a structure helps avoiding erosion at the culvert's outlet. The design of those structures depends on the culvert's outlet velocity, which allows to calculate the rockfill medium characteristic diameter.

5 Concluding Remarks

In the present study, the results of the hydrological and hydraulic analyses of different road culverts in southern Portugal were investigated. Flow rate capacity was evaluated considering revised design flows for a return period of 100 years, based on annual maximum daily precipitation records of two meteorological stations. The main objective of the present study was to verify if existing culverts would still have a suitable flow rate capacity under a possible climate change scenario.

From a sample of eight culverts, it was possible to observe that about 40% of such samples would not have enough flow rate capacity, which is a high percentage.

Although it would not be correct, from a scientific point of view, to assign climate change with those culverts malfunction, it seems reasonable to expect that a considerable number of culverts may no longer be suitable in terms of flow rate capacity, resulting in a potentially high number of critical points in the Portuguese road network.

A further investigation of this preliminary study should consider an extended sample of culverts, covering other regions of the Portuguese territory, along with a review of the hydrological analysis and hydraulic design methods.

Acknowledgements The authors would like to thank to the Public Portuguese Infrastructures Administration (IP) and to COBA—Engineering and Environmental Consultants—in particular to André Colaço, for all the support given throughout this study.

References

Brandão C, Rodrigues R, Costa, JP (2001) Análise de fenómenos extremos. Precipitações intensas em Portugal Continental. Direcção dos Serviços de Recursos Hídricos (in Portuguese)

IPCC (2013) Climate change 2013: the physical science basis. In Stocker TF, Qin D, Plattner G-K, Tignor M, Allen SK, Boschung J, Nauels A, Xia Y, Bex V, Midgley PM (eds) Contribution of working group I to the fifth assessment report of the intergovernmental panel on climate change. Cambridge University Press, Cambridge, United Kingdom and New York

Martins F (2000) Dimensionamento Hidrológico e Hidráulico de Passagens Inferiores Rodoviárias para Águas Pluviais. Faculdade de Ciências e Tecnologia da Universidade de Coimbra, Coimbra (M.Sc. thesis) (in Portuguese)

Nordhaus W (2018) Projections and uncertainties about climate change in an era of minimal climate policies. Am Econ J Econ Policy 10(3): 333–360

Portela MM (2006) Estimação de Precipitações Intensas em Bacias Hidrográficas de Portugal Continental. Associação Port dos Recursos Hídricos 27(1):15–32 (in Portuguese)

Ramos C (2005) Drenagem em Infra-Estruturas de Transportes e Hidráulica de Pontes, 1st edn. LNEC, Lisbon (in Portuguese)

Schall JD, Thompson PL, Zerges SM, Kilgore RT, Morris JL (2012) Hydraulic design of highway culverts, 3rd edn. U.S Department of Transportation, Federal Highway Administration, USA

Wilhere GF, Atha JB, Quinn T, Tohver I, Helbrecht L (2017) Incorporating climate change into culvert design in Washington State. USA. Ecol Eng 104(A):67–79

Around of Pico Island Geology: Meeting Natural Hazards

Geology and Volcanology of Pico Island (Azores, Portugal): A Field Guide

João Carlos Nunes

Abstract

Pico is the biggest island of the Central Group, the second-largest (445 km^2) of the Azores archipelago and exhibits the highest point of Portugal (2351 m), the Pico Mountain volcano, which is the 3rd highest volcano of the North Atlantic Ocean. In relation to the surrounding sea-floor Pico Mountain is a 3500 m high polygenetic volcano, characterized by dominant effusive basaltic volcanism episodes of Hawaiian-type eruptions and extensive *pahoehoe* lava fields. The main geological features of the island are related to the fact that Pico is the youngest island of the archipelago (about 300,000 years old) and its volcanism is almost exclusively of basaltic nature, evident either in the Pico Mountain volcano, the Topo shield volcano or in the volcanic ridge of the Planalto da Achada. This volcanic ridge (e.g., São Roque-Piedade Volcanic Complex) has a length of 30 km and is composed of about 190 scoriae and spatter cones and eruptive fissures. This work outlines the field trip on the Pico Island volcanoes and eruptive history, with emphasis on its main morphotectonics features and geological hazards.

Keywords

Pico Island • Eruptive history • Polygenetic • Volcanic ridge • Geological hazard

1 Introduction

Pico Island is located on the Faial-Pico Fracture Zone, a 350 km long leaky transform with a general WNW-ESE trend between the Mid-Atlantic Ridge and an area south of the Hirondelle basin, that generally marks the southern border of the so-called “Azores Block” (Fig. 1). The elongated shape of Pico Island along that tectonic alignment demonstrates the strong tectonic control of its volcanism (e.g., Nunes et al. 2006; Nunes 2004; França et al. 2003; Pacheco et al. 2013; Kueppers and Beier 2018).

Three main groups of tectonic lineaments are observed on Pico Island (Fig. 2): (i) WNW-ESE lineaments, the most common ones; (ii) NNW-SSE lineaments, mostly as left strike-slip faults, with a dip-slip component, and (iii) NE-SW lineaments, which is also the main trend of most of the dykes in the island (Nunes 1999). In terms of frequency and length of tectonic features the WNW-ESE to E-W directions dominates, some with clear displacements of the topography (e.g., fault scarps and local, small *graben*-like structures). The majority of these tectonic features are right strike slip faults (Madeira 1998).

Morphologically the Pico Island is characterized by three main units (Nunes 1999): (i) the 2350 m high basaltic stratovolcano of the Pico Mountain, with a subaerial volume of 97 km^3, that dominates the western part of the island, (ii) a shield-like volcanic structure, the Topo Volcano, located on the south, central part of the island and (iii) a 30 km long, WNW-ESE to E-W volcanic ridge, the Planalto da Achada, between the central part of the island and its eastern end, characterized by about 190 scoria and spatter cones and associated lava flows (Fig. 3).

The present work is based on the field guide produced to support the Technical Visit established under the framework of the NATHAZ19 (2nd International Workshop on Natural Hazards), organized by the LREC—Regional Civil Engineering Laboratory, that took place in Pico Island, Azores. The itinerary of the visit is presented in Fig. 4.

J. C. Nunes (✉)
University of Azores, Ponta Delgada, S. Miguel Island, Portugal
e-mail: joao.cc.nunes@uac.pt

© Springer Nature Switzerland AG 2020
F. Fernandes et al. (eds.), *Advances in Natural Hazards and Hydrological Risks: Meeting the Challenge*, Advances in Science, Technology & Innovation,
https://doi.org/10.1007/978-3-030-34397-2_35

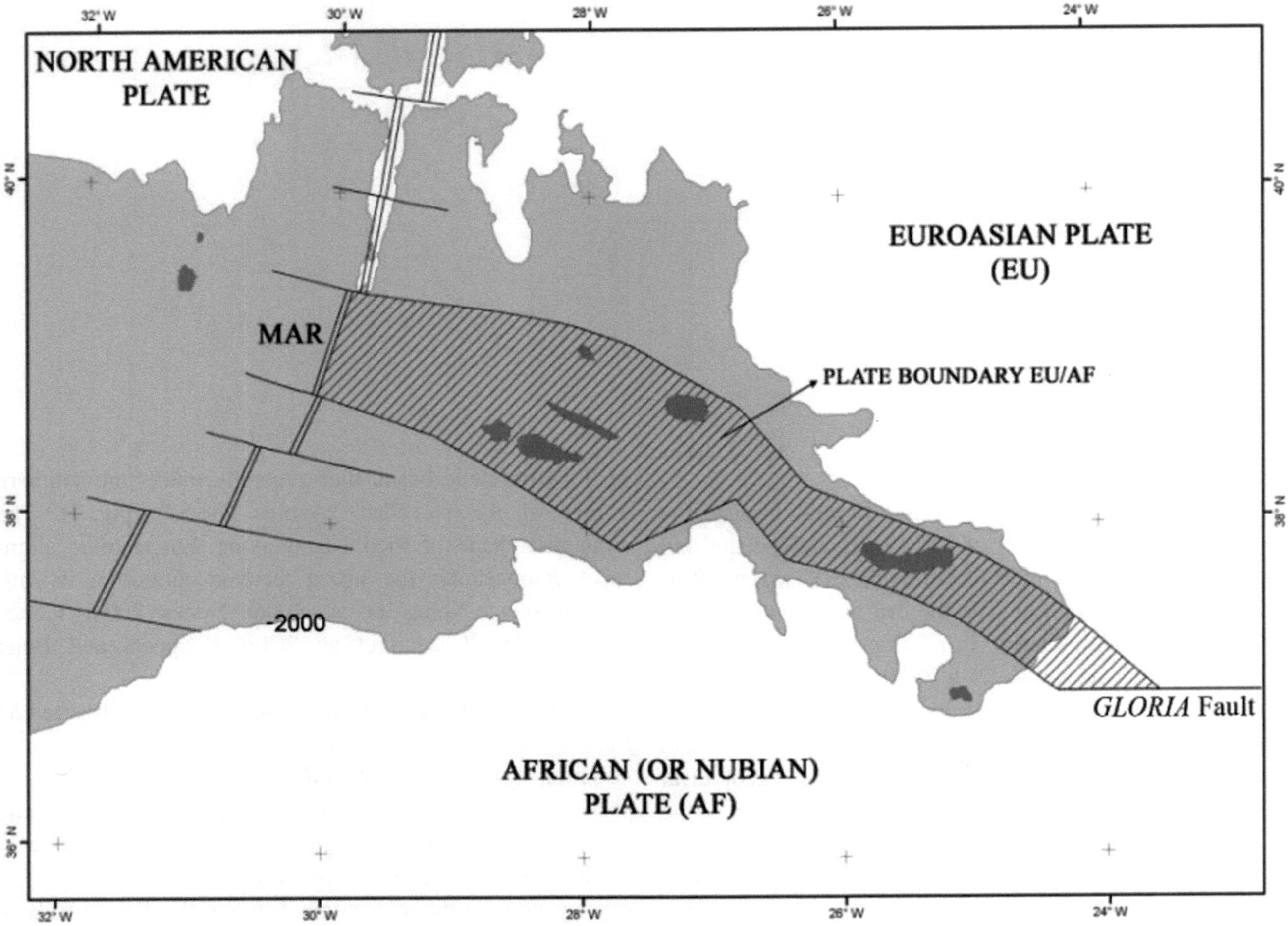

Fig. 1 General geotectonic framework of the Azores Triple Junction (ATJ). *MAR*: Mid-Atlantic Ridge, the western linear/discrete boundary of the ATJ; shaded area: the "Azores Block or Domain", the diffuse plate boundary between EU and AF plates; gray area: the "Azores Plateau", defined by the 2000 m bathymetric contour line. © J.C. Nunes

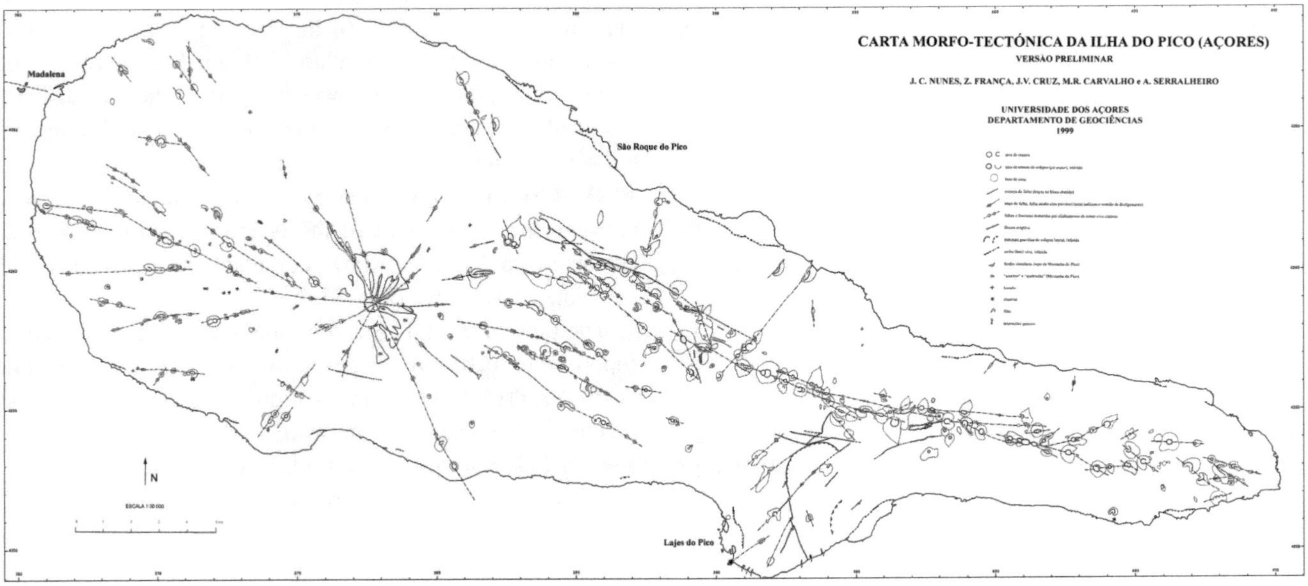

Fig. 2 Morphotectonic map of Pico Island (Nunes 1999)

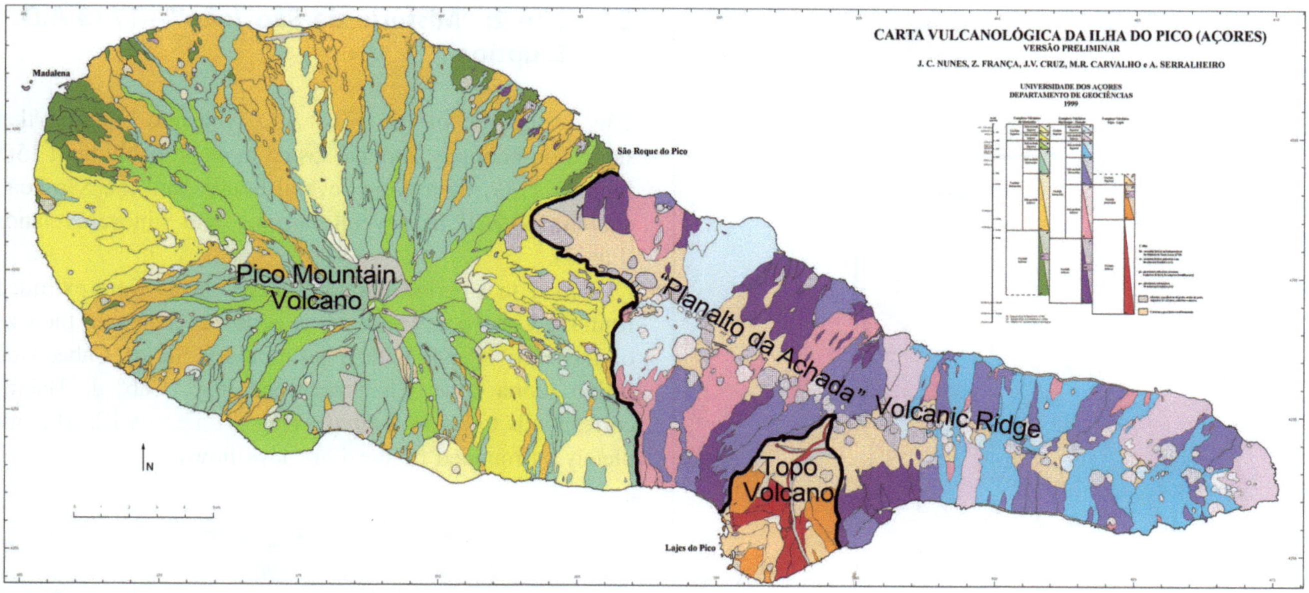

Fig. 3 Volcanological map of Pico Island (Nunes 1999)

Fig. 4 Itinerary of the NATHAZ19 Technical Visit (IGeoE topographic map)

2 Description of Visit Sites

2.1 Site 1: Lajes do Pico Lava Delta

The Lajes do Pico village is emplaced on a lava delta (lava "fajã", as locally named), formed by *pahoehoe* type lava flows extruded from a small spatter cone located near the Cabeço do Geraldo scoria cone (Fig. 5). Both the spatter and the scoria cones are monogenetic volcanoes emplaced on the SW flanks of the Topo shield volcano.

The old sea cliff that was spilled over by the Lajes do Pico basaltic lava flow is well preserved to the East of the village, e.g., along the main road, and South of the lava delta is located the volcanic neck of Castelete, the inner remains of an old scoria cone (Fig. 6).

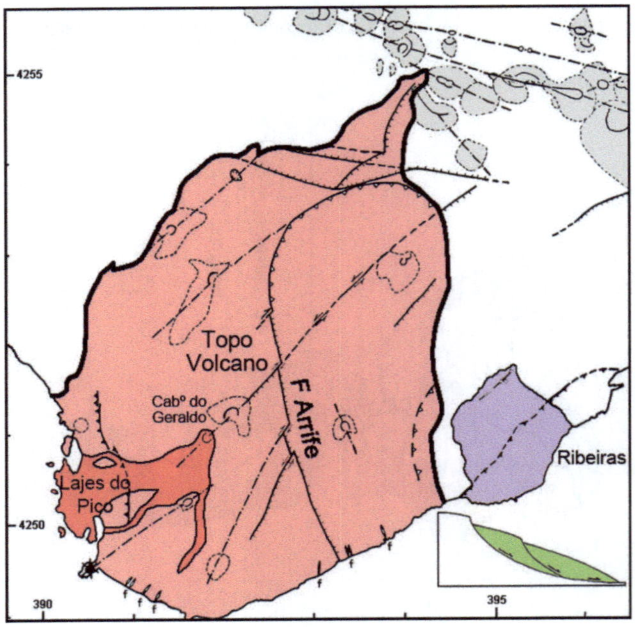

Fig. 5 Morphotectonic map of Topo shield volcano. The spatter cone and associated lava flow that originated the Lajes do Pico lava delta are highlighted in red. The NE-SW eruptive fissure and lava flow of the Ribeiras lava delta are highlighted in blue. Adapted from Nunes (1999)

Fig. 6 Aerial view of the Lajes do Pico lava delta and the volcanic neck of Castelete, south of the lava delta. *Photo* P. H. Silva (SIARAM)

This low altitude coastal platform is vulnerable to flooding episodes namely associated with storms ("enchentes de mar" as named in old records): to mitigate such risk an Antifer blocks breakwater was built on its northern sector (Fig. 6).

2.2 Site 2: "Mistério de São João"—1718 A.D. Eruption

This is the site of the 1718 A.D. volcanic eruption that started on February 2nd and lasted most probably until 15th January 1719, destroying several houses and the São João church, as well as several farming and vineyards lands (Fig. 7).

During that period basaltic *aa*-type lavas were extruded from the Cabeço de Cima scoria cone that increased the size of the Pico Island, with a promontory SE of Cabeço de Baixo scoria cone. The later cone (e.g., Cabº de Baixo) corresponds today to a *kipuka* (Fig. 8), i.e., an island completely surrounded by the 1718 lava flows.

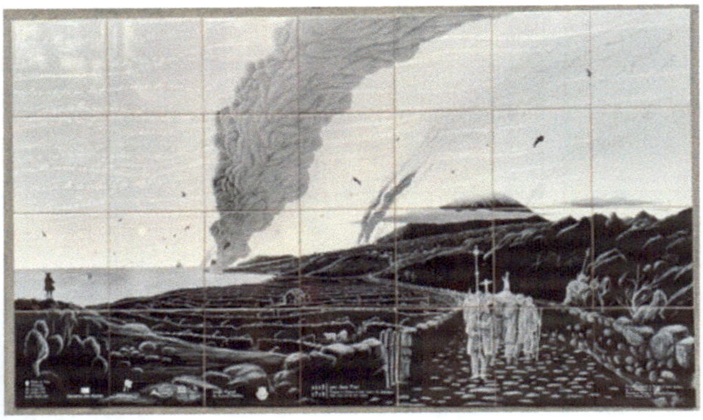

Fig. 7 Tiles panel put in place on the evocation of the 300th years of the 1718 A.D. historical eruption. *Photo* M. P. Costa

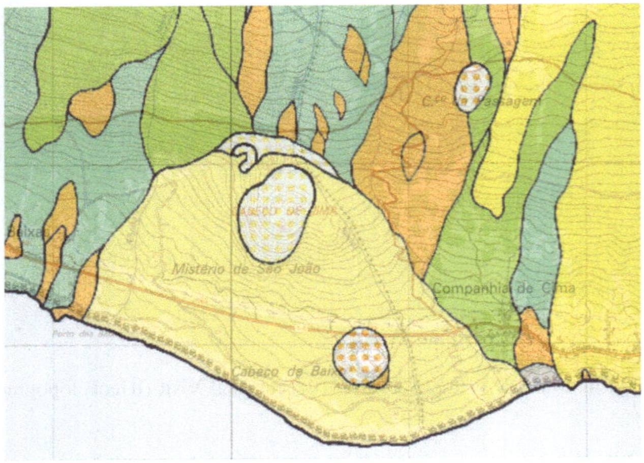

Fig. 8 Lava field of the 1718 A.D. eruption, with eruptive center on the Cabeço de Cima scoria cone. Cabeço de Baixo scoria cone is nowadays a *kipuka* (Nunes 1999)

The "mistério" wording: scared and lacking understanding about those natural phenomena, local people named after "mistério" ("mystery") the lava flows they observed during the volcanic eruptions.

2.3 Site 3: Alluvial Deposits of São Caetano-Santa Margarida

While the upper northern and eastern flanks of the Pico Mountain volcano are dominated by impressive epiclastic deposits—resulting from rockfall and avalanche-type mass movements (the designated "arieiros" and "quebradas", Fig. 9)—the low altitude and coastal southern slopes of the Pico Mountain volcano are dominated by massive alluvial deposits associated with important streams in the area, namely the Ribeira Grande, Ribeira Nova and Ribeira do Dilúvio streams, with torrential regime.

Thus, the geological formations that characterize an extensive area between São Caetano and Santa Margarida urban areas are thick and impressive alluvial deposits (cf. Pico Island volcanological map, Figs. 3 and 10), often mixed with colluvial deposits, driven by the steep slopes of Pico Mountain upwards and also the São Caetano Fault.

This way, the adequate land-use management of that area imposes this as a *non aedificandi* area and justifies the hydraulic infrastructures recently implemented in the area, as a relevant measure to mitigate the often river floods occurring in the area, many times jeopardizing namely the main road that cross it.

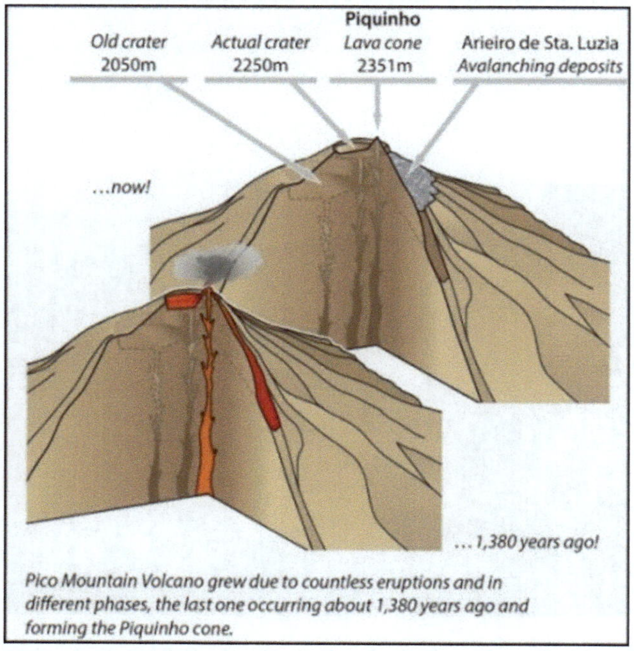

Fig. 9 Schematic evolution of Pico Mountain stratovolcano, with a general northwards migration of its plumbing system (cf. old crater, actual crater and Piquinho entrails *pahoehoe* lavas cone)

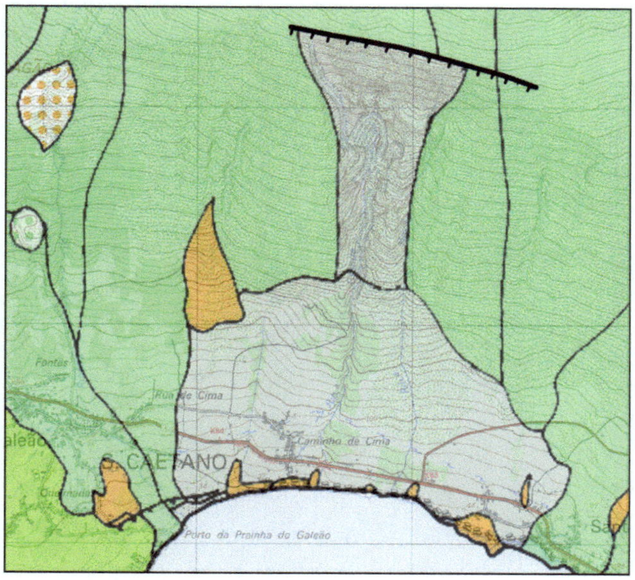

Fig. 10 Massive alluvial deposits often mixed with colluvial deposits (in gray) driven by the steep slopes of Pico Mountain upwards and also the São Caetano Fault (e.g., hachure line). Nunes (1999)

2.4 Site 4: Gruta das Torres Lava Cave

The Gruta das Torres lava cave was formed by *pahoehoe*-type basaltic lava flows extruded from the Cabeço Bravo scoria cone (Fig. 11) and is the largest lava tube cave known in the Azores: it has a total length of about 5150 m and 15 m of maximum height. It is composed of one main large-sized tube and several secondary lateral and upper channels, which are smaller in size but display a greater variety of geological structures.

The cave is rich in speleological formations, namely different kinds of lava stalactites and stalagmites, lateral benches, lava balls, grooved walls, and ropy lavas. The floor is formed by *aa* and *pahoehoe* lavas and is well preserved in most parts of the cave (Fig. 12). In some areas, the cave's walls are covered by secondary mineral deposits (e.g., silica) and microbiological activity deposits (e.g., bacteria and molds).

This volcanic cave is a protected area and its Visitors Centre was open to the public on May 24th, 2005.

2.5 Site 5: "Lajido da Criação Velha" *Pahoehoe* Lava Field

The word "lajido" applies to the flatten, smooth and regular surface lava flows of basaltic nature that characterizes

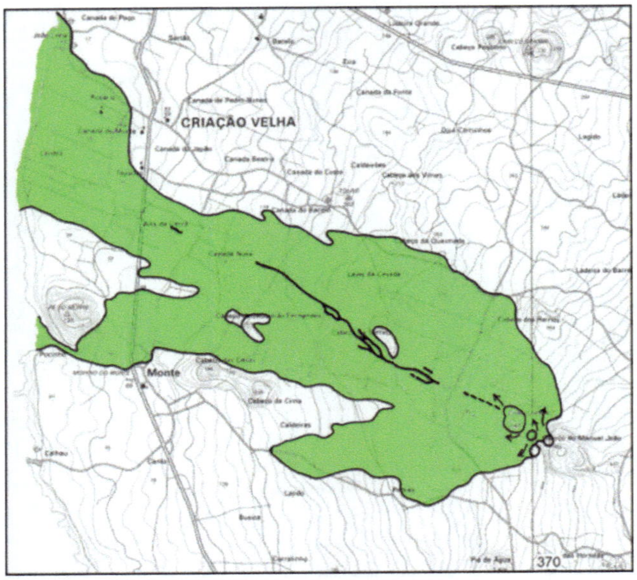

Fig. 11 Lava field of the Cabeço Bravo—Gruta das Torres volcanic episode, on the western flanks of Pico Mountain volcano. Nunes (1999)

Fig. 12 Gruta das Torres lava cave main section. *Photo* P. H. Silva (SIARAM)

several areas of Pico Island, mostly in the Pico Mountain volcano area: thus "lajido" can be considered a synonymous of *pahoehoe*-type lava flows.

Tumulus, pressure ridge, lava trench, *pahoehoe toe*, ropy lava, lava blister, and lava tree mold, are some of the structures and micro-landforms types that can be observed in this lava field (Fig. 13).

This is also an important Azorean geo-cultural landscape, part of the Landscape of the Pico Island Vineyard Culture WHS—World Heritage Site, nominated by UNESCO in 2004. It is the case of the "rola-pipas", man-cut aperture in the lava flows to allow the transport of the wine barrels to nearby anchored small ships (Fig. 14).

2.6 Site L (Lunch): "Mistério de Santa Luzia"—1718 A.D. Eruption

The "Mistério de Santa Luzia" corresponds to the benmoreite/mugearite-type lava flows of the 1718 A.D. eruption that started at 6 a.m. on February 1st, at the northern flanks of Pico Mountain volcano (Fig. 15).

Together with the "Mistério de São João" and the submarine vent offshore the south coast, the 1718 A.D. eruption puts into evidence a major NW-SE volcano-tectonic lineament that crosses the Pico Mountain volcano (cf. Pico Island morphotectonic map, Fig. 2).

Tiles panels were put in place on the evocation of the 300th years of the 1718 A.D. historical eruption that greatly affected the Pico Island inhabitants and was the seed for relevant cultural events at Pico and Faial islands.

Fig. 13 Ropy lavas in the Lajido da Criação Velha *pahoehoe* lavas field. *Photo* P. H. Silva (SIARAM)

Fig. 14 "Rola-pipas" carved in the *pahoehoe* lavas field. *Photo* P. H. Silva (SIARAM)

2.7 Site 6: "Lajido de Santa Luzia" *Pahoehoe* Lava Field and UNESCO WHS

Like the Criação Velha field, this "lajido" area is composed by very fluid basaltic lava flows, most of them extruded from the uppermost part of the Pico Mountain volcano, and sometimes flowing for more than 10 km before reaching the sea, as in the Arcos area (Fig. 16). Those *pahoehoe* lava flows exhibits the diversity of structures and micro-landforms that were mentioned before (cf. Site 5), and to the west they contrast and overlaps *aa*-type lava flows.

This area is the main core of the Pico Island Vineyard Culture World Heritage, including its interpretation center, the site of the Azores UNESCO Global Geopark island delegation and of the future "Casa dos Vulcões" (Volcanoes House), where the links between the Man and the geodiversity are promoted and valued. As it is the case of the "relheiras" (Fig. 17), linear and parallel grooves carved in the massive, hard *pahoehoe* lavas by the friction of the bowls cars wheels, decade after decade!

2.8 Site 7: Santo António Old Sea Cliff

Between Santo António and São Roque do Pico villages, and for about 4.5 km length, there is an old sea cliff often as a rocky steep slope that marks the old coastline of the Pico Island in this area (Fig. 18).

In several places younger basaltic lava flows spill over the edge of this old sea cliff, often as very fluid pahoehoe type lava flows extruded from the Pico Mountain Volcano that gave origin to spectacular lava cascades (Fig. 19), as those that can be seen at the Furna—camping area, and at Ginjal.

In the nearby old quarry of "Nariz de Ferro" outcrops the ankaramite lava flow, which blocks were used to build the São Roque do Pico harbor breakwater.

2.9 Site 8: Lagoa do Capitão Lake and Fault Scarp

The Capitão lake is emplaced in the westernmost part of the São Roque—Piedade volcanic ridge, on a tectonic sag pond (Fig. 20). Thus, the lake is bordered by the Capitão fault

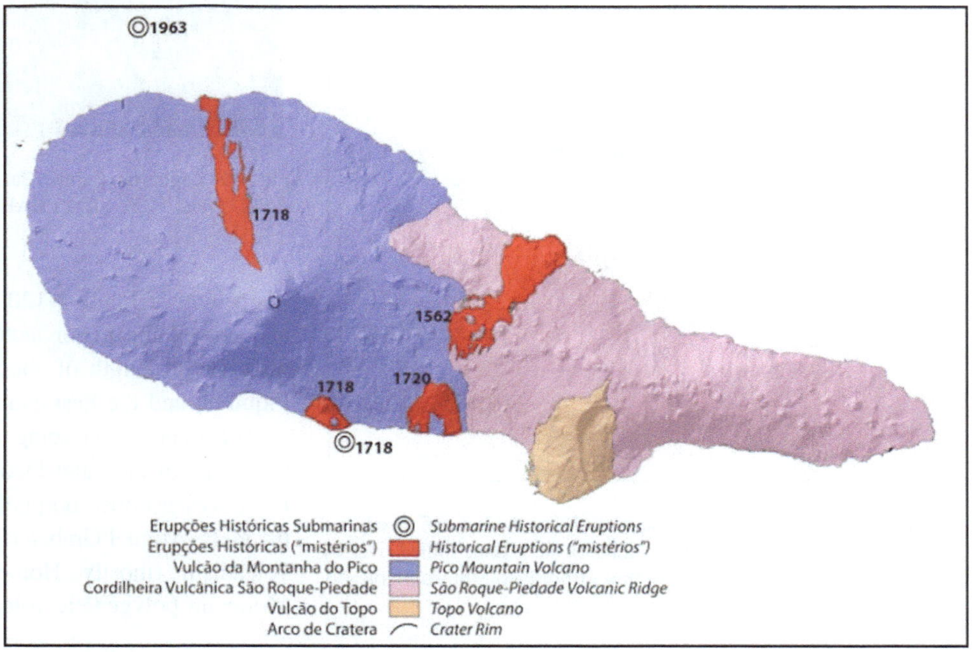

Fig. 15 Simplified volcanological map of Pico Island, including the historical eruptions ("mistérios"), in red. Nunes (2004)

Fig. 16 "Lajido de Santa Luzia" *pahoehoe* lavas field. *Photo* M. P. Costa

Fig. 17 "Relheiras" carved in the massive, hard *pahoehoe* lavas field by the bowls cars wheels. *Photo* M. P. Costa

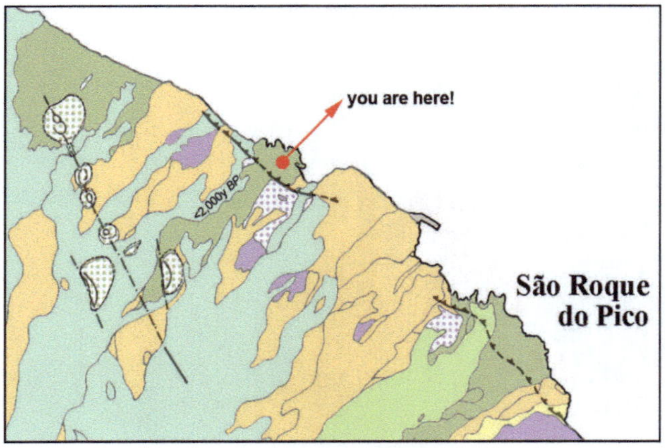

Fig. 18 Santo António—São Roque do Pico old sea cliff (hachured line), the signature of the past coastline of the island, spilled over by Holocene (e.g., <2000 y B.P.) lava flows from Pico Mountain volcano. Nunes (1999)

Fig. 19 Old sea cliff (cf. steep slope on the right) of Santo António. *Photo* M. P. Costa

Fig. 20 Lagoa do Capitão lake and fault scarp (steep slope on the right). *Photo* P. H. Silva (SIARAM)

scarp, with azimuth N120°, maximum high of about 20 m and one of the major tectonic features of the Pico Island, with a total length of about 9 km, between the Cabeço do Piquinho and the Passagem/Caiado lake areas (Fig. 21).

This fault scarp (facing towards south) is a morphological testimony of the Faial-Pico Fracture Zone in this island and is the westernmost morphological signature in Pico island of the Pedro Miguel Graben (Faial Island), hidden by the young volcanism (mostly Holocene lava flows) of the Pico Mountain polygenetic volcano.

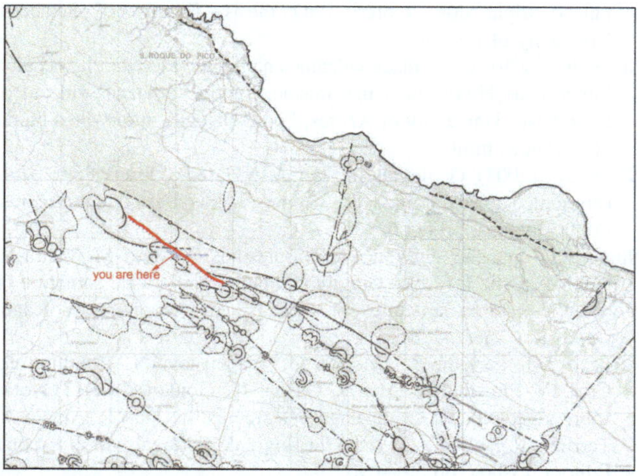

Fig. 21 Morphotectonic map of Pico Island with Lagoa do Capitão fault scarp highlighted in red. Adapted from Nunes (1999). Topographic base map by IGeoE

Fig. 22 Aerial view of the Arrife fault. Note the NE-SW fault scarp crosscutting the Arrife fault (see also Fig. 5). *Photo* P. H. Silva (SIARAM)

2.10 Site 9: Arrife Fault Scarp and Topo Volcano Gravitational Collapses

The Arrife fault scarp marks the western border of a massive gravitational collapse structure that affects the south flanks of the Topo shield volcano (Figs. 5 and 22). This polygenetic volcano (built about 95% by basaltic lava flows), comprises the oldest geological formations of Pico Island, with an estimated maximum age of about 300 ka, that outcrop only at the base of the shield volcano pile, at the Arrife shoreline (Fig. 23).

Located east of Arrife, the Ribeiras lava delta was formed in several episodes, being its youngest sector (e.g., south) associated with a NE-SW eruptive fissure active about 3500 years ago (Fig. 5).

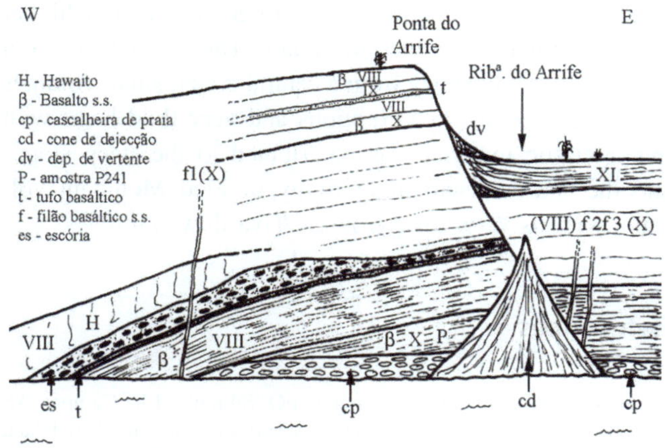

Fig. 23 Schematic cross-section on the older sector of Topo shield volcano, including Arrife fault. Nunes (1999)

3 Concluding Remarks

Previous studies (e.g., Nunes 1999) based on the absolute ages available on Pico Island eruptions, coupled with field relations determined in connection with detailed geological mapping were able to provide a semi-quantitative assessment of volcanic productivity during the recent eruptive history of Pico Island. The figures obtained for Pico Mountain volcano over the past 1500 years give average productivity of 0.040 km^3/century, and for the São Roque-Piedade (e.g., Planalto da Achada) volcanic ridge over the past 2000 years the average productivity is 0.023 km^3/century. This gives an overall figure for the productivity of the island of 0.063 km^3/century.

Assuming that those productivity figures have remained approximately constant during the formation of the subaerial part of Pico Mountain volcano (e.g., 2350 m high and total volume of 97 km^3), the age of the subaerial basement of that polygenetic volcano would be about 240,000 years. Similarly, the maximum age expected for the all Pico Island (e.g., 207 km^3) would be about 330,000 years, well within the error limit of the K/Ar dates available at the time and in good agreement with absolute ages of old rocks exposed in the island (cf. Fig. 23). Nevertheless, recent papers dealing with new ages for Pico Island (e.g., Costa et al. 2015; Silva et al. 2018) report some inconsistencies between sampled formations location, ages obtained, and volcanostratigraphic relationships observed in the field and already mapped.

As a final note, it is worth mentioning the fact that the higher volcanic productivity of Pico Mountain volcano than that of São Roque-Piedade volcanic ridge is compatible and strengthened with the observation that lava flows emitted from Pico Mountain are smaller in volume than those erupted on the São Roque-Piedade ridge, indicating a higher eruption frequency in Pico Mountain. The combination of a high eruption frequency from a central crater, small volumes produced by individual eruptions and very fluid lavas with minor clastic material, has contributed to the steep slopes and the stratovolcano morphology of Pico Mountain volcano, with its very low pyroclasts/lava flows ratio.

References

Costa ACG, Hildenbrand A, Marques FO, Sibrant ALR, Campos AS (2015) Catastrophic flank collapses and slumping in Pico Island during the last 130 kyr (Pico-Faial ridge, Azores Triple Junction). J Volcanol Geoth Res 302:33–46

França Z, Cruz JV, Nunes JC, Vorjaz VH (2003) Geologia dos Açores: uma perspectiva actual. Açoreana 10(1):11–140

Kueppers U, Beier C (eds) (2018) Volcanoes of the Azores: revealing the geological secrets of the Central Northern Atlantic Islands. Springer, Berlin

Madeira J (1998) Estudos de neotectónica nas ilhas do Faial, Pico e S. Jorge: uma contribuição para o conhecimento geodinâmico da junção tripla dos Açores. PhD thesis, Faculty of Sciences, University of Lisbon

Nunes JC (1999) A actividade vulcânica na Ilha do Pico do Plistocénico Superior ao Holocénico: mecanismo eruptivo e hazard vulcânico. PhD thesis, University of Azores, Ponta Delgada. http://dited.bn.pt/30404/index.html

Nunes JC (2004) Geologia. In: Forjaz VH (ed) Observatório Vulcanológico e Geotérmico dos Açores. Atlas Básico dos Açores. Ponta Delgada, pp 60–62

Nunes JC, Camacho A, França Z, Montesinos F, Alves M, Vieira R, Velez E, Ortiz E (2006) Gravity anomalies and crustal signature of volcano-tectonic structures of Pico Island (Azores). J Volcanol Geoth Res 156(1–2):55–70

Pacheco JM, Ferreira T, Queiroz G, Wallenstein N, Coutinho R, Cruz JV, Pimentel A, Silva R, Gaspar JL, Goulart C (2013) Notas sobre a geologia do arquipélago dos Açores. In: Dias R, Araújo A, Terrinha P, Kullberg JC (eds) Geologia de Portugal, vol 2. Escolar Editora, Lisboa, pp 595–690

Silva PF, Henry B, Marques FO, Hildenbrand A, Lopes A, Madureira P, Madeira J, Nunes JC, Roxerová Z (2018) Volcano-tectonic evolution of a linear volcanic ridge (Pico-Faial Ridge, Azores Triple Junction) assessed by paleomagnetic studies. J Volcanol Geoth Res 352:78–91

MIX
Papier aus verantwortungsvollen Quellen
Paper from responsible sources
FSC® C105338

If you have any concerns about our products,
you can contact us on
ProductSafety@springernature.com

In case Publisher is established outside the EU,
the EU authorized representative is:
Springer Nature Customer Service Center GmbH
Europaplatz 3, 69115 Heidelberg, Germany

Printed by Libri Plureos GmbH
in Hamburg, Germany